PROCEEDINGS
SPIE—The International Society for Optical Engineering

Optical Scatter: Applications, Measurement, and Theory

John C. Stover
Chair/Editor

24-26 July 1991
San Diego, California

Sponsored and Published by
SPIE—The International Society for Optical Engineering

Volume 1530

SPIE (Society of Photo-Optical Instrumentation Engineers) is a nonprofit society dedicated to the advancement
of optical and optoelectronic applied science and technology.

The papers appearing in this book comprise the proceedings of the meeting mentioned on the cover and title page. They reflect the authors' opinions and are published as presented and without change, in the interests of timely dissemination. Their inclusion in this publication does not necessarily constitute endorsement by the editors or by SPIE.

Please use the following format to cite material from this book:
    Author(s), ''Title of paper,'' *Optical Scatter: Applications, Measurement, and Theory,* John C. Stover, Editor, Proc. SPIE 1530, page numbers (1991).

Library of Congress Catalog Card No. 91-62702
ISBN 0-8194-0658-9

Published by
SPIE—The International Society for Optical Engineering
P.O. Box 10, Bellingham, Washington 98227-0010 USA
Telephone 206/676-3290 (Pacific Time) ● Fax 206/647-1445

Printed in the United States of America.

OPTICAL SCATTER:
APPLICATIONS, MEASUREMENT, AND THEORY

Volume 1530

CONTENTS

(continued)

OPTICAL SCATTER:
APPLICATIONS, MEASUREMENT, AND THEORY

Volume 1530

OPTICAL SCATTER:
APPLICATIONS, MEASUREMENT, AND THEORY

Volume 1530

OPTICAL SCATTER:
APPLICATIONS, MEASUREMENT, AND THEORY

Volume 1530

CONFERENCE COMMITTEE

Conference Chair
John C. Stover, TMA Technologies, Inc.

Cochairs
Jeffrey W. Garrett, Lockheed Missiles & Space Company, Inc.
Donald J. Janeczko, Martin Marietta Electronic Systems
H. Philip Stahl, Rose-Hulman Institute of Technology and Stahl Optical Systems
Cynthia L. Vernold, Hughes Danbury Optical Systems, Inc.

Session Chairs
Session 1—Analysis and Theory
John C. Stover, TMA Technologies, Inc.

Session 2—Surface Scatter
H. Philip Stahl, Rose-Hulman Institute of Technology and Stahl Optical Systems

Session 3—Scatter from Be Mirrors
Donald J. Janeczko, Martin Marietta Electronic Systems

Session 4—Instruments and Techniques
Cynthia L. Vernold, Hughes Danbury Optical Systems, Inc.

Session 5—Scatter Measurements
Jeffrey W. Garrett, Lockheed Missiles & Space Company, Inc.

Conference 1530, *Optical Scatter: Applications, Measurement, and Theory*, was part of a five-conference program on Optical Design and Fabrication held at SPIE's 1991 International Symposium on Optical Applied Science and Engineering, 21–26 July 1991, in San Diego, California. The other conferences were:

Conference 1527, *Current Developments in Optical Design and Optical Engineering*
Conference 1528, *Nonimaging Optics: Maximum Efficiency Light Transfer*
Conference 1529, *Ophthalmic Lens Design and Fabrication*
Conference 1531, *Advanced Optical Manufacturing and Testing II.*

Program Chair: **Robert E. Fischer,** OPTICS I, Inc.

INTRODUCTION

Optical scatter remains an important issue for designers and users of many optical systems. Much effort has been expended in reducing the causes of optical scatter, since it limits resolution, reduces system throughput, and is a source of optical noise. On the other hand, optical scatter can be used as a sensitive measure of component quality. Outside the optics industry (where scatter itself is not the primary problem), it can be used to monitor other changes in product attributes such as roughness, position, and contamination. This volume reports progress over the last two years in this field of metrology.

Researchers continue working on comparing surface roughness measurements obtained from profiling instruments to those found from surface BRDF. The comparisons are not always easy. The issue is further complicated when surface scatter is mixed with nontopographic scatter, as in the case of beryllium mirrors used in the infrared. Understanding these effects is still important, because nontopographic scatter may be found to dominate in many materials at many wavelengths as the amount of topographic scatter is reduced through superpolishing techniques. Work continues on scatter standards, both samples and practices. New instrumentation has been introduced over the last two years. In particular, vacuum cryogenic scatterometers, out-of-plane measurements, small dedicated systems, and even hand-held units have all been built. Controlling polarization of both source and receiver is expected to add to the information that can be extracted from scatter measurements and to increase the use of scatter measurement outside the optics industry. Measurements from a wide variety of materials at many wavelengths were reported at the conference.

I would like to thank the speakers for their well-prepared talks, the session chairs for their time (and recruiting), and the SPIE staff for their usual professional support. Your contributions made my job easy—and almost unnecessary.

John C. Stover
TMA Technologies, Inc.

SESSION 1

Analysis and Theory

Chair
John C. Stover
TMA Technologies, Inc.

Optical scatter: an overview

John C. Stover
TMA Technologies Inc.
Bozeman MT 59715

ABSTRACT

Optical scatter is a bothersome source of optical noise, limits image resolution and reduces system throughput. However, it is also an extremely sensitive metrology tool. It is employed in a wide variety of applications in the optics industry (where direct scatter measurement is of concern) and is becoming a popular indirect measurement in other industries where its measurement in some form is an indicator of another component property - like roughness, contamination or position . This paper presents a brief review of the current state of this technology as it emerges from university and government laboratories into more general industry use. The bidirectional scatter distribution function (or BSDF) has become the common format for expressing scatter data and is now used almost universally. Measurements made at dozens of laboratories around the country cover the spectrum from the uv to the mid-IR. Data analysis of optical component scatter has progressed to the point where a variety of analysis tools are becoming available for discriminating between the various sources of scatter. Work has progressed on the analysis of rough surface scatter and the application of these techniques to some challenging problems outside the optical industry. Scatter metrology is acquiring standards and formal test procedures. The available scatter data base is rapidly expanding as the number and sophistication of measurement facilities increases. Scatter from contaminants is continuing to be a major area of work as scatterometers appear in vacuum chambers at various laboratories across the country. Another area of research driven by space applications is understanding the non-topographic sources of mid-IR scatter that are associated with Beryllium and other materials. The current flurry of work in this growing area of metrology can be expected to continue for several more years and to further expand to applications in other industries.

2. INTRODUCTION

In general, scatter from optical components can propagate in any direction over the entire observation sphere centered about the sample. The distribution of light within the scatter pattern is a function of incident angle and wavelength as well as sample parameters such as orientation, transmittance, reflectance, absorptance, surface finish, index of refraction, bulk homogeneity, contamination, etc. The bidirectional scatter distribution function (BSDF) is the generic term commonly used to describe scattered light patterns. The BRDF, BTDF and BVDF are subsets of the BSDF, used specifically for reflective, transmissive and volume scatter sources, respectively. Because the BSDF format is a common form of scatter characterization and can be used to generate quantitative scatter specifications, it is worthwhile to understand its simple mathematical definition and to become familiar with its variations and limitations.

The derivation and notation for BRDF is credited to F. E. Nicodemus[1] who, with his co-workers, expended considerable effort examining the problems of measuring and defining the reflectance of optics that exhibit both diffuse or specular properties (ie: all reflective optics). Since the objective here is to understand the expression, its complete derivation is not provided. The defining geometry is shown in Figure 1 where the subscripts i and s are used to denote incident and scattered quantities respectively. Nicodemus started with the general case of light reflected from a surface and made several logical approximations to arrive at a simple, manageable form for BRDF. He examined the relatively simple case of a collimated beam of light of uniform cross-section reflecting from a surface. The situation is further simplified by also assuming that the illuminated surface area, A, is isotropic, and that all scatter comes from the surface and none from the bulk. The BRDF is then defined in radiometric terms as the scattered surface radiance at some point on the hemisphere in front of the reflective sample, divided by the incident surface irradiance. After several simplifying assumptions are made, a manageable expression results that is used by almost all scatter measurement laboratories.

0-8194-0658-9/91/$4.00

$$BRDF = F = \frac{P_s/\Omega}{P_i \ COS \ \theta_s} \qquad \qquad (1)$$

The quantity Ω_s is the solid angle through which the light power P_s is scattered as shown in Figure 1. This equation is appropriate for all angles of incidence and all angles of scatter. The $\cos(\theta_s)$ factor can be viewed as a correction to adjust the illuminated area, A, to its apparent size when viewed from the direction of scatter. The value of BRDF is "bidirectional" in that it depends on both the incident direction (θ_i, ϕ_i) and the scatter direction (θ_s, ϕ_s), and may be viewed as directional reflectance per unit steradian. Notice that BRDF has units of inverse steradians and, depending on the relative sizes of P_s and Ω_s, can vary from very large to very small. For example, the ratio (P_s/P_i) is nearly unity when the specular reflection is measured so the BRDF is approximately $(1/\Omega_s)$, usually very large. Scatter measurements well removed from specular generally encounter very small values of P_s and require larger apertures. The differential form $(dP_s/d\Omega_s)$ is more correct and is only approximated when measurements are taken with a finite diameter detector aperture. This approximation is very good when the flux density is reasonably constant over the measuring aperture, and is very poor when using a large aperture to measure focused specular beams. Although Nicodemus worked with reflective samples, the same equation is also used to calculate BSDF for transmissive and volume (bulk) samples.

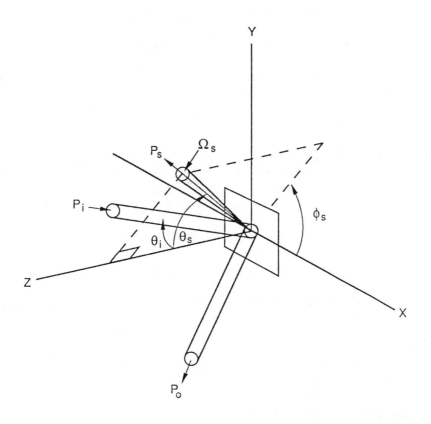

Figure 1. BSDF Geometry

The assumptions made in the Nicodemus derivation are seldom completely true in real measurement situations. A slightly converging incident laser beam with a Gaussian intensity cross-section is often used instead of one that is collimated and uniform. And, the sample area is often anisotropic (instead of isotropic) and never 100% front surface reflective. For the case of a transparent sample, where two surfaces and the bulk are scattering, the idea of an illuminated surface area becomes a little fuzzy at best. These differences mean that, although we can measure it as defined in Equation 2, the BSDF is not truly equal to the scattered radiance per unit incident irradiance, as defined by Nicodemus. However, these reservations apply only to the original definition in terms of differential radiance and irradiance; it is still sensible to measure the BSDF, as defined in Equation (2), and to use these measurements as component specifications.

Light scattered from a particular optic into any given solid angle, from any hypothetical source, can be estimated by multiplying the appropriate value of the BSDF by the incident power, the cosine of the scatter angle, and solid angle of interest. Thus a designer with a library of typical BSDF data can study system scatter issues and then assign meaningful BSDF specifications to components. Optics can be accepted or rejected, on the basis of BSDF measurements, just as interferometer measurements are used to confirm specified surface contour. Software that will make these computations through an optical system is now available for operation on both PC and VAX computers.

The following sections review key points in the general areas of study explored by our conference papers. A final section suggests several areas of work that are expected to be of continued interest over the next two years.

3. ANALYTICAL DEVELOPMENTS

A key analytical technique that is emerging is the use of polarization sensitive instrumentation to extract more sample information from BSDF measurements. A linear systems approach has been used where the light is characterized by a 1x4 Stokes vectors that completely describes the polarization state of both the incident and scatter signals. Once these are known, a 4x4 Mueller matrix can be calculated that represents the sample action on the incident light. In many cases, one or more Mueller elements are far more sensitive to sample changes than an overall BSDF measurement.

Another hot topic in the last two years has been the surface induced scatter peaks that appear back in the incident beam direction. Termed retro-scatter, or enhanced back scatter, these unexpected signals are of intense interest to workers in the fields of ring laser gyros and laser RADAR. The theoretical explanations and the instrumentation required for measurement are equally complicated. Scatter from multilayer coatings and from X-ray optics are also receiving a lot of attention.

4. SURFACE SCATTER

The analysis of scatter from surface roughness has been the subject of a huge amount of work. The work on angle of incidence scaling (shift invariance), wavelength scaling and diffraction grating results, confirm that the Rayleigh-Rice vector perturbation theory is indeed the yellow brick road connecting the BRDF of smooth, clean, front surface mirrors to their surface statistics. A number of authors have compared surface statistics obtained with surface profiling instrumentation to that found form the measured scatter with excellent results as long as the correct prescriptions are followed for computing the surface power spectral density function (PSD). In some sense, this is more easily done from scatter data than from profile data, because the PSD and the BRDF are nearly proportional. The complexities of the profile to PSD conversion, which involves a bandwidth limited Fourier Transform and consideration of the differences between a surface and a sampled one dimensional profile, are being addressed by an ASTM committee studying terminology. Once published, this should reduce some of the difficulties that roughness/scatter comparisons pose.

5. BERYLLIUM MIRROR SCATTER

Scatter from clean, reflective optics can not always be related to just surface roughness. In particular beryllium mirrors used in the infrared often scatter as much as an order of magnitude more light than expected from their surface roughness measurements. This caused some initial confusion as these optics were originally manufactured to roughness specifications - not scatter specifications. Several studies are currently in progress to find a way to produce low scatter Be mirrors and to explain the cause of the unwanted anomalous (non-topographic) scatter. In fact, an entire session of this conference has been devoted to reporting their experimental and theoretical results. The bottom line appears to be that although some low scatter Be mirrors have now been produced, we still do not understand the origin(s) of non-topographic scatter. Possible candidates are surface non-uniformities in the optical associated with grain boundaries and the relatively large number of small pits (or inclusions) that are found on these mirrors. The data obtained in the last year appear to favor the former over the latter; however, a clear cut correlation between experiment and theory that explains the effect has not yet been reported. Polarization measurements may play a role in solving this problem.

It may turn out that anomalous scatter is present to some degree in many materials at many wavelengths. If this is the case, it could well pose a problem for those producing very low scatter optics. A scatter specification, instead of a roughness specification, may be the better choice.

6. INSTRUMENTATION AND TECHNIQUES

Instrumentation is getting increasingly more sophisticated. In addition to the problems associated with polarization measurement and retro-scatter, other hardware issues have been faced during the last two years. A number of vacuum cryogenic scatterometers have been built and used successfully. Scatter measurement is routinely leaving the plane of incidence. Broadband measurements, from 0.25 to 14 micrometers, are being made. The degree of automation is increasing - operators are free to leave the lab during measurement sequences that may take several hours to complete.

On the other hand, instrumentation is also getting small and simple. Dedicated instrumentation has been designed to measure scatter characteristics of jet engine exhaust materials - in situ. Hand held units are used to measure IR sensor window degradation. Scatter measurements at selected absorption bands are used to find and quantify thin film contamination. A host of general industry applications awaits application of existing technology.

7. MEASUREMENTS

The ASTM Standard BRDF Measurement Practice is now in place. Now that measurements at the major labs are in good agreement, requirements for increased accuracy are appearing - as low as 1%. NIST is working on functional measurement standards (samples of known BRDF) which will further improve industry accuracy. Measurements are being made on a huge variety of samples - everything from freshly coated laser optics to surgical specimens. Papers in this conference report measurements on curved optics, semi-conductor wafers, coated paper, space shuttle tiles, and intra-ocular lens material as well as a variety of optical materials and systems.

8. THE NEXT TWO YEARS

The number of scatter related conferences, publications and measurement facilities has grown dramatically in the last two years. Based on the work reported at this conference, it would be reasonable to expect a high level of activity over the next two years. Polarization analysis is expected to become of increasing importance as both hardware and software improve. Requirements for BRDF measurement to "1% or better" are currently being discussed for instrumentation that is required in the coming two years. Work continues on determining the sources of anomalous (or non-topographic) scatter from beryllium (and other materials) in the infra red. In fact Be mirrors exhibiting very little of this effect have now been reported. Scatter measurement applications are expanding into areas outside the optics industry. Expect the next two years in our industry to be both exciting and rewarding.

9. REFERENCES

1.   Nicodemus, F.E., J.C. Richmond, J.J. Hsia, I.W. Ginsberg, T. Limperis. <u>Geometric Considerations and Nomenclature for Reflectance</u>. U.S. Dept. of Commerce, NBS Monograph 160 (1977).

Stokes vectors, Mueller matrices and polarized scattered light: experimental applications to optical surfaces and all other scatterers

William S. Bickel and Gorden Videen

University of Arizona, Department of Physics
Tucson, Arizona 85721

ABSTRACT

We discuss scattering in the context of the Stokes vectors and Mueller matrices that characterize the interaction. In order to study surface structures using light-scattering techniques it is useful to examine the nature of light scattered from perfect and perturbed mirror surfaces.

1. INTRODUCTION

A highly motivated, systematic, and fundamental approach to surface scattering requires that the initial surfaces be fundamental and that the contamination to produce the surface scattering be known and controlled. With this in mind any rough surface can be considered to be a perturbed perfect surface that has reached its final condition through some continuum contamination process.

The powerful Stokes vector, Mueller matrix light-scattering techniques can be applied directly to study surfaces. Scattering from perfect surfaces can be theoretically predicted and experimentally measured. Experimentalists have an advantage in that they can measure what theorists cannot calculate. If the data are to be used to check theory, the experimentalist must relate the light scattering signal to the exact structure and orientation of the surface that scattered the light.

Consider how a perfect optical surface (lens) is treated by taking into account its geometrical and diffraction-limited properties. Geometrical optics is used to predict where a point on the object will be focused to a point on the image. Diffraction theory will predict how the image point is actually a diffraction pattern; i. e., not all light from the object point ends up at the image point. Nevertheless, the diffraction-limited image is exactly predicted by theory and its intensity distribution is related to fundamental constants. The problem arises when the perfect diffraction-limited optic and its image is perturbed by a defect in or on the optic. This defect scatters light into all directions — out of the paths so well defined by geometrical and diffraction optics.

The question is where does this scattered light go and what does it do to the image (information). There will be a loss of image intensity, but more troublesome is the light initially intended for one part of the image that is scattered to another part of the image causing loss of contrast and image definition. We can write that the object (point A) is transformed into a diffuse diffraction-limited geometrical image (point A') by a transformation matrix [L, D, S] where L is the geometrical function of the lens, D is the diffraction function and S is the scattering function which contains all of the scattering parameters of the defect.

2. POLARIZED LIGHT SCATTERING

What is S? If the defect were a perfect sphere the scatter from it is nothing more than the diffraction by a spherical object — a problem solved by Gustav Mie in 1908.[1] As the sphere becomes more irregular in shape and inhomogeneous in optical constants, exactly solveable sphere diffraction goes over to the statistical aspects of irregular particle scatter.[2] A scatterer is said to be characterized by its scattering properties when we know how it will rearrange the properties of light incident on it. For example, consider perfect sphere scattering. A sphere with radius r, refractive index n and aborption μ has a scattering matrix [S]. When it is illuminated with light of intensity I_0, wavelength λ_0, polarization Π_0, at angle ϑ_0, it creates a scattered field of I_s, λ_s, Π_s, and ϑ_s. We can write $(I_s, \lambda_s, \Pi_s, \vartheta_s) = [S](I_0, \lambda_0, \Pi_0, \vartheta_0)$ where [S] is exactly known for spheres. The matrix [S] can be calculated only for certain highly symmetric cases. It can be approximated for slightly distorted systems, but can only be measured for truly irregular particles. Therefore, in order to learn how a defect on or in a surface (an irregular, imperfect surface) will affect all aspects of the incident light, it is better to measure what happens exactly instead of relying on approximation and theoretical models. Generally two things are required of scattering experiments: 1. produce a defect and measure its scatter; or 2. measure the scatter and predict the defect — the inversion process. Before we extract information from light-scattering data it is important to be sure that we have all the information available.

3. POLARIZED SCATTERING NEPHELOMETER

The experimental set-up that can measure the polarized intensities scattered by the scatterer [S] into the angles ϑ and φ is shown in Fig. 1. The input optics can be selected to be an open hole [o], or horizontal linear polarizer [h], or +45° linear polarizer [+] or a right-handed circular polarizer [r]. The exit optics choices are the same and can be chosen independently of the input optics and can swing with the detector through the scattering angle ϑ from 0° to 180°.[3]

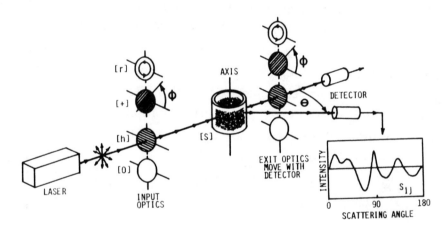

Fig. 1. Arrangement of entrance-exit polarizers to measure the matrix elements of a scatterer.

The following example shows how to determine v. ..at scattering matrix elements S_{ij} are involved when a particular set of input-output polarizers are used to prepare and analyze the scattered light.[4] For our first choice we assume that the arbitrary scatterer (defect) is illuminated with horizontally-polarized light [h]. The scattered Stokes vector will be $[V_s] = [S]*[h]$ or

$$\begin{pmatrix} S_{11} & S_{12} & S_{13} & S_{14} \\ S_{21} & S_{22} & S_{23} & S_{24} \\ S_{31} & S_{32} & S_{33} & S_{34} \\ S_{41} & S_{42} & S_{43} & S_{44} \end{pmatrix} \begin{pmatrix} 1 \\ 1 \\ 0 \\ 0 \end{pmatrix} = \begin{pmatrix} S_{11} + S_{12} \\ S_{21} + S_{22} \\ S_{31} + S_{32} \\ S_{41} + S_{42} \end{pmatrix}.$$

We see that the scatterer [S] mixes the initial pure polarization state [h] to produce a scattered Stokes vector with mixed polarizations. The first component (S_{11} + S_{12}) is the total intensity and is a sum of two matrix elements. If this scattered light is now passed through a +45° linear polarizer [+] we get [V_f] = [+][V_s]. Specifically we have

$$\begin{pmatrix} S_{11} + S_{12} + S_{31} + S_{32} \\ 0 \\ S_{11} + S_{12} + S_{31} + S_{32} \\ 0 \end{pmatrix} = \begin{pmatrix} 1 & 0 & 1 & 0 \\ 0 & 0 & 0 & 0 \\ 1 & 0 & 1 & 0 \\ 0 & 0 & 0 & 0 \end{pmatrix} \begin{pmatrix} S_{11} + S_{12} \\ S_{21} + S_{22} \\ S_{31} + S_{32} \\ S_{41} + S_{42} \end{pmatrix}.$$

The first component of the final Stokes vector is now a mixture of four matrix elements. The element sum (S_{11} + S_{12} + S_{31} + S_{32}) is the total intensity that will be measured by the detector. We put the result of all such calculations from all 16 Stokes vector combinations into a final array shown in Fig. 2. The main point of Fig. 2 is that there are 4 X 4 = 16 measurements to be made to completely characterize the polarized light scattered from the defect. These measurements can be routinely made by nephelometers that incorporate the various polarizers in their entrance-exit optics. The 16 matrix elements are needed to completely characterize the scattered light or to use in the inversion process to determine the properties of the scatterer.

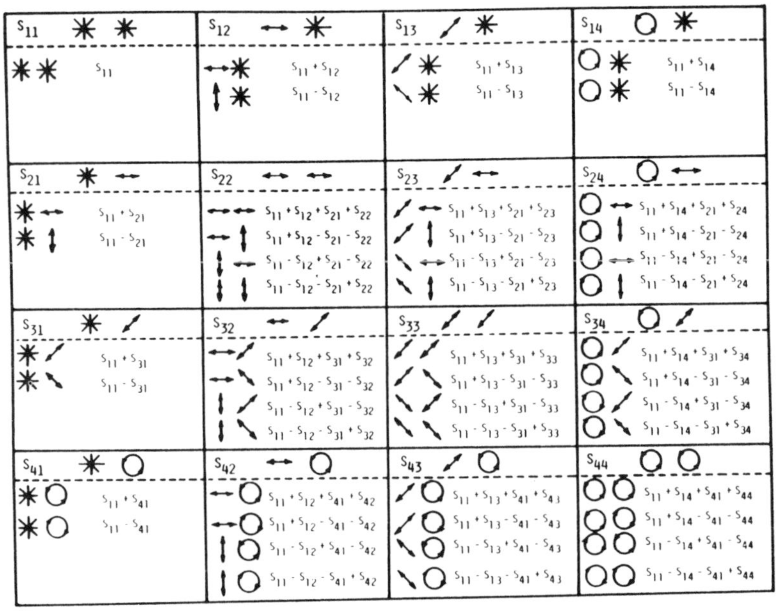

Fig. 2. Matrix array showing matrix element combinations measured with various arrangements of input and output polarizers.

4. SURFACE ORIENTATION

The geometrical orientation of the rough surface with respect to the entrance-exit beams must be addressed early because there are an infinite number of possible orientations, all of which might give a different S_{ij} for the same surface. Figure 3 shows some of the geometrical parameters involved in surface scattering studies. The laser beam, after preparation into a definite polarization state, strikes the surface at angle α and is scattered into all 4π. The light scattered into angle ϑ is analyzed with the polarization exit optics and detected by the photomultiplier. The angle τ is a surface tilt measured in the surface plane. Our work has shown that geometry is important and that no universal scaling factor exists and that no best orientation exists for all cases.

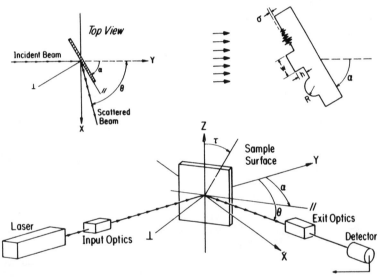

Fig. 3. Optical and sample arrangement for surface scattering measurements.

5. SURFACE SCATTERING

We now show some general results of scattering from some surfaces to show how light-scattering nephelometry and BRDF are related.[5,6] Figure 4 shows four matrix elements for a reflective aluminum surface. The 4 order of magnitude angular decrease of the total intensity matrix element S_{11} is an indication of the quality of the reflecting surface. Note that even though the total scattered intensity is down by over 5 orders of magnitude near $\vartheta = 150°$ (back-scatter) the polarization of the scattered light is significant as demonstrated by matrix S_{33} and S_{34}.

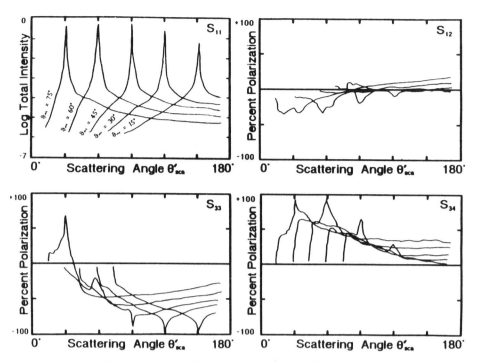

Fig. 4. Four Mueller matrix elements for a reflecting aluminum surface.

Figure 5 shows the matrix elements for the same surface but now roughened to saturation. Further roughening will not change the surface character. All hints of the location of the specular peak are gone. S_{12} as well as S_{34} are independent of surface orientation; and polarizations, as indicated by all matrix elements, remain large.

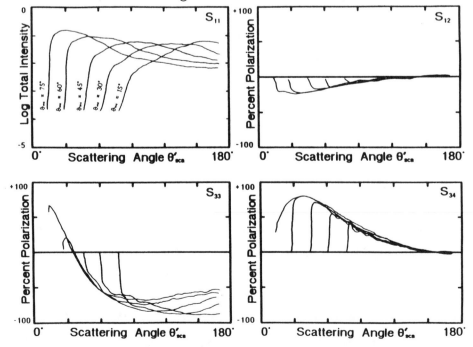

Fig. 5. Four Mueller matrix elements for a saturated rough aluminum surface.

Now we show how dramatically the S_{ij} from surface defects depend on the angle at which the surface (and its defect) is illuminated. Figure 6 shows the matrix element S_{33} for a rectangular line (h = 0.46μm, w = 1.10μm) on a smooth aluminum surface illuminated at near-grazing incidence and at normal incidence. Normal incidence not only restricts the angular range over which the data is received but it also wipes out all the phase information needed to characterize the defect.

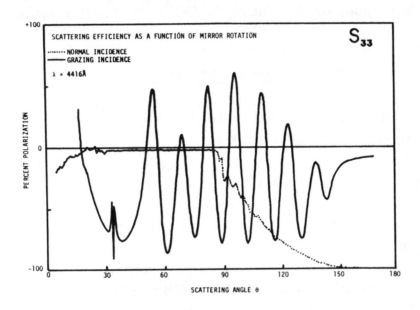

Fig. 6. Matrix element S_{33} for a rectangular line on a mirror surface measured at two different angles of incidence.

8. CONCLUSIONS

One general goal of light scattering is to develop an algorithm to put into a computer which will predict either the scattered field (from particle properties) or particle properties (from the scattered field). The data (particle properties or scattered field) must be determined experimentally. We must determine how much scattering data is needed, how good it must be and how well it describes the scatterer in a practical way. We also must determine the amount of information contained in the various matrix elements and whether signal changes can be related to changes in the optical or geometrical properties. We see that polar nephelometry gives sixteen matrix element signals. The record shows that polar nephelometry, ellipsometry, BRDF, and other optical techniques can all complement each other and yield important information about surface scattering when used with care, keeping in mind their limitations and range of validity.

9. ACKNOWLEDGMENTS

This research was funded by the Army Armament Research and Development Command, the U. S. Air Force Office of Scientific Research (AFSC) and the Itek Corporation. We thank Vincent J. Iafelice for many helpful discussions.

10. REFERENCES

1. G. Mie, "Beitrage zer Optik trüber Meiden speziell kolloidaler Metallosüngen," Ann. Physik, $\underline{25}$, 377-445 (1908).

2. W. S. Bickel, H. A. Yousif, and W. M Bailey, "Masking of information in light scattering signals from complex scatterers," Aerosol Sci. and Technol. $\underline{1}$, 329-335 (1982).

3. A. J. Hunt and D. R. Huffman, "A new polarization modulated light scattering instrument," Rev. Sci. Instrum., $\underline{44}$, 1753-1762 (1973).

4. W. S. Bickel and W. M. Bailey, "Stokes vectors, Mueller matrices, and polarized scattered light," Am. J. Phys. $\underline{53}$, 468-478 (1985).

5. V. J. Iafelice, The Polarized Light Scattering Matrix Elements for Select Perfect and Perturbed Optical Surfaces, M. S. thesis, University of Arizona 1985.

6. R. Anderson, "Matrix description of radiometric quantities, " Appl. Opt. $\underline{30}$, 858-867 (1991).

Sinusoidal surfaces as standards for BRDF instruments

Egon Marx, Thomas R. Lettieri, and Theodore V. Vorburger

National Institute of Standards and Technology
Gaithersburg, MD 20899

and

Malcolm McIntosh

Martin Marietta Energy Systems
P. O. Box 2000, Oak Ridge, TN 37831

ABSTRACT

This study of light scattered by sinusoidal surfaces shows that such a configuration can be used as a material standard to help calibrate instruments that measure the BRDF of arbitrary surfaces. Measured and computed values of the power scattered into the diffraction peaks show good agreement, and such calculations can be further improved and used to verify the standards.

1. INTRODUCTION

A round robin study of the bidirectional reflectance distribution function (BRDF), in which a number of laboratories made measurements of the BRDF of a set of four different surfaces, has disclosed significant discrepancies between the results.[1,2] The wide variation between these results clearly suggests that material standards are needed to calibrate the many instruments in use today. One such standard should provide a way to determine the accuracy of BRDF instruments over several decades of intensity.

We propose that sinusoidal surfaces be used as standards to verify the relative power measurements in BRDF instruments. A conducting sinusoidal surface illuminated by a light beam produces a set of peaks in the angular distribution of the scattered power which can vary over seven or more orders of magnitude when the appropriate grating and source parameters are used. This wide dynamic range in scattered power and the well known angular distribution of these peaks may be used to calibrate the linearity response and geometric configuration of BRDF instruments. For instance, CO_2 laser radiation with a wavelength of 10.6 μm incident on a sinusoidal surface of amplitude ~0.5 μm and spatial wavelength ~100 μm can be used for this purpose. Here we compare the measured values of the angular distribution of the power scattered by this surface with computed ones. We also measured the scattered intensity for He-Ne laser light with a wavelength of 0.6328 μm incident on a sinusoid of amplitude 0.050 μm and spatial wavelength 6.67 μm. Measurements with different collection apertures show the shapes of the diffraction peaks and also the total power in each peak. We computed the power scattered into each diffraction peak by using approximate results for perfect sinusoids, and we also computed the angular distribution of power using an actual profile measured with a stylus instrument. The two types of calculations are described in Section 2. In Section 3 we compare measured and computed results that correspond to the CO_2 laser, and in Section 4 we do the same for results obtained with the He-Ne laser.

2. SCATTERING FROM SINUSOIDAL SURFACES

The electromagnetic energy scattered by a sinusoidal surface[3,4] is concentrated in a number of directions θ_n with respect to the normal of the average surface. These angles are given by

$$\sin\theta_n = -\sin\theta_i + n\lambda/\Lambda, \quad n = 0, \pm1, \pm2, \ldots \tag{1}$$

where θ_i is the angle of incidence with respect to the normal, λ is the wavelength of the incident light, and Λ is the spatial wavelength of the surface. We use the same sign convention for θ_i and θ_n, so that the specular direction corresponds to $\theta_0 = -\theta_i$. The power associated with the diffraction peak of order n is then approximately given by[4]

$$P_n = P_0[J_n(\Delta_n)\cos\theta_n]^2, \tag{2}$$

where J_n is the Bessel function of order n,

$$\Delta_n = ka(\cos\theta_i + \cos\theta_n), \quad k = 2\pi/\lambda, \tag{3}$$

and a is the amplitude of the sinusoid. For small values of the argument, Δ,

$$J_n(\Delta) \approx (1/n!)(\Delta/2)^n, \tag{4}$$

so that, assuming that $\cos\theta_i \approx \cos\theta_n \approx 1$,

$$P_n/P_{n-1} \approx (ka/n)^2. \tag{5}$$

This formula shows that this ratio becomes smaller with increasing n; for example, $(P_6/P_5)/(P_1/P_0) \approx 1/36$. Thus, the sinusoidal grating does not provide a constant ratio between the powers in successive diffraction peaks.

To calculate the intensity of the scattered light for a given surface that has a one-dimensional profile, $z = \zeta(x)$, we assume that the direction of incidence is in the xy-plane and use the Kirchhoff approximation to find the intensity,[3]

$$I(\theta) \propto [1 + \cos(\theta_i - \theta)/(\cos\theta_i + \cos\theta)]^2|\int\exp(i\vec{v}\cdot\vec{r})dx|^2, \tag{6}$$

where θ is the scattering angle,

$$\vec{v} = -k[(\sin\theta_i + \sin\theta)\hat{e}_x + (\cos\theta_i + \cos\theta)\hat{e}_z], \tag{7}$$

$$\vec{r} = x\hat{e}_x + \zeta(x)\hat{e}_z. \tag{8}$$

We omitted constants such as the incident power and the illuminated area in Eq. (6). The intensity must be integrated over the aperture to provide the values of the power measured as a function of the scattering angle.

3. MEASUREMENTS WITH A CO_2 LASER

We measured the radiation scattered by a sample with a diamond-turned, nickel-plated, sinusoidal surface of wavelength $\Lambda = 100.13$ μm and amplitude a = 0.520 μm with incident radiation from a CO_2 laser ($\lambda = 10.6$ μm). These values of Λ and a were obtained with a stylus instrument. The BRDF instrument we used was the Oak Ridge complete angle scattering instrument (CASI).[*,5] In Fig. 1 we show the computed power obtained from Eq. (6) using the measured stylus profile of the actual surface and integrating the intensity over an aperture, $\Delta\theta$, of 1.43°. The angles shown as abscissas in all the figures are measured from the specular direction, that is, they correspond to $\theta + \theta_i$. A set of ten profiles was obtained using a Talystep instrument with a stylus of radius ~1 μm and a force of ~4 mgf. The computed power distribution is an average of the ten computed distributions. In Figs. 2 and 3 we show the measured BRDF, multiplied by $\cos\theta$ to yield a quantity proportional to the power measured by the detector, as a function of scattering angle with collection apertures of 1.43° and 0.055°, respectively, for an angle of incidence $\theta_i = -5°$. Measurements were made with the larger aperture to integrate over the peaks to provide the total powers, while the smaller aperture gives more precise shapes and locations for the diffraction peaks. The flat tops of the peaks shown in Fig. 2, to be compared with those in Fig. 1, indicate that the intensities have been effectively integrated to give the powers in the diffraction peaks. The beam incident on the sample in CASI is focused onto the detector when it is located at the specular beam. As the detector scans, it does not follow the locus of focal points, which gives rise to an angular broadening of the diffraction peaks at large angles from the specular direction. This effect can be seen in Fig. 2: for the eighth-order peak, the angular spread was calculated to be about 1.2°, which has to be combined with the 1.43-degree aperture to give the measured halfwidth of 1.5°. The flat tops of the peaks in Fig. 2 are narrower than expected due to this defocusing effect.

In the second and third columns of Table 1 we show the values of the angles for

Order	$\theta°$ (calc.) Eq. (1)	$\theta°$ (meas.)	Power (c.) Eq. (2)	Power (c.) Eq. (6)	Power (m.)
0	5.0	5.0	1.00	1.0	1.0
1	11.1	11.1	9.95×10^{-2}	8.3×10^{-2}	1.10×10^{-1}
2	17.4	17.4	2.13×10^{-3}	1.7×10^{-3}	1.4×10^{-3}
3	23.9	23.9	1.77×10^{-5}	1.3×10^{-4}	9.5×10^{-5}
4	30.7	30.7	6.80×10^{-8}	3.2×10^{-5}	1.6×10^{-5}
5	38.1	38.1	1.26×10^{-10}	6.4×10^{-5}	1.8×10^{-5}
6	46.2	46.3	1.06×10^{-13}	2.8×10^{-5}	6.7×10^{-6}
7	55.9	55.9	3.12×10^{-17}	1.7×10^{-5}	4.0×10^{-6}
8	69.1	69.2	1.22×10^{-21}	1.4×10^{-5}	9.9×10^{-7}

Table 1. Calculated and measured angles and power of the diffraction peaks for CO_2 laser radiation incident on a sinusoidal surface.

---

*Certain commercial equipment is identified in this report to specify adequately the experimental procedure. In no case does such identification imply recommendation or endorsement by NIST, nor does it imply that the equipment identified is necessarily the best available for the purpose.

the diffraction peaks θ_n computed from Eq. (1) and obtained from the measurements carried out with CASI and shown in Fig. 3. In the fourth column we show the power computed from Eq. (2) relative to the power of the 0-order peak. The fifth and sixth columns show the relative power of the peaks, minus the background, obtained from Figs. 1 and 2, respectively. The small power values calculated for the higher-order diffraction peaks of an ideal sinusoidal surface are not valid for real surfaces, where the surface imperfections dominate the scattering at large angles. For this surface and source we have $P_1/P_0 \approx (ka)^2 \approx 0.095$. The ratio $(P_6/P_5)/(P_1/P_0) \approx 1/118$, computed from column four, is even smaller than the approximate value determined from Eq. (5).

4. MEASUREMENTS WITH A HE-NE LASER

We measured the other sample, a holographically produced, aluminum-coated sinusoidal surface with $\Lambda = 6.67$ μm and $a = 0.050$ μm using the detector array for laser light angular scattering (DALLAS)[6] at NIST. We show the results obtained for this sample in Table 2. With DALLAS we get only the first five peaks of light scattered by this sample, since the dynamic range of the instrument is limited to about five orders of magnitude. The incident beam was normal to the surface, $\theta_i = 0°$, and the intensity readings were taken every 0.2°. The angles of the diffraction peaks were determined with the aperture of the detector reduced to 0.2° by means of a slit. We estimate that the values of the measured angles, shown in column three, have an error of ~0.3°. We measured the power using the full aperture of the detector, 1.5°, to integrate the peak intensity. The 0-order peak was obtained by going 2° off normal incidence in the azimuth angle to avoid one of the mirrors. The incident beam had to be attenuated with a filter to avoid saturating the detector for this peak. The intensity value of the fourth-order peak was obtained without the filter, and the two sets of measurements were then combined by matching the power in a peak that both sets have in common. The measured power, shown in column five, is again normalized to the value of the 0-order peak. We computed the power using Eq. (2) and not yet from a measured profile, which was not available in digital form for this sample.

Order	$\theta°$ (calc.) Eq. (1)	$\theta°$ (meas.)	Power (calc.) Eq. (2)	Power (meas.)
0	0	–	1.00	1.00
1	5.4	5.0	3.21×10^{-1}	3.3×10^{-1}
2	10.9	10.6	2.03×10^{-2}	2.2×10^{-2}
3	16.5	16.2	5.07×10^{-4}	5.2×10^{-4}
4	22.3	22.2	6.25×10^{-6}	5.2×10^{-6}

Table 2. Calculated and measured angles and power of the diffraction peaks for He-Ne laser light incident on a sinusoidal surface.

The manufacturer of the holographically produced sample specified a nominal value of the amplitude of the sinusoid of $a = 0.04$ μm, but comparison of the measured and computed peak intensities indicated that $a = 0.05$ μm gave the best agreement. Measurements of the surface using a Talystep[7] stylus instrument and a WYKO interferometric microscope[8] showed that the larger value of the amplitude was indeed correct. The agreement between calculated and measured angles indicates that the nominal value of the spatial wavelength is accurate.

5. CONCLUSIONS

Examination of the measured values of the power shows that a sinusoidal surface produces an angular distribution of scattered light which can be used to verify the calibration of BRDF instruments and other instruments that measure related quantities, such as BTDF or intensity values. The peaks used to measure the relevant power are well separated in angle, and the range of intensity values is about 6 decades, which is adequate to calibrate a variety of instruments.

Comparison of measured and calculated values shows good agreement for the lower-order diffraction peaks, although the roughness of the actual surfaces creates a background noise of scattered light that is significant at the larger scattering angles both in the measured values and in those calculated from the actual surface profile. The noise of the calculated distribution is higher than the measured noise, which may be due to the errors introduced in the measurement of the profile and numerical errors in the calculations. The agreement between the powers in the peaks calculated for the perfect sinusoidal surface and the measured powers shows that the holographically produced sample is closer to an ideal sinusoid than the diamond-turned sample. Comparisons such as these can also be used to determine to what extent the calculations of the location and power of the diffraction peaks, using Eqs. (2) and (6), specify the spatial wavelengths and amplitudes of sinusoidal surfaces to be used as standards for each wavelength of incident light used in BRDF instruments.

6. ACKNOWLEDGEMENTS

We wish to thank A. Hartman and J.-F. Song for their assistance in carrying out some of the measurements discussed in this paper. We are also grateful to C. Asmail, J. Hylton, and R. Hinebaugh for valuable discussions during the formulation of this research. This work was supported in part by the SDI Office through the Aerospace Guidance and Metrology Center and the Survivable Optics MODIL.

7. REFERENCES

1. T. A. Leonard and M. Pantoliano, "BRDF Round Robin," in *Stray Light and Contamination in Optical Systems*, SPIE vol. 967, pp. 226-235, 1988.
2. T. A. Leonard, M. Pantoliano, and J. Reilly, "Results of a CO2 BRDF Round Robin," in *Scatter from Optical Components*, SPIE vol. 1165, pp. 444-449, 1989.
3. P. Beckmann and A. Spizzichino, *The Scattering of Electromagnetic Waves from Rough Surfaces*, Pergamon, New York (1963).
4. John C. Stover, *Optical Scattering: Measurement and Analysis*, McGraw-Hill, New York (1990).
5. D. R. Cheever, F. M. Cady, K. A. Klicker, and J. C. Stover, "Design Review of a Unique Complete Angle-Scatter Instrument (CASI)," in *Current Developments in Optical Engineering II*, SPIE vol. 818, pp. 13-21, 1987.
6. Egon Marx and T. V. Vorburger, "Direct and Inverse Problems for Light Scattered by Rough Surfaces," *Appl. Opt.*, vol. 29, pp. 3613-3626, 1990.
7. J. F. Song and Theodore V. Vorburger, "Stylus profiling at high resolution and low force," *Appl. Opt.*, vol. 30, pp. 42-50, 1991.
8. Katherine Creath and James C. Wyant, "Interferometric measurement of the

roughness of machined parts," in *Optical Testing and Metrology II*, SPIE vol. 954, pp. 246-251, 1988.

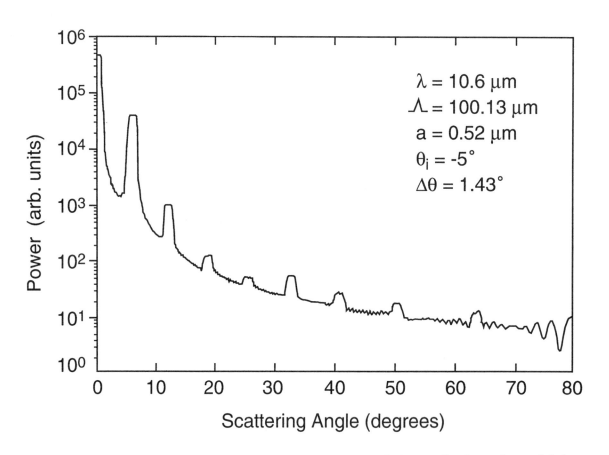

Fig. 1. Power computed from the stylus-measured profiles of the sinusoidal sample used with a CO_2 laser.

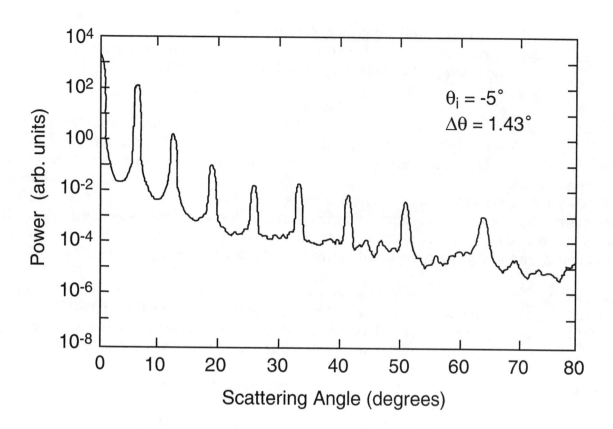

Fig. 2. Measured BRDF×cosθ, incident radiation from a CO_2 laser, large aperture.

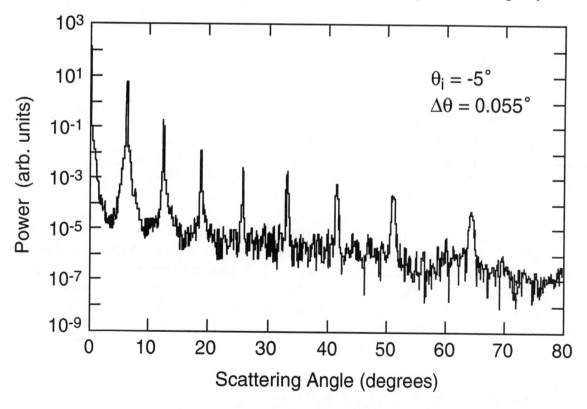

Fig. 3. Measured BRDF×cosθ, incident radiation from a CO_2 laser, small aperture.

Stray Light Implications of Scratch/Dig Specifications[*]

by Isabella T. Lewis, Arno G. Ledebuhr (Lawrence Livermore National Laboratory),
and Marvin L. Bernt (Toomay Mathis and Associates, Inc.)

Abstract

The bidirectional transmittance distribution function (BTDF) of two sets of scratch/dig standard sets were measured. These sets were representative of the inspection standards used in the optical industry to characterize polished surface defects. Measurements were taken with a small (1 mm diameter) illumination beam to maximize signal. The increase in average BTDF that results from a single scratch or dig over the MIL-STD 20 mm diameter surface was then calculated to determine what overall impact a defect will have on system stray light above base surface scattering due to surface micro-roughness. A BTDF measurement was taken with the illumination beam centered on the defect, then with it centered on a smooth section of the reference sample to find the increase in scattering caused by the defect.

Results show that dig scattering, when normalized to account for the single dig per 20 mm MIL-STD inspection area criteria, did not catastrophically increase the 0.05 B_0 (B_0 is the BTDF at 0.57°) at 633 nm characteristic of a high quality optical surface. As intuitively expected, dig scattering was angularly symmetric. Scratches, however, scattered highly directionally. Normalized BTDF is substantially increased from a smooth surface's typical 0.05 B_0 perpendicular to the scratch axis, but is unaffected in other angles. On average, the scratches may not have increased net surface scattering.

Scattering from the defects on the surfaces below the 40-20 scratch/dig level was found to not cause a catastrophic increase in scattering over the level os a well-polished optic (typically 4Å rms roughness). Since comparisons with scratch/dig samples only serve to provide a measure of the localized defects, and fail to be useful in determining the low-level scattering from the surface microroughness, one should not assume that a "40-20" surface is necessarily a low-scattering optic.

Introduction

Sensor systems designers have always had to worry about overall performance margin in the presence of stray light. In order to achieve a signal to noise ratio, the amount of scattering from bright out-of-field radiation sources onto the focal plane (or film plane) must be controlled. In many systems, the scattering off the optical elements, which is a function of the optical surface quality or microroughness, is the limiting factor in out-of-field stray light transmission to the focal plane.

The best way to measure surface scatter is a direct test of the scattering, such as bidirectional reflectance distribution function (BRDF) or bidirectional transmittance distribution function (BTDF). The method preferred in the absense of scattering measurement capability is surface roughness data, which generally correlates well with BR/TDF. Unfortunately, the cost in capital equipment and time associated with both of these tests (especially the BR/TDF tests) is often prohibitive. Most optical houses cannot perform either test. The optical industry standard is the scratch/dig reference. The scattering from these standards is visually compared to the scattering of the optical surface in question. While this is a good technique for

---

[*] Work performed under the auspices of the U.S. Department of Energy by the Lawrence Livermore National Laboratory under contract No. W-7405-ENG-48.

determining scatter from the prominent defects, it does not allow information on the average scattering over the entire surface.

Past work has been published about the confusion surrounding standard sets[1]. (Common scattering community perception is that quantitative BR/TDF can vary by up to a factor of ten between so called standards). While reference articles with attempts to malign the standard were readily available, no articles that characterize the scattering from the standards were found. No data was found relating the average scattering of a scratch or dig to an overall surface BRDF. This study was undertaken to quantify the typical amount of scattering from surface flaws and to find the resulting impact on average surface BR/TDF.

Experiment.

Two sets of scratch/dig standards were purchased from Beale Optical. Since the purpose of this project was to find typical scattering values of defects, variation between the two sets was not a concern. To the contrary, by inspecting 2 sets of reference marks, it was hoped some insight into the variation between "reference" standards could be gleaned. Our end goal was to determine how the surface quality specification of an optic purchased from the optical manufacturing community related to scattering from the entire surface.

The first step in the characterization of the two standard sets was a visual comparison with OCA's quality control standards, which were received from and have been annually recalibrated by the Frankfort Arsenal.[2] Apparent brightness was evaluated both in transmission and in reflection. Two evaluators compared the Beale sets to the OCA standard. One evaluator was a quality control inspector, familiar with the standard visual comparison through 15 years of optical QC experience; the other evaluator was the primary author of this paper. The second step in the characterization was scattering measurements

The general trend: the newer Beale standards of a particular flaw appear dimmer than reference flaws from the Frankfort Arsenal set. For example, one of the #80 scratches from the newer sets is visually equivalent to the #60 scratch from the Frankfort Arsenal reference set.

BTDF Measurements.

BTDF data was taken with an elliptical gaussian illumination beam, with 1/e points at ±0.23 mm vertical by ± 0.30 mm horizontal. Instrument profile of the scatterometer was performed for this beam diameter, and was saved for comparison.

The illumination spot was centered on the defect with centration search precision of 1/10 spot diameter.

---

[1] Young, Matt. "The scratch standard is only a cosmetic standard", SPIE Vol. 1164 Surface Charaterization and Testing II (1989), pp. 185 - 189.

[2] Historically, scratched and digs were fabricated solely by and rated against a reference at the Frankfort Arsenal. Later, scratch/dig widths were given in MIL-0-13830A in an attempt create a universal reference that could be fabricated at remote facilities. This is generally considered to yield only marginally correct comparison flaws, with yearly Arsenal re-calibrations *de rigueur* procedure to ensure uniformity among scratch/dig QC sets.

Centering is determined by maximizing the observed scatter. Scattering from digs was found to be angularly invariant, so a single rotation position was measured for each sample. Scratches, however, scatter quite preferentially in the direction perpendicular to their axes. Scans are performed in this maximum scatter direction, and in the perpendicular direction. To determine how constant the stray light is along the scratch, measurements were taken at 2 points. Scratch alignment is performed visually by observing the scatter from the defect, rotating the sample first for maximum scattering in the horizontal scan axis, then for maximum scattering in the vertical axis. (The scatterometer always sweeps in the horizontal axis.) For reference of the sample scattering surface in the absence of the defect, the BTDF is scanned with the illumination spot centered on an area that is visually free of defects.

Data Analysis.

The goal of these experiments is to quantify the increase in scattering caused by a given defect level over a uniformly illuminated surface. Intuitively, it can be seen that the scattering of a surface under uniform illumination, I_0, is the sum of the scattering over smaller segments when each is illuminated with I_0. Figure 1 illustrates this principle.

BTDF is the energy scattered (per solid angle), divided by the energy impinging on a surface. Consider a surface broken into two sections, of areas A_1 and A_2, with a uniform illumination of total energy E impinging on the surface. The light scattered from A_1 at an angle q, $E_{s1}(q)$, is:

$$E_{s_1} = \frac{E \cdot A_1}{(A_1 + A_2)} \cdot BTDF_1 (q) \tag{1}$$

and the energy scattered from area A_2 is similarly:

$$E_{s_2} = \frac{A \cdot A_2}{(A_1 + A_2)} \cdot BTDF_2 (q) \tag{2}$$

The BTDF of the entire surface is $(E_{s1} + E_{s2}) / E$:

$$BTDF (q) = \frac{A_1 \cdot BTDF_1 (q) + A_2 \cdot BTDF_2}{A_1 + A_2} \tag{3}$$

Since scratches and digs occupy only a small area fraction of the optical surface, by definition of the cosmetic standard which limits the number of defects in any 20 mm observation diameter, the average BR/TDF over the entire surface cannot be significantly affected unless the scattering is quite high from the local defect.

The amount of light scattered from a defect is proportional to the light incident on the defect. The BR/TDF, equal to scattered light divided by total incident light, is proportional to the light incident on the defect divided by the total illumination spot diameter. In effect, larger measurement beams dilute the influence of the defect with average scattering from the nominal surface. To maximize the BTDF measurement from a defect, illumination should be done with the minimum reasonable beam diameter. This ensures that the fraction of scattered light divided by incident light is a maximum, increasing the local

BTDF measurement to hopefully higher than that of the average surface.

In the CASI-C2 scatterometer, the illumination beam is not a constant: it is a gaussian. This alters the previous equations (Eqs. 1 - 3). With the largest scratch nominally 8 μm wide, no concern exists that the illumination varies over the scattering width. Normalization is, however, required for the variations along the beam diameter over the length of a scratch, and for the higher flux density of the central beam portion over the diameter of a dig. At the larger dig diameters, up to 0.5 mm, the illumination density does vary across the defect, making interpretation of the results somewhat ambiguous: scattering from the center/edge of the dig will be based on varying illumination intensity.

Figure 2 illustrates the averaging of the effective scattering over a 20 mm diameter surface area from a BTDF measurement with a smaller gaussian illumination beam. Variance in illumination intensity is seen along the length of the scratch. Considering the definition of cosmetic finish, one 20 mm long scratch can exist within a 20 mm diameter clear aperture. If a scratch were illuminated with a flux density source, I_0 {W/cm^2}, the amount of light scattered into a particular viewing solid angle would be BTDF $\cdot I_0 \cdot W \cdot L$, where W and L are the width and length of the scratch. This is the integral along the length of the scratch of:

$$E_s = \int_l BTDF \cdot I_o \cdot W \, dl \tag{4}$$

With gaussian illumination, I is a function of L, but not W as mentioned earlier. Note that the experimental beam was an elliptical gaussian, so vertical and horizontal scratch orientations must be treated separately. For beam 1/e radius w_0 in the direction of the scratch width, and 1/e radius of l_0 along the length of the defect, and a total energy of E_0, the beam illumination intensity function is given by:

$$I(w,l) = \frac{E_o}{P \, w_o l_o} \cdot e^{-\left\{\left(\frac{w^2}{w_o}\right)+\left(\frac{l}{l_o}\right)^2\right\}} \tag{5}$$

Integrating along the length of the scratch over a 20 mm maximum length to find the average flux density on the scratch:

$$\frac{1}{WL} \cdot \int_{-w_s}^{w_s}\int_{-10\,mm}^{+10\,mm} I(w,l) \, dl \, dw$$

$$\frac{1}{P \, w_o \, l_o \, WL} \cdot E_o \, P \, w_o \, l_o \cdot erf\left(\frac{w_s}{w_o}\right) \cdot erf\left(\frac{10 \, mm}{l_o}\right)$$

$$\frac{E_o}{WL} \cdot erf\left(\frac{w_s}{w_o}\right) \cdot erf\left(\frac{10 \, mm}{l_o}\right)$$

$$@\frac{E_o}{WL} \cdot erf\left(\frac{w_s}{w_o}\right) \tag{6}$$

This simplification of erf { 10 mm/ l_0 } is possible for the small gaussian beam radii (l_0) used in our experiment.

The energy impinging on the defect is:

$$E_o \cdot \text{erf}\left(\frac{W_s}{W_o}\right)$$

(7)

From Eqs. 1 - 3, recall that the total scattering from a surface is a function of the energy impinging on the surface. The scatterometer instrumentation registers the ratio of scattered light to total light incident on the measurement piece- not just the total light incident on the defect. The measured BTDF values must thus be multiplied by a correction factor that accounts for the ratio of energy on the defect to total illumination energy to find the true, or corrected, BTDF, before incorporation in equations 1, 2, and 3.

$$BTDF_{corrected} = \frac{BTDF_{meas}}{\text{erf}\left(\frac{W_s}{W_o}\right)}$$

(8)

For a constant illumination intensity, I_s, the energy impinging on the scratch, E_{scr}, would be $I_s \cdot WL$, and the amount of scattered energy would be $E_{scr} \cdot BTDF_{corrected}$. The scattered energy from the normal surface finish over the 20 mm defect definition area is :

$$BTDF_{surf} \cdot I_s \cdot \pi/4 \cdot (20 \text{ mm})^2$$

(9)

where BTDF is the BTDF of the base surface without defects. The percentage of increase in scattered light is:

$$\% \text{ increase BTDF} = \frac{100 \cdot I_s \cdot BTDF_{meas} \cdot WL}{\text{erf}\left(\frac{W_s}{W_o}\right) \cdot BTDF_{surf} \cdot I_s \cdot \frac{P}{4} \cdot (20 \text{ mm})^2}$$

$$= \frac{100 \cdot BTDF_{meas} \cdot WL}{BTDF_{surf} \cdot \text{erf}\left(\frac{W_s}{W_o}\right) \cdot \frac{P}{4} \cdot (20 \text{ mm})^2}$$

(10)

Figure 3 shows the scattering from a #80 scratch perpendicular to the scratch axis. The width, W, of this defect is defined to be 8 μm, for a w_s of 4 μm. In the experimental set up, the gaussian beam half width is 230 μm vertically and is 300 μm horizontally. It can be seen from the figure that the BTDF slope is actually steeper than -2. This is characteristic of scratch measurements. A second example is given in figure 4. Digs tend to have BTDF slope shallower than -2, as shown in the examples of Figures 5 and 6. The ideal comparison (equation 10) would present impact on BRDF as function of angle. The data presentation in Table 1 summarizes the scattering by selecting a small, fixed range of angles, and extrapolating the BTDF value to the 0.57° scattering angle. This allows a single "$BTDF_{meas}$", reported as B_0 at 0.57°, to compress data. The value is a conservative fit (highest reasonable value) to scattering data in the 5° to 40° angle range.

For a well polished surface at the measurement wavelength of 633 nm, the B_0 value is generally taken as 0.05. Thus, the percentage scattering increase of the defect and base over a 20 mm dia. is the measured B_0 value of the scratch times 69. This is an increase of average surface scatter of up to 4200 percent (42X increase) for one of the #80 scratch references. Table 2 gives the corrected BTDF of each scratch, then the percentage of increase in scattering over a "high quality" 0.05 B_0 surface for one 20 mm long scratch per 20 mm dia. Note that data in Table 2 is given only for scattering perpendicular to the scratch axis. Measurements with a concentrated spot outside a few degrees from perpendicular showed no appreciable increase over the base surface scatter.

For digs, a different set of equations is needed to correlate the overall increase in surface scattering with measured BTDFs. Since the 2-axis gaussian test beam widths are very similar, the illumination function can be approximated by a symmetric gaussian with the average beam radius with little loss of precision. The illumination intensity is modeled by:

$$I(r) = \frac{E_o \cdot e^{-\left(\frac{r}{r_o}\right)^2}}{P \, r_o^2} \tag{11}$$

(Variations between references have more impact than this illumination simplification. This allows simpler math.) The average illumination intensity over a dig is then:

$$I_{ave} = \frac{1}{P \, r_d^2} \cdot \int_o^{r_d} \int_o^{2P} 2Pr \, I(r,q) \, dq \, dr \tag{12}$$

The energy incident on the dig is $I_{ave} \cdot p \, r_d^2 = E_o \cdot \left\{1 - e^{-\left(\frac{r_d}{r_o}\right)^2}\right\}$

To correct BTDF, as with the scratch measurements, the measured values must be multiplied by the ratio of the total light incident on the test piece divided by the light incident on the defect. This yields:

$$BTDF_{corrected} = \frac{BTDF_{meas}}{1 - e^{-\left(\frac{r_d}{r_o}\right)^2}} \tag{14}$$

The percentage of increase in scattered light for a dig defect over a 20 mm diameter area is:

$$\% \text{ increase BTDF} = \frac{100 \cdot I_s \cdot p \, r_d^2 \cdot BTDF_{meas}}{\left(1 - e^{-\left(\frac{r_d}{r_o}\right)^2}\right) \cdot BTDF_{surf} \cdot I_s \cdot \frac{p}{4} \cdot (20 \text{ mm})^2}$$

$$= \frac{400 \, BTDF_{meas} \cdot r_d^2}{BTDF_{surf} \cdot \left(1 - e^{-\left(\frac{r_d}{r_o}\right)^2}\right) \cdot (20 \text{ mm})^2} \tag{15}$$

As mentioned before, the illumination density varies substantially over the diameter of the larger digs. With the mean 1/e radius of the illumination equal to 260 μm, the illumination at the outer diameter of a

#50 dig is 40% of the intensity at the center of the dig. Under such circumstances, it would be desirable to know the empirical mechanism for scattering. Specifically, we would want to know how the edge's contribution compared to the center's. This might be done in the future by measuring a single dig with varying gaussian illumination profiles. Correlation between scattering and both average illumination density and illumination density at 250 µm radius can be done. These correlations were not made in this series of tests. In lieu of this, one could examine the variance between digs to determine whether scattering varies with the periphery of the dig and the illumination density at the dig diameter, or with the dig area and the average illumination over the dig. Equations 12 - 14 are developed based on average intensity over the dig area. If this approach is correct, one would expect the corrected BTDF to be relatively constant between dig samples. Table 3 shows the corrected BTDF (based on average illumination density over the dig area), and the percentage of average scatter increase compared to a "high quality" surface without defects. Note that $BTDF_{corrected}$, while not constant, has no readily apparent correlation with dig area.

Alternatively, the scattering can be correlated to the illumination density at the periphery and the length of the dig periphery. In such a case, the illumination density at the periphery:

$$I_{periphery} = \frac{E_o}{p\, r_d^2} \bullet e^{-\left(\frac{r_d}{r_o}\right)^2}$$

(16)

and the corrected BTDF would be the scattered light divided by the total light incident on the dig if $I_{periphery}$ were the uniform illumination density over the dig:

$$BTDF_{corrected} = BTDF_{meas} \bullet \left(\frac{r_o}{r_d}\right)^2 \bullet e^{\left(\frac{r_d}{r_o}\right)^2}$$

(17)

Table 4 lists $BTDF_{corrected}$ based on periphery illumination domination of scatter and the percentage of average scatter increase compared to a "high quality" surface without defects. Note that $BTDF_{corrected}$ varies quite similarly with the calculations on Table 3. Since this table's analysis assumes an edge function, the BTDF should not be expected to be constant for the various dig diameters, as would an area-scattering dig. The corrected BTDF would be expected to vary inversely with dig radius, since the BTDF definition is based on the total energy incident on the defect (varying with diameter^2) while scattering would actually be varying with periphery, which varies linearly with diameter.

Comparing the data correlation in Tables 3 and 4 for the methods of anticipating dig scatter, it is unclear that either of the two dig scattering assumptions is superior to the other. Apparently, scattering variation between the digs is quite large. Definitive empirical scattering mechanism studies would have to be done on a single dig, varying the average to edge illumination ratio. For now, it can be stated that either analytic assumption will yield equally acceptable results.

Conclusions.

The single #50 dig within a 20 mm diameter aperture will increase the overall scattering level of an otherwise 0.05 B_0 {ster^{-1}} smooth surface by approximately a factor of 2 to 3. A #40 dig induces an increase of 2X; a #20 dig 1X; a #10 dig 10%; and a #5 dig 2%. A single large dig in 20 mm diameter will noticeably increase the average scattering of an otherwise high quality surface. The smaller digs do not severely impact the scattering of an optical surface.

Scratches scatter directionally, perpendicular to the scratch axis. A single 20 mm long #80 scratch will increase overall "smooth" surface 0.05 B_0 by 20X to 40X in the maximum scattering direction. No noticeable increase is seen except within a few degrees of perpendicular to the scratch. A #60 scratch increases scatter by approximately 20X; a #40 by 10X; a #20 by 3X, and a #10 by 5X.

Whether the scratch scattering adversely affects system stray light suppression is unclear. While BTDF is certainly increased in one direction, the increase in total scattering may be a more significant measure of system impact, pending stray light propagation path geometry. A suggestion for future experiments would be to conduct TIS measurements with the scratches, and similarly over a "smooth" section of the reference surface, then normalize the effect over the 20 mm diameter inspection criteria as a lower bound for scattering impact.

Variations between reference standards are clearly evident, with the scattering from the #10 scratches actually measuring higher than from the #20 scratches.

Typical data presented in this paper does allow general conclusions about the area-averaged scattering impacts of cosmetic defects. Thus, the significance of a scatch/dig on overall scatter can now be made. On the other hand, comparisons of a surface with a scratch dig standard do not give microroughness data. Meeting a scratch/dig comparison only means that the surface may be as low as the scattering levels measured on the isolated defects (averaged over the surface), but could be much higher if the microroughness is large.

Please note: Since the raw data from the scratches/digs does not follow a $\{ \sin(theta) \}^{-2}$ mapping, simple comparisons with the "B_0" of a nominal low-scatter surface should ideally be adjusted for the scattering angle of greatest concern. To generate the summary tables in this paper, the scattering data in the 10° to 30° from specular region was extrapolated at a "-2" slope.[3]

---

[3] "-2" refers to the exponent n in $BTDF = B_0 \cdot \{ [\sin(theta) - \sin(theta_0)] / 0.01 \}^n$

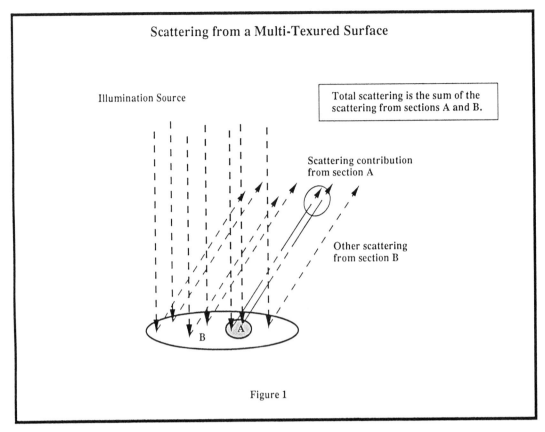

Figure 1

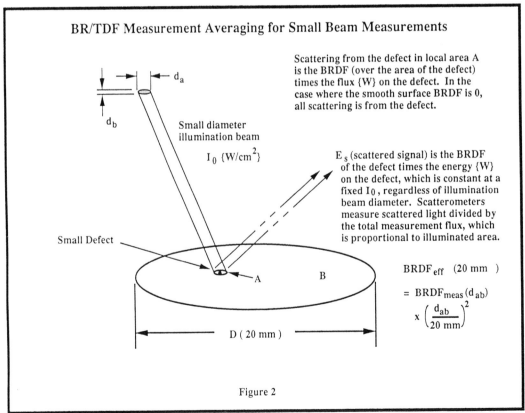

Figure 2

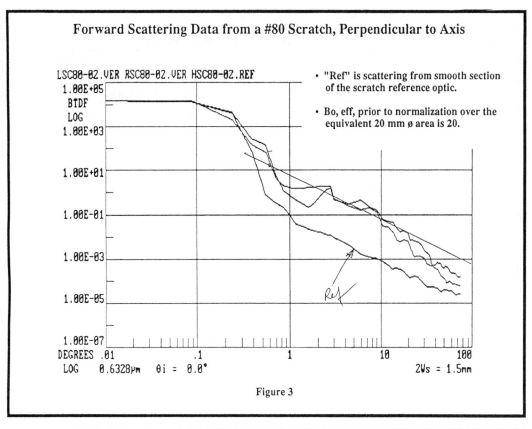

Forward Scattering Data from a #80 Scratch, Perpendicular to Axis

Figure 3

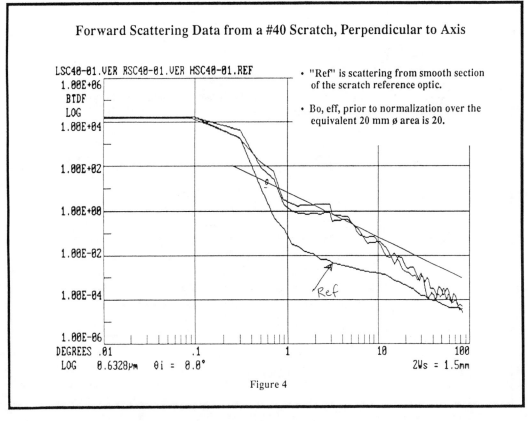

Forward Scattering Data from a #40 Scratch, Perpendicular to Axis

Figure 4

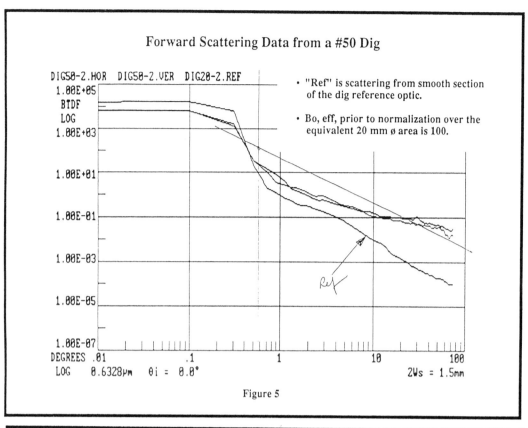

Forward Scattering Data from a #50 Dig

DIG50-2.HOR DIG50-2.VER DIG20-2.REF

- "Ref" is scattering from smooth section of the dig reference optic.

- Bo, eff, prior to normalization over the equivalent 20 mm ø area is 100.

Figure 5

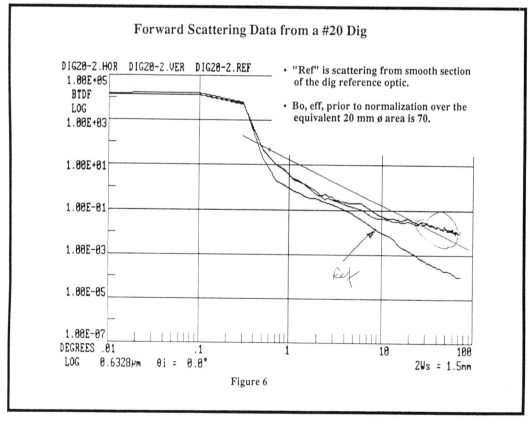

Forward Scattering Data from a #20 Dig

DIG20-2.HOR DIG20-2.VER DIG20-2.REF

- "Ref" is scattering from smooth section of the dig reference optic.

- Bo, eff, prior to normalization over the equivalent 20 mm ø area is 70.

Figure 6

Table 1
Visual Inspection Matrix of the Scratch/Dig Sets

Beale Optical sets were compared to OCA's Frankfort Arsenal reference set. The left column (defects) are the stated values of the Beale sets. The data given in the columns is the Frankfort Arsenal reference mark that is visually equivalent. In some of the Beale scratches, variation was seen along the defect. In these instances, the minimum and the maximum are given. Defects were compared to the reference both in transmission and in reflection; transmissive/reflective differences are given only as noted.

Defect	Inspection	Set #1 Equivalent	Set #2 Equivalent
#50 Dig:	reflective (visual)	50	50
	transmissive (visual)	between 40 & 50	between 40 & 50
#40 Dig	reflective/transmissive	40	40
#20 Dig	reflectivetransmissive	20	20
#10 Dig	reflective/transmissive	between 5 & 10	10
#05 Dig	reflective/transmissive	5	<5 (barely visible)
#80 Scratch	reflective	between 60 & 80	60
	transmissive	between 40 & 80	60
#60 Scratch	reflective/transmissive	60	40
#40 Scratch	reflective/transmissive	40	40
#20 Scratch	reflective	10	20
	transmissive	10	between 10 & 20
#10 Scratch	reflective/transmissive	between 10 & 20	20

Table 2
BTDF of Scratches, Perpendicular to Scratch Axis

This table contains raw and corrected B_0 data for each scratch of 2 sets of standards. The first column, $BTDF_{meas}$, reports the B_0 as printed form the CASI scatterometer. The second column, $BTDF_{corrected}$, lists the BTDF that would have been measured if the scratch were exactly filled with constant illumination flux. The third column is the increase in BTDF over a 20 mm surface of nominal $B_0 = 0.05$ of the base surface (absent from defects) when a single 20 mm long scratch is located within the illuminated clear aperture.

Scratch	$BTDF_{meas}$	$BTDF_{corrected}$	% average B_0 increase
#80, set 1	60	4100	4200
#80, set 2	20	1400	1400
#60, set 1	30	2700	2100
#60, set 2	30	2700	2100
#40, set 1	20	2700	1400
#40, set 2	10	1400	700
#20, set 1	4	1100	280
#20, set 2	5	1400	350
#10, set 1	4	2200	280
#10, set 2	10	5500	700

Table 3
BTDF of Digs, Corrected for Area Scattering

This table contains raw and corrected B_0 data for each dig of 2 sets of standards. The first column, $BTDF_{meas}$, reports the B_0 as printed form the CASI scatterometer. The second column, $BTDF_{corrected}$, lists the BTDF that would have been measured if the scratch were exactly filled with constant illumination flux, assuming that the mechanism for scattering is functionally equivalent to scattering from the area of the dig. The third column is the increase in BTDF over a 20 mm surface of nominal $B_0 = 0.05$ of the base surface (absent from defects) when a single dig is located within the illuminated clear aperture.

Scratch	$BTDF_{meas}$	$BTDF_{corrected}$	% average B_0 increase
#50, set 1	60	100	130
#50, set 2	100	170	210
#40, set 1	100	230	180
#40, set 2	100	230	180
#20, set 1	70	520	100
#20, set 2	40	300	60
#10, set 1	5	140	7
#10, set 2	7	200	10
#5, set 1	2	220	2.8
#5, set 2	0.7	80	1.0

Table 4
BTDF of Digs, Corrected for Edge Scattering

This table contains raw and corrected B_0 data for each dig of 2 sets of standards. The first column, $BTDF_{meas}$, reports the B_0 as printed form the CASI scatterometer. The second column, $BTDF_{corrected}$, lists the BTDF that would have been measured if the scratch were exactly filled with constant illumination flux, assuming that the mechanism for scattering is functionally equivalent to scattering from the edge of the dig. The third column is the increase in BTDF over a 20 mm surface of nominal $B_0 = 0.05$ of the base surface (absent from defects) when a single dig is located within the illuminated clear aperture.

Scratch	$BTDF_{meas}$	$BTDF_{corrected}$	% average B_0 increase
#50, set 1	60	160	210
#50, set 2	100	270	340
#40, set 1	100	310	240
#40, set 2	100	310	240
#20, set 1	70	560	110
#20, set 2	40	320	65
#10, set 1	5	140	7
#10, set 2	7	200	10
#5, set 1	2	220	2.8
#5, set 2	0.7	80	1.0

Scattering from multilayer coatings: a linear systems model

James E. Harvey and Kristin L. Lewotsky

The Center for Research in Electro-optics and Lasers (CREOL)
The University of Central Florida
12424 Research Parkway
Orlando, Florida 32826

ABSTRACT

It is finally being recognized that residual surface roughness over the *entire range of relevant spatial frequencies* must be specified and controlled in many precision optical systems. This includes the "mid" spatial frequency surface errors that span the gap between the traditional "figure" and "finish" errors. This is particularly true for enhanced reflectance multilayers if both high reflectance and high spatial resolution are desired. If we assume that the interfaces making up a multilayer coating are uncorrelated at high spatial frequencies (microroughness) and perfectly correlated at low and mid spatial frequencies, then the multilayer can be thought of as a surface power spectral density (PSD) filter function. Multilayer coatings thus behave as a low-pass spatial frequency filter acting upon the substrate PSD, with the exact location and shape of this cut-off being material and process dependent. This concept allows us to apply conventional linear systems techniques to the evaluation of image quality, and to the derivation of optical fabrication tolerances, for applications utilizing multilayer coatings.

1. INTRODUCTION

When light is reflected from an imperfect optical surface, the reflected radiation consists of a specularly reflected component and a diffusely reflected component as illustrated in Figure 1. The light scattered from optical surface irregularities degrades optical performance in several different ways: a.) it reduces optical throughput since some of the scattered radiation will not even reach the focal plane, b.) the wide-angle scatter will produce a veiling glare which reduces image contrast or signal-to-noise ratio, and c.) the small-angle scatter will decrease resolution by producing an image blur.

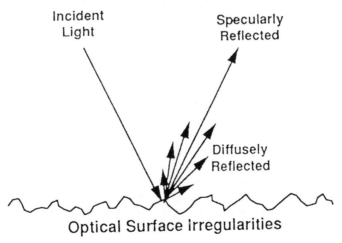

Figure 1. Optical surface irregularities produce a specularly reflected beam with a diffusely reflected component that can degrade optical performance in several different ways.

Although any one of the above effects may dominate the optical performance degradation in a given application, it is finally being recognized that residual surface roughness over the *entire range of relevant spatial frequencies* must be specified and controlled in many precision optical systems. This includes the "mid" spatial frequency surface errors that span the gap between the traditional "figure" and "finish" errors and is particularly true for enhanced reflectance multilayers if both high reflectance and high spatial resolution are desired. Figure 2 illustrates an optical surface profile with rms roughness σ, the spatial frequency spectrum of that surface obtained by Fourier transforming the surface profile, the surface power spectrum or PSD function, and the surface autocovariance (ACV) function. This schematic diagram also shows the relationship between each of these functions (the autocorrelation operation is indicated by the symbol "$\bigstar$", and the Fourier transform operation by the symbol "$\Leftrightarrow$"). Note that the peak value of the surface autocovariance function is σ^2, and from the central ordinate theorem of Fourier transform theory the volume under the surface PSD is also σ^2. In this paper we will try to develop a linear systems model of surface scatter phenomena in enhanced reflectance multilayer coatings

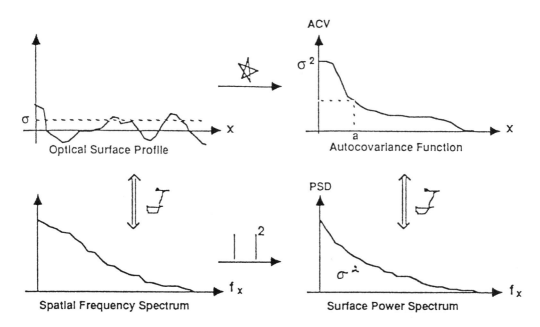

Figure 2. Illustration showing the relationship between the optical surface profile, the surface power spectral density (PSD) function and the surface autocovariance (ACV) function.

2. MULTILAYER SCATTERING CHARACTERISTICS

The reflectance, at normal incidence, from a single interface (as illustrated in Figure 1) can be calculated using the following Fresnel reflectance formula, if the refractive indices of the respective media separated by the interface are known.

$$R = |r|^2 = (n_2 - n_1)^2 / (n_2 + n_1)^2 \tag{1}$$

This value for reflectance is valid whether the interface is rough or not. In the case of a rough interface, the reflected wavefront takes on a disturbance of the same form as the interface but of twice the amplitude. As a result, some fraction of the light will be scattered away from the specular direction. We represent this fact by writing the specularly reflected power as

$$P_s = R A P_o \tag{2}$$

and the diffusely scattered power as

$$P_d = R B P_o \tag{3}$$

where R is the Fresnel reflectance, P_o is the incident power, and

$$A = \exp[-(4\pi\sigma/\lambda)^2] \qquad (4)$$

and

$$B = \text{total integrated scatter} = 1 - A \qquad (5)$$

are the fraction of reflected power in the specular and scattered beams, respectively.[1,2] Here σ is the rms surface roughness over the entire range of relevant spatial frequencies, from high spatial frequency microroughness, through mid spatial frequency surface irregularities, and including low spatial frequency figure errors

$$\sigma^2 = \sigma_L^2 + \sigma_M^2 + \sigma_H^2 \qquad (6)$$

Similarly, the radiation reflected by an enhanced reflectance multilayer will consist of a specular and diffuse part as shown in Figure 3. The specularly reflected and diffusely reflected components are still given by equations of the form of Eqs. (2) and (3); however, the relative strength of the specular beam will now depend upon the 'effective' roughness, σ', which in turn depends upon the degree of *correlation* between the various interfaces. Hence, for the multilayer

$$A' = \exp[-(4\pi\sigma'/\lambda)^2] \qquad (7)$$

and

$$B' = \text{total integrated scatter} = 1 - A' \qquad (8)$$

where

$$\sigma'^2 = \sigma'_L{}^2 + \sigma'_M{}^2 + \sigma'_H{}^2 \qquad (9)$$

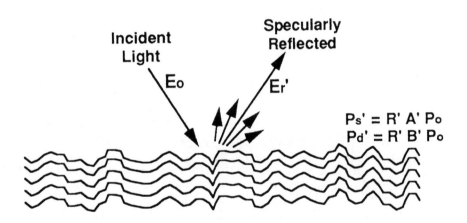

Figure 3. An enhanced reflectance multilayer also produces a specularly reflected beam and a diffusely reflected component whose relative strengths depend upon the degree of correlation between the various interfaces.

Eastman (1974), Carniglia (1979), Elson (1980), Spiller (1986), Sterns (1989), and Amra (1991) have all contributed to our understanding of scattering from optical surfaces with multilayer coatings.[3-8] However, in most cases they have dealt only with microroughness and have assumed that the interfaces are either perfectly correlated or perfectly uncorrelated. Much of the recent interest in scattering from multilayer structures has been with regard to the rapidly emerging technology of enhanced reflectance coatings for soft x-ray applications.[7,9] It has been pointed out by Spiller that spatially uncorrelated microroughness in X-ray/EUV multilayers will yield an "effective" rms surface roughness reduced by a factor of $1/\sqrt{N}$, where N is the number of layer pairs.[6] The wide-angle scatter produced by this microroughness will decimate absolute reflectance before significantly degrading image quality. The small-angle scatter produced by correlated mid spatial frequency surface irregularities does not significantly reduce the absolute reflectance; however, it can severely degrade image quality (depending upon the correlation width of the surface irregularities).[10]

3. THE MULTILAYER AS A SURFACE PSD FILTER FUNCTION

Let us assume that the surface PSD of the substrate upon which a multilayer is to be deposited is described by an inverse power law as illustrated in Figure 4

$$PSD(f) = \frac{h f_o^{\alpha}}{(f + f_o)^{-\alpha}}$$

(10)

Note that this is a radial profile of a two-dimensional PSD plotted as Log Power expressed in waves squared per spatial frequency squared versus spatial frequency. The units are thus consistent with a volume under the PSD of σ^2 as required. The low, mid, and high spatial frequency domains are indicated in the figure. Furthermore, knowing the functional form of the surface PSD now enables one to calculate the total rms surface error σ from a band-limited measurement of the microroughness σ_{μ} with an instrument such as a micro phase-measuring interferometer.

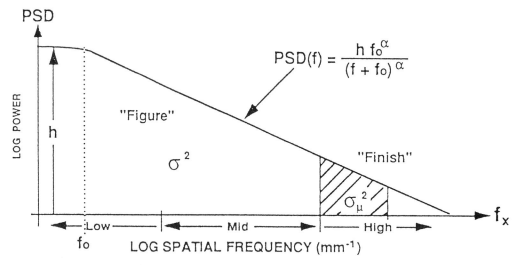

Figure 4. The substrate surface PSD is assumed to obey an inverse power law which spans the entire range of spatial frequencies from low spatial frequency "figure" errors to high spatial frequency "finish".

Since we are considering the surface irregularities over the entire range of relevant spatial frequencies from the very low spatial frequency "figure" errors to the very high spatial frequency "finish" errors. It seems rather intuitive that for any reasonable thin-film deposition process, the low spatial frequency figure errors will *print through* and be correlated from layer to layer, while the high spatial frequency microroughness inherent to the deposition process itself will be uncorrelated from layer to layer. Let us also assume that each interface making up the multilayer has the same surface statistics.

From the previous description of uncorrelated microroughness on multilayer interfaces resulting in an 'effective' roughness reduced by the factor of $1/\sqrt{N}$, and the knowledge that the interfaces will be highly correlated at mid and low spatial frequencies, it is clear that the multilayer will act as a low-pass spatial frequency filter which has a value of unity for correlated low spatial frequencies and drops to a value of $1/N$ for uncorrelated high spatial frequencies. The exact location and shape of this cut-off depends upon the material and the deposition process. This PSD filter function is illustrated qualitatively in Figure 5(a). Figure 5(b) illustrates the effective multilayer PSD obtained by multiplying the filter function by the interface PSD. Note that the microroughness (and hence the wide-angle scatter) has been substantially reduced.

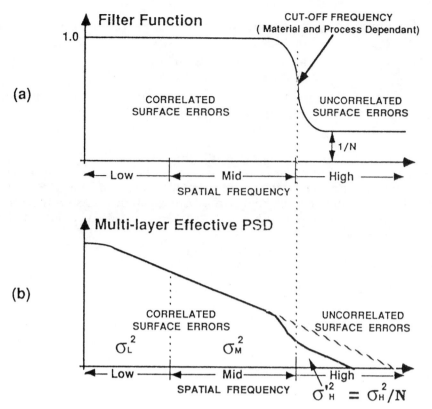

Figure 5. (a) Enhanced reflectance multilayer coatings behave as a low-pass spatial frequency filter acting upon the interface PSD. (b) The filtered interface PSD is the effective PSD of the multilayer.

4. IMAGE QUALITY PREDICTIONS

Assuming that the scattering in the multilayer is not so bad as to severely degrade the reflectance, let us turn our attention to the quality of the image produced. The image of a point source, or point spread function (PSF), will consist of a central image core surrounded by a scattered light halo produced by the surface irregularities (residual optical fabrication errors) of the optical components comprising the imaging system.

When the smooth surface approximation is valid ($\sigma \ll \lambda$), the scattering function is proportional to the effective multilayer PSD as illustrated in Figure 6.

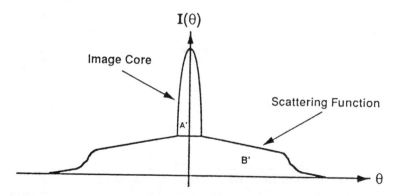

Figure 6. In the smooth surface approximation, the PSF is made up of an image core and a scattering function proportional to the filtered PSD. The fractional encircled energy is also indicated.

The image intensity distribution or point spread function (PSF) can therefore be written as

$$I(\theta) = A' I_c(\theta) + (B'/\sigma'^2) \, PSD(\theta) \tag{11}$$

where $I_c(\theta)$ is the image core intensity distribution as determined by diffraction, geometrical aberrations, misalignment errors, and any environmental errors such as thermal distortion or jitter.[10]

If the smooth surface approximation is not valid (which is very likely for extreme ultraviolet and soft x-ray applications), a scalar diffraction approach to surface scattering phenomena can still be applied but the scattering function is no longer proportional to the surface PSD. Instead, the image characteristics are now illustrated in Figure 7 where the system transfer function is given by the product of the transfer function of the image core with the transfer function of the effective optical surface[11]

$$H(\hat{x},\hat{y}) = H_c(\hat{x},\hat{y}) \; H_s(\hat{x},\hat{y}) \tag{12}$$

This surface transfer function is given by

$$H_s(\hat{x},\hat{y}) = \exp\{-(2\pi\hat{\sigma}_w)^2 \, [1 - C_w(\hat{x},\hat{y})/\sigma_w^2]\} \tag{13}$$

where the wavefront autocovariance function for normal incidence is just twice the Fourier transform of the effective PSD of the multilayer illustrated in Figure 4 (b).[2,12]

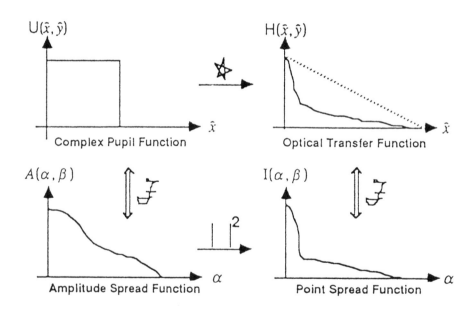

Figure 7. Illustration showing the relationship between the complex pupil function, the point spread function, and the optical transfer function.

Different image quality criteria might be imposed for different applications. For example, an x-ray astronomy application might have a requirement specified in terms of fractional encircled energy whereas a soft x-ray microlithography application might have a requirement specified in terms of the square wave response (modulation in the image of a three-bar target) at a particular wavelength and spatial frequency. These different image quality criteria can obviously be modeled once the PSF is known.

5. SUMMARY

We have reviewed multilayer scattering characteristics and have cited references which state that reflectance and image quality (spatial resolution) are degraded by optical fabrication errors in totally different spatial frequency domains. By assuming that the interfaces making up a multilayer coating are uncorrelated at the high spatial frequencies (microroughness) and perfectly correlated at the low and mid spatial frequencies, we have shown that a multilayer can be thought of as a surface PSD filter function. Multilayer coatings thus behave as a low-pass spatial frequency filter acting upon the interface PSD, with the exact location and shape of this cut-off being material and process dependent. By multiplying the PSD of a typical interface in a multilayer coating by this PSD filter function, the effective PSD of the multilayer is obtained. When the smooth surface approximation is valid, the image characteristics follow directly from the specular reflectance and the shape of this effective multilayer PSD. For very short wavelength applications where the smooth surface approximation is not valid, determining the image characteristics is slightly more involved. The effective PSD must now be Fourier transformed to obtain the effective surface autocovariance function. This autocovariance function is then used to calculate the effective surface transfer function, which is then multiplied by the conventional transfer function characterizing the diffraction and aberration properties of the imaging system. This product then comprises the system transfer function which is Fourier transformed to yield the generalized point spread function which includes scattering effects. The appropriate image quality criterion for a given application can then be evaluated as a function of the optical fabrication parameters and compared to optical performance requirements. This process will allow appropriate optical fabrication tolerances to be derived.

6. REFERENCES

1. H. E. Bennett and J. O. Porteus, "Relation between Surface Roughness and Specular Reflectance at Normal Incidence", J. Opt. Soc. Am. **51**, 123 (1961).

2. J. E. Harvey, "Light-Scattering Characteristics of Optical Surfaces," Ph.D Dissertation, University of Arizona (1976), (Available from University Microfilms, Ann Arbor, MI 48106).

3. J. M. Eastman, "Surface Scattering in Optical Interface Coatings", Ph.D. Dissertation, University of Rochester, Rochester, NY (1974), (Available from University Microfilms, Ann Arbor, MI 48106).

4. C. K. Carniglia, "Scalar Scattering Theory for Multilayer Optical Coatings", Opt. Eng. **18**, 104 (1979).

5. J. M. Elson, J. P. Rahn, and J. M. Bennett, "Light Scattering from Multilayer Optics: Comparison of Theory and Experiment", Appl. Opt. **19**, 669 (1980).

6. E. Spiller, A. E. Rosenbluth, "Determination of Thickness Errors and Boundary Roughness from the Measured Performance of a Multilayer Coating," Opt. Eng. **25**, 898 (1986).

7. D. G. Stearns, "The Scattering of X-rays From Nonideal Multilayer Structures", J. Appl. Phys. **65** (2), 498 (15 Jan. 1989).

8. C. Amra, J. H. Apfel, and E. Pelletier, "The Role of Interface Correlation in Light Scattering by a Multilayer", submitted to Appl. Opt.

9. T. W. Barbee, Jr., "Multilayers for X-ray Optics", Opt. Eng. 25, 954 (1986).

10. J. E. Harvey, W. P. Zmek, and C. Ftaclas, "Imaging Capabilities of Normal-incidence X-ray Telescopes", Opt. Eng. **29**, 603 (1990).

11. J. E. Harvey, E. C. Moran and W. P. Zmek, "Transfer function characterization of grazing incidence optical systems", Appl. Opt. **27** (8), 1527-1533 (1988).

12. J. E. Harvey, "Surface scatter phenomena: a linear, shift-invariant process", Proc. SPIE 1165 (1989).

IR window damage measured by reflective scatter

Marvin Bernt
John C. Stover
TMA Technologies Inc.
Bozeman MT 59715
(406) 586-7684

ABSTRACT

It has been demonstrated that erosion of sensor window surfaces can be measured on the flight line with reflective scatter instrumentation.   At issue is the relationship between these damage sensitive measurements and the corresponding loss of system performance. Rain and sand erosion of IR sensor windows can limit system performance in three ways.   In the case of IRST's, background scatter from window defects increases the system noise floor which limits range.   Image resolution degrades in FLIR instrumentation as window erosion increases. Finally for both systems, severe damage can cause window breakage resulting in loss of the sensor system and possibly the aircraft. This paper reports the results of initial studies that correlate reflective scatter measurements to the loss of mid-IR performance. High angle and near angle transmissive scatter from window damage are responsible for different types of system degradation. Both are studied and related to reflective scatter measurements.

2. INTRODUCTION

Two types of optical surveillance systems are commonly found on modern military aircraft. These are the Infrared Search and Track (IRST) and the Forward Looking Infrared Receiver (FLIR). The former is used both night and day to find and track objects with a relatively high IR signal at long distances. Images consist of a bright point on a dark field.   The latter is used as a means of night vision so image detail is required. Background noise in these detection systems is caused by optical scatter, as well as by electronic noise. The effects of scatter can often be ignored if optics in the sensor field of view are relatively clean and defect free. Unfortunately, this is not always the situation with IRST's and FLIR's, because the system windows are subjected to an environment that includes rain, ice, sand, etc. Scatter from the resulting damage and contamination will degrade system operation.

The degree to which a window scatters is characterized by the bidirectional transmittance distribution function (BTDF)[1,2] as shown in Figure 2. It is simply the scatter density (watts/steradian) normalized by the incident power and is given in units of inverse steradians. The scatter signal is also divided by the cosine of the polar scatter angle (measured from surface normal). The BTDF of optical components can be measured in laboratories on scatterometers.

$$BTDF = \frac{P_s/\Omega}{P_i \cos\theta_s} \tag{1}$$

Once the BTDF is known at the wavelength of interest, the scatter signal into any solid angle, in any direction, can be found by simply solving Equation 1 for P_s.

The measured BTDF of damaged windows has been show to increase by two or three orders of magnitude in the IR[3]. An example is shown in Figure 1 where the 10.6 micrometer BTDF of damaged and undamaged ZnSe windows are compared. Notice that the near specular scatter is much larger than high angle scatter. High angle scatter of about 10^{-5} sr^{-1} at 10.6 micrometers is typical of undamaged windows. The

0-8194-0658-9/91/$4.00

increased BTDF can produce two undesirable effects. Near specular light scattered out of the target signal degrades the resulting image, and light scattered through large angles from non-target sources, such as the sun, increases the background noise floor. Window specifications are generally written around IR transmission - not BTDF. Neither of these can be measured without removing the window which often means removing the entire pod encasing the sensor - time and labor consuming process. This paper is concerned with measuring the quality of these windows using reflective scatter techniques that do not require removal of either the detection system or the window.

Following sections review the issues for IRST's and FLIR's, and present correlation data between reflective scatter and IR transmission specifications.

3. IRST's and FLIR's

The IRST geometry is shown in Figure 2, where the system is represented by a lens, field stop and sensor behind a window. The lens and field stop act to define the sensor field of view (FOV). The relatively narrow FOV is raster scanned over most of a hemisphere much like a radar. Points of high signal are located and displayed. As the window degrades, either through contamination or impact damage, the signal to noise ratio is effected in two ways. First, the transmission of signal to the sensor drops slightly, because small amounts of signal are scattered out of the field of view. This is not a serious effect - signal degradation is usually less than a few percent, and the image is generally a near point source on a dark field. The second effect is more serious. Bright sources of radiation outside of the field of view (such as the sun) that illuminate the window section inside the field of view, now scatter within the sensor field of view. Because the sun is much brighter than the target signal, the potential exists for a dramatic drop in signal to noise as window scatter increases. The effect is similar to driving your car with a dirty windshield. Close (high signal) objects are still easy to see. Objects farther away (low signal) are lost in the background light created from the sun scattering from the windshield. If the sensor FOV is moved closer to the sun, the resulting background noise increases. Thus the effect of window damage on the sensor system is to increase the minimum angle from the sun that a target can be tracked. The minimum angle between the tracked object and the sun depends on the FOV cone size, signal strength, and electronic noise floor, as well as the window BTDF. Details of this calculation, which have already been reported[3], reveal that a 0.1 degree FOV looking through a heavily damaged window will have a signal to noise ratio of less than one inside a 20° cone about the sun.

FLIR systems are designed to produce an image on an array camera and look into a much larger FOV. The objective is night-time vision, not search and track, thus the sun is not a problem. Image resolution slowly degrades due to near specular scatter caused by increasing window damage. The effect is similar to driving your car on a misty day. If the weather continues, the wipers eventually have to be turned on. In the case of the FLIR, the question is when - not if - because the window damage always gets progressively worse.

4. MEASUREMENT ISSUES

The advantage of reflective measurements is that the window can be inspected without removing it from the aircraft. This provides a large advantage in time and labor. The problem is relating reflective measurements to transmissive degradation. There are three key measurement issues.

Surface or Surface/Bulk Measurements?

When the windows damage, both the surface and the bulk material degrade. The relative effect on surface and bulk depends on the materials used for both the window and the coating. Both surface and bulk damage cause transmissive scatter and result in the types of system degradation previously described. For a given window material, a key choice is whether or not to choose an illumination wavelength that is absorbed by the window or transmitted by the window. The resulting signal in the first case is just from the surface damage. In the second case from a combination of bulk and surface damage.

Scatter or Specular Measurements?

The total reflective signal is composed of both specular and scatter contributions. The specular signals come from the front (and possibly back) surfaces. The scatter signal comes from the front surface and possibly the bulk. Little scatter signal comes from the back surface, even in the transmissive case because it does not damage. Changes in scatter are generally a more sensitive measure of changes in damage because as damage occurs the scatter signal can increase several orders of magnitude. Specular signals generally decrease by much smaller amounts. On the other hand the scatter signals are much smaller and more difficult to detect. As a result, in the mid-IR it may be impractical to measure scatter.

Partial or Complete Measurement Coverage?

The window initially degrades in discrete spots. Eventually the surface takes on a sand paper texture. At this point essentially the same BRDF value is found regardless of measurement location. The question is whether sample uniformity occurs before or after system degradation becomes an issue. If system degradation is not a problem until after uniform damage is present then a few measurements can be used as being representative of the window. If one or more discrete defects can cause the system to fail then the measurements must cover the entire window.

The answers to these questions must be decided on a case by case basis. For example, we have determined that AR coated ZnSe windows can be adequately characterized by measuring scatter from a visible laser in a few representative locations. This may not be the case if the same windows are protected with a hard carbon coating, and does not appear to be the case for carbon coated germanium windows. Each set of choices needs to be made on the basis of a correlation study. Initial data from two such studies are reviewed in the next section.

5. REFLECTIVE AND TRANSMISSIVE CORRELATIONS

Study efforts were undertaken to determine correlation between transmissive window behavior in the IR and reflective window behavior at visible wavelengths. Two window types tested were zinc sulfide on zinc selenide FLIR windows, and IRST windows of carbon coated germanium. The FLIR windows suffered primarily sand erosion from use in desert environments. The damage was in general a uniform sandblast with a minimum of outstanding local defects. The surface lent itself very well to reflective scatter measurements at a few representative locations. Correlation was made between visible reflective scatter at .670μm with a handheld scatterometer and specular transmission at 10.6μm. Correlation for the initial set of seven windows is shown in plot #3. The IRST windows were damaged by exposure to various environments while mounted on a whirling arm to simulate flight. Damage to the germanium samples was much less uniform than on the softer ZnS/ZnSe material, and consisted of larger local surface defects spaced randomly on a still smooth surface. The correlation shown in plot #4 shows 10.6μm scatter at 50° from specular plotted against .633μm scatter at 10° from specular. Two substrate finishes are compared, polished and diamond turned germanium. The data was taken using a illuminating spot of about 5mm, sufficiently large to average over several defects.

6. SUMMARY

Preliminary data shows good correlation between IR transmissive behavior and visible reflective scatter. The data presented suggests that a thorough study of each material type, finish type, and different damage mechanisms is required to develop a reliable fit. Presumably subsurface damage could be investigated in a similar fashion by proper choice of wavelength and polarization selection. Motivation for the aforementioned effort is the possibility of flightline or in-situ testing of windows that currently require removal from the aircraft for testing.

7. REFERENCES

1. Nicodemus, F.E., J.C. Richmond, J.J. Hsia, I.W. Ginsberg, T. Limperis. Geometric Considerations and Nomenclature for Reflectance. U.S. Dept. of Commerce, NBS Monograph 160 (1977).

2. Stover, J.C.. Optical Scattering: Measurement and Analysis; McGraw-Hill, NY (1990).

3. Stover, J.C., "Practical measurement of rain erosion and scatter from IR windows," Proc. SPIE 1325-16, July 1990.

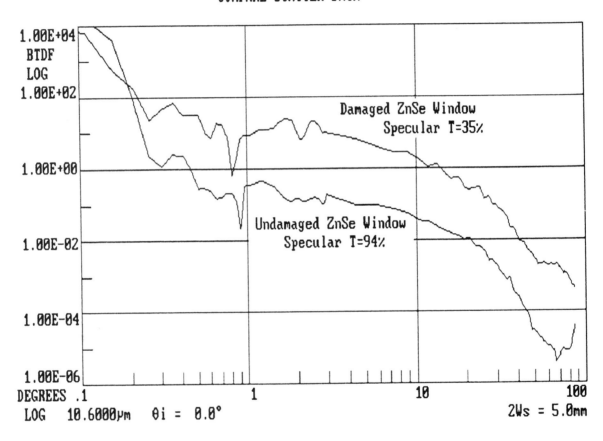

COMPARE SCATTER DATA

Damaged ZnSe Window
Specular T=35%

Undamaged ZnSe Window
Specular T=94%

BTDF LOG: 1.00E+04, 1.00E+02, 1.00E+00, 1.00E-02, 1.00E-04, 1.00E-06

DEGREES .1   1   10   100
LOG   10.6000μm   θi = 0.0°   2Ws = 5.0mm

Figure 1

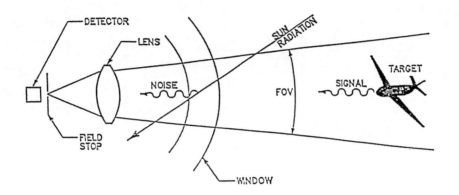

FIGURE 1. RADIATION FROM THE SUN CONTRIBUTES TO BACKGROUND NOISE AFTER IT SCATTERS FROM THE WINDOW.

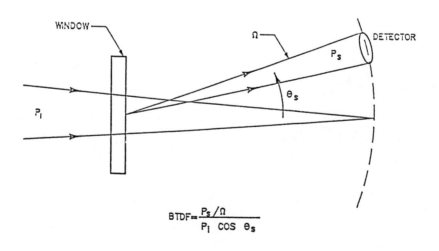

$$BTDF = \frac{P_s / \Omega}{P_l \cos \theta_s}$$

Figure 2

10.6μm Transmission : Visible Scatter

Polished ZnS on ZnSe

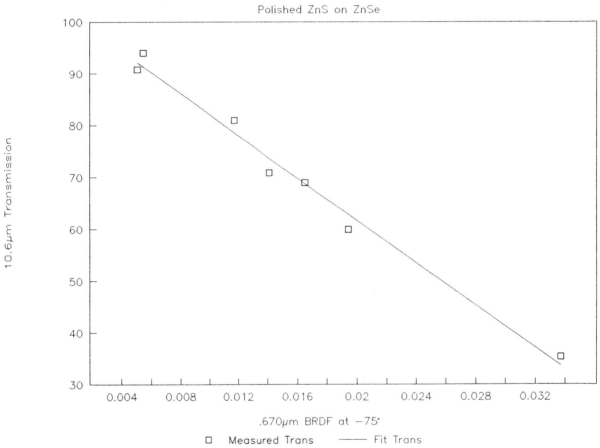

Figure 3

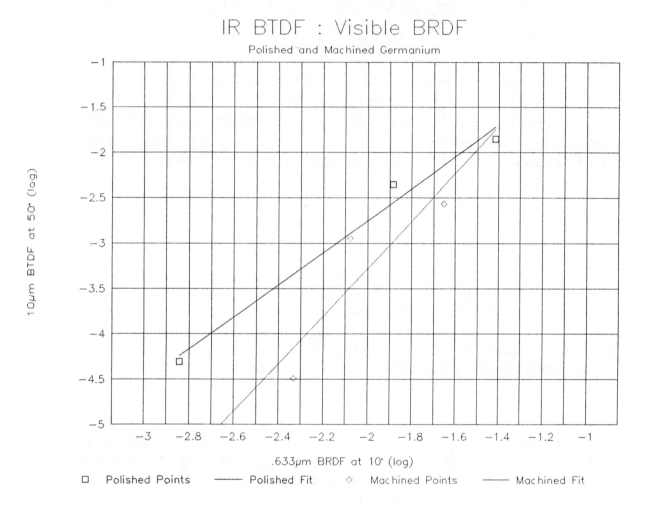

Figure 4

BACKSCATTERING IMAGE RESOLUTION AS A FUNCTION OF PARTICLE DENSITY

Paul Rochon and Daniel Bissonnette

Physics Department, Royal Military College of Canada
Kingston, Ontario, Canada K7K 5L0

ABSTRACT

Light which is emitted from a point source and is randomly scattered by an ensemble of particles is found to form an image of the point source. Results are presented for light which is emitted from a 25 *micrometer* pin hole and is then scattered by a suspension of .296 *micrometer* polystyrene spheres. This phenomenon is interpreted in terms of coherent multiple backscattering. The present study investigates the effects of particle density on the image resolution.

INTRODUCTION

There has been considerable interest in recent years in the phenomenon of enhanced backscattering of light by random media and rough surfaces[1-8]. It is found that when a light beam undergoes multiple scattering constructive interference effects give rise to an increase in the intensity of the light which is scattered in the backward direciton. Much progress has been made in the theoretical models for multiple scattering which include the above effect. Of particular interest are the effects of the properties of the scattering medium of the backscattered light since the later may then be used to characterize the medium.

We have recently shown[9,10] that the backscattering phenomenon can also be used to produce an image of the light source. This imaging process was predicted in a number of studies involving the double passage of light trough a random screen. The observation of the formation of a pseudo-image by retro-reflection from diffusing surfaces has also been reported although the explanation for the phenomenon was not fully developed and did not include multiple scattering[8].

In the present study we consider the effect of particle density on the image resolution. The experimental configuration is illustrated in figure 1.

0-8194-0658-9/91/$4.00

An unpolarized 10-*mW He-Ne* laser beam is passed through a spatial filter that has a 25 *micron* pinhole. The laser wavelength is 632.8 *nm*. The pinhole can now be considered as the source of a diverging beam of light that is

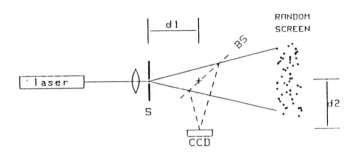

<u>Figure 1</u>

sent through a beam splitter and falls upon a random medium. A charge-couple device (CCD) array acts as the detector for the image and is placed at the equivalent backward-source position. The array records the light which is scattered from the random screen. Care was taken to ensure that no direct specular reflections were incident upon the detector array from the beam splitter surface or from the flat surfaces of the screen. In the present experiments the random screen consisted of a number of suspensions of (0.296 *micron* diameter) latex spheres in a 10 *cm* x 10 *cm* x 0.1 *cm* container. The Brownian motion of the spheres helped to remove the speckle which would otherwise mask the imaging effect. The image of the pinhole was recorded for a range of particle densities in the suspension and the results are compared to theoretical predictions.

THEORETICAL MODEL

To characterize the image formed by multiple backscattering we consider the following model. As illustrated in figure 2, we concentrated on two particles

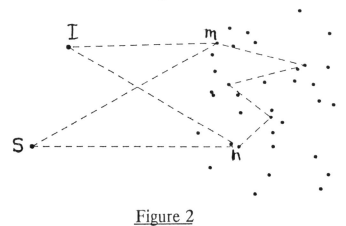

Figure 2

n and m of the random ensemble. Light is emitted from the source S and is first scattered by particle n, it is then randomly scattered in the medium to finally be scattered towards the image point I by particle m. There is an alternative path in which the light is first scattered by particle m, follows the same random path in the medium but in reverse, and is finally scattered towards the imaging point by particle n. The net phase difference ϕ between these paths is

$$\phi = \frac{2\pi}{\lambda} \left(|r_n - r_S| + |r_m - r_I| - |r_m - r_S| - |r_n - r_I| \right) \qquad (1)$$

where λ is the wavelenght, and r_S, r_I, r_n and r_m are the positions of the source, image, particle n and particle m, respectively. This phase difference leads to an interference term when these two waves are added at the image position. We now assume that the interparticle distance $|r_n - r_m|$ and yhe source to image distance $|r_s - r_I|$ are much less than the source to screen distance D. As well we assume that in this random screen situation only the fields which are related by the above reverse paths contribute to interference terms. The net irradiance measured at the image position is therefore the sum of the individual irradiances due to particular multiple scattering events these can be written as

$$I_{nmi} = c\epsilon_o \frac{E_n^2}{D^4} |P_{nmi}|^2 \left\{ 1 + \cos\left(\frac{k}{2D}(r_n - r_m) \cdot (r_S - r_m)\right)\right\} \qquad (2)$$

where $|E_n| \approx |E_m|$ is the field at n and m and where P_{nmi} is the amplitude probability for selecting particles n and m and path i between these.

We define a set of coordinates by placing the $y\ axis$ in the direction from the source and the screen. For an image point in the x-z plane we define the $z\ axis$ to be from S to I. We assume a homogeneous distribution of spheres and use the diffusion approximation to estimate P_{nmi}. The probability that n and m are the particles which emit the light to the image and receive the light from the source is proportional to

$$e^{-Yn/\ell} \quad e^{-Ym/\ell} \qquad (3)$$

where "ℓ" is the diffusion length of the light in the scattering medium and y_n (y_m) is the distance from the front surface of the screen to particle n (m).

The probability that the light is diffused from n to m in the medium is taken to be proportional to

$$\frac{1}{\rho^2} \qquad (4)$$

where ρ is the distance from n to m. The net intensity can then be written as

$$I = C\int_o^\infty d\rho \int_o^\pi d\phi \int_o^\pi d\phi \ e^{-\rho \ \sin\theta \ \sin\phi/\ell} \qquad *$$

$$\left(1 + \cos\left(\frac{kr\rho}{2D}\cos\theta\right)\right)\sin\theta \qquad (5)$$

where C is a constant and r is the distance from the source to the image. The above integral is simplified to

$$I = C \int_0^{\pi/2} \int_0^{\pi/2} \left(\frac{1}{\sin \theta} + \frac{\sin^2\theta\sin\phi}{\sin^2\theta\sin^2\phi + \zeta^2 \cos^2\theta} \right) d\theta d\phi \qquad (6)$$

where $\zeta = kr\ell/2D$. This integral is then numerically evaluated as a function of the parameter ζ and the comparison is made with the experimental results.

EXPERIMENTAL RESULTS AND ANALYSIS

The apparatus was set-up as shown in figure 1 and the image recorded by the CCD array was transfered to a computer for further processing. A typical image is presented in figure 3. The image of the pinhole appears as a bright

Figure 3

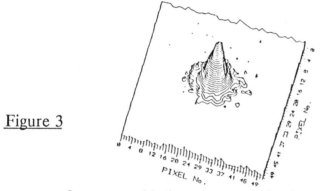

spot on top of a reasonably homogeneous background. The background signal is produced by light which is single and multiply scattered and the pinhole image is typically 10% above the background. Figure 3 presents the recorded image in each pixel (23 x 23 μm^2) after most of the background has been removed.

The exact position of the image along the *y axis* is obtained by plotting its fullwidth at half maximum as a function of the array position. Typical results are presented in figure 4. This shows that the pinhole is imaged at a reasonably precise location.

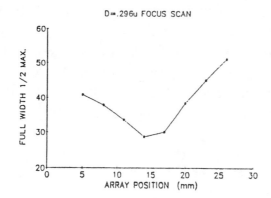

Figure 4

The minimum in figure 4 give us the best image position where the images are recorded as a function of particle density in the random screen. In the present set-up the distance between source and screen was $D \approx 50\ cm$.

In Figure 5 we present typical results of the enhancement above the background signal for three particle concentrations. A qualitative analysis indicates that both relative amplitude and the line width are a function of the particle concentration. In the present study we concentrate on comparing these

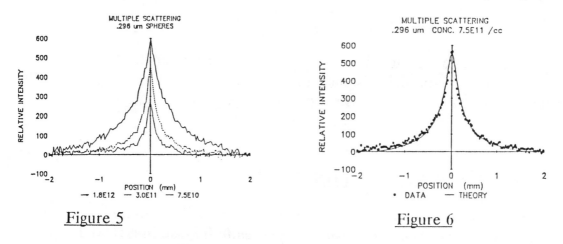

Figure 5   Figure 6

results to the model reprented in equation 5 and evaluate the diffusion length "ℓ" which will best fit the results since all other variables in the parameter ζ are known. In figure 6 we present a representative fit to the data which is obtained by adjusting the diffusion length "ℓ" and the amplitude constant "C". The theory represents the data quite well and has features which are similar to that which was presented by Akkermans et al.[5]. However we note an important difference, the present theory includes the three dimensional distribution of the particles in the ensemble, while the theory by Akkermans et al. considered that particles n and m lie on the same XZ plane. This approximation ignores the wide angle multiple scattering enhancement which occurs when $(r_n - r_m)$ and are not parallel due to the dot product in equation 2. As a consequence, if we normalize the multiple scattering at a scattering angle of 2° for example, the enhancement due to multiple scattering at the peak is approximately a factor of 1.8 to 1.9 and not 2.0, a fact which had been experimentally observed but remained unexplained.

Finally in figure 7 we present the diffusion length "ℓ" as a function of interparticle distance which is taken to be the cube root of the particle density.

Figure 7

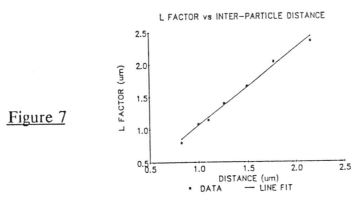

The results indicate that, for these particles, the diffusion length is slightly greater that the interparticle distance and is directly proportinal to it as long the interparticle distance is kept above two to three particle diamaters.

CONCLUSION

We have shown that a pinhole acting as light source can be imaged onto a CCD array by the phenomenon of multiple scattering within a random ensemble. The image width, a measure of its resolution, is found to vary with the particle density in the scattering medium. The light diffusion length, ℓ, in the theoretical model is seen to be directly proportional to the interparticle distance. The theoretical model also predicts an enhancement factor of 1.8 to

1.9 depending on the choice of normalization point since it is found that, in three dimensions, the enhancement due to multiple scattering can occur over wide angles.

REFERENCES

1.   P.E. Wolf and G. Maret, "Weak Localization and Coherent Backscattering of Photons in Disordered Media", Phys. Rev. Lett 55, 2696-2699, 1985.
2.   M. Kaveh, M. Rosenbluh, I. Edrei, and I. Freund, "Weak Localization and Light Scattering from Disordered Solids", Phys. Rev. Lett 57, 2049-2052, 1987.
3.   E. Akkermans, P.E. Wolf, and R. Maynard, "Coherent Backscattering of Light by Disordered Media: Analysis of the Peak Line Shape", Phys. Rev. Lett 56, 1471-1474, 1986.
4.   S. Etemad, R.Thompson and M.J. Andrejco, "Weak Localization of Photons; Universal Fluctuations and Ensemble Averaging", Phys. Rev. Lett. 57, 575-578, 1986.
5.   P.E. Wolf, G. Maret, E. Akkermans and R. Maynard, "Optical Coherent Backscattering by Random Media" J. Phys. France 49, 63-98, 1988.
6.   Y.A. Kravtsov and A.I. Saichev, "Effects of Partial Wave-front Reversal During the Reflection of Waves in Randomly Inhomogencous Media, Sov. Phys. JETP 56, 291-294, 1982.
7.   Y.A. Kravtsov and A.I. Saichev, "Properties of Coherent Waves Reflected in a Turbulent Medium", J.O.S.A. A2, 2100-2105, 1985.
8.   J.S. Preston, "Retro-reflexion by Diffusing Surfaces", Nature 213, 1007-1008, 1967.
9.   P. Rochon, "Image Production Due to Backscattering", Opt. Lett 15, 1334-1385, 1990.
10.   P. Rochon and D. Bissonnette, "Lenless Imaging Due to Backscattering", Nature 348, 708-710, 1990.

OPTICAL SCATTER:
APPLICATIONS, MEASUREMENT, AND THEORY

Volume 1530

SESSION 2

Surface Scatter

Chair
H. Philip Stahl
Rose-Hulman Institute of Technology and Stahl Optical Systems

Coherence in single and multiple scattering of light from randomly rough surfaces

Zu-Han Gu

Surface Optics Corporation, P.O. Box 261602, San Diego, CA, 92196, U.S.A.

Alexei A. Maradudin

University of California, Physics Department, Irvine, CA, 92717, U.S.A.

E. R. Mendez

CICESE, Apdo. Postal 2732, Ensenada, B.C., Mexico.

ABSTRACT

One of the most interesting phenomena associated with the scattering of light from a randomly rough surface is that of enhanced backscattering. This is the presence of a well-defined peak in the retroreflection direction in the angular distribution of the incoherent component of mean scattered intensity of the light scattered from such a surface which is due primarily to the coherent interference of each multiple reflected optical path with its time-reversed partner. It is an example of a broader class of multiple scattering phenomena that goes under the name of weak localization.

Not all of the manifestations of weak localization in the interaction of light with a randomly rough surface are in backscattering. It was recently shown that the average diffuse intensity from randomly rough surfaces with even symmetry can be enhanced or reduced in the specular direction due to the constructive interference between correlated pairs of scatters.

In this paper we will present recent theoretical analysis and experimental results that cover four kinds of enhancement: the enhanced backscattering, the enhanced transmission, the enhanced specular, and the enhanced refraction for 1-d and 2-d surfaces. These are manifestations of coherent effects that remain after ensemble averaging.

1. INTRODUCTION

One of the most interesting phenomena associated with the scattering of light from a randomly rough surface is that of enhanced backscattering. The phenomenon is manifested by the presence of a well-defined peak in the retroreflection direction in the angular distribution of the intensity of the incoherent component of the scattered light. The exact physical mechanism giving rise to this effect is still unclear, but it is now known that the effect is due to multiple scattering of waves (either optical or in the form of surface polaritons), giving rise to constructive interference effects in the retroreflection direction; that is the interference of each multiply scattered optical path with its time-reversed partner. The effect is an example of a broader class of multiple scattering phenomena that goes under the name of weak localization.[1-6]

Not all of the manifestations of weak localization in the interaction of light with randomly rough surfaces are in backscattering. There have been reports of enhanced transmission in the antispecular direction.[7,8] Also, the results of the numerical simulations of the scattering of light from randomly rough surfaces with even symmetry conducted by Nieto-Vesperinas and Soto-Crespo,[9-10] have shown the presence of a very sharp peak in the specular direction. This peak is part of the incoherent or diffuse component. In previous work this effect has been termed, in reflection, specular enhancement, and in transmission, enhanced refraction. For simplicity in the present paper we shall refer to this effect as the specular enhancement. It must be understood, however, that in the transmission geometry we mean enhanced refraction. More recently, it has been shown that the light scattered from surfaces with symmetries can be either enhanced or reduced in the specular direction due to the interference between correlated pairs of scatterers.[11-14] So, this phenomenon is also due to coherent effects that remain, and in fact only emerge, after ensemble averaging.

In the work reported here, the specular enhancement is presented in Section 2. The analysis shows, among other things, that the effect is primarily due to single scattering. The enhanced backscattering is presented in Section 3 and the enhanced transmission in Section 4. It is generally believed that these two enhancements are due to the multiple scattering effect. Finally Section 5 shows the summary and conclusion.

2. SPECULAR ENHANCEMENT

Coherent effects in the scattering of light from random surfaces with symmetry have recently been investigated. The diffused specular enhancement including enhanced specular and enhanced refraction, which is part of the diffuse component of the scattered light, can only be detected with even symmetrical random surfaces which must to be rough enough to extinguish the non-enhanced specular or coherent component of the scattered light.

Since symmetric surfaces do not, in general, occur naturally, they had to be fabricated. For the work reported here, the symmetric surfaces were fabricated by randomly exposing photoresist-coated plates under computer control, in the manner described in Reference 15. Figure 1 shows the measured surface profile of a gold coated 1-d grating with even symmetry fabricated with this technique which is composed of several (~ 150) segments, each with a center of symmetry of random roughness. A large number of segments is necessary to obtain a good estimate of the ensemble average. The center of symmetry is indicated by the dotted line. Each segment with even symmetry is 200 μm long. A Dektak 3030 stylus machine was used to contactly measure the surface profile. The rms roughness over the average of all 2 mm profiles is $\sigma = 1.4 \pm 0.16$ μm, and the 1/e correlation length is 8.0 μm.

The observation of these effects present some difficulty. The main problem lies in the estimation of the ensemble average. Normally, in angle-resolved scattering instruments, the mean intensity is estimated by performing a spatial average with the detector, which typically subtends a solid angle much greater than the speckle size. This spatial average will only provide a good estimation of the ensemble average when the mean intensity changes slowly, as compared with the angular subtense of the detector. This is not the case in the present situation, since the angular width of the enhancement, or reduction in the intensity, has been shown to be comparable to the speckle size. Evidently, a spatial average is not equivalent to an ensemble average in this situation.

Despite these problems, enhancements in the light scattered by a specially prepared photoresist surface have been measured. The profiles of the surfaces employed, fabricated in the manner described above, are composed of several (~ 150) segments, each with a center of symmetry (a large number of segments is necessary in order to obtain a good estimate of the ensemble average). The average complex amplitude in the specular direction is nearly zero (~ exp {$-(\sigma/\lambda)^2$}), which means that the coherent component is extinguished by the randomness of the surface. Figure 2 is a plot of the transmission measurement of the gold-coated 1-d grating with even symmetry. The 0.6328 micron laser beam is normally incident on the random surface. A very sharp peak is observed on the specular transmission direction which is the enhanced diffuse intensity of a randomly rough surface with even symmetry in the vicinity of the specular transmission direction.

Figure 3 shows another surface profile of a gold-coated randomly rough 1-d grating with even symmetry and constant steps. Although there are only two segments here, totally there are 150 segments for a 3 cm fabricated surface. Figure 4 is a plot of the transmission measurement of the gold-coated 1-d grating with even symmetry and steps. The incident beam is normal to the surface. A sharp reduction in intensity is observed in the specular direction which is due to destructive interference of the diffuse intensity of a randomly rough surface with even symmetry and height steps. Figure 5 is the analytical results of mean intensity as a function of the scattering angle based on the Kirchhoff single scattering approximation.[13,14] There is no specular enhancement of random phase surface in Figure 5(a), however there is specular enhancement of random phase surface with even symmetry in Figure 5(b), and specular depletion of random phase surface with even symmetry and steps in Figure 5(c), which are similar to the experimental results.

Specular enhancement in reflection has also been observed experimentally. In Figure 6 we present the reflection measurement of a gold-coated 1-d grating with even symmetry deposited on a glass substrate. The measured rms roughness over the average of all 2 mm profiles is $\sigma = 2.3$ μm, and the 1/e correlation length $a = 9.5$ μm. The surface consists of 150 different symmetrical pairs, each of which is 200 μm long, so that the measurement is equivalent to the result of averaging over 150 different realizations of a random surface. A well-defined peak is observed in the specular direction, when incident angle $\theta_r = 10°$.

3. ENHANCED BACKSCATTERING

Recent investigations have shown that for optical rough surfaces, the high-sloped nature of the surface leads to multiple scattering and this contribution enhances the scattered light. We found that the enhancement was already significant in a double-scattering approximation.[14] This was done by writing the integral equation for the surface field or its normal derivative (in the case of p- or s-polarization, respectively) on a perfectly conducting, one-dimensional random rough surface in the form of an inhomogeneous Fredholm equation of the second kind. In this equation the inhomogeneous term corresponds to the Kirchhoff, or single-scattering, approximation for the surface field, or its normal derivative, depending on the polarization. When this equation is solved iteratively, the nth term in the resulting expansion describes an n-fold scattering process. The first few terms in such a solution have been calculated and averaged with a Monte Carlo computer simulation approach. The angular dependent contribution to the mean differential reflection coefficient from the incoherent component of the scattered light in the single-scattering approximation displays no enhanced backscattering; the contribution from pure double-scattering processes shows a well-defined enhanced backscattering peak. The inclusion of the contribution from triple-scattering processes modifies the sum of the other two contributions from the single- and double-scattering by only a small amount in the vicinity of the enhanced backscattering peak, and is most significant in the region of large scattering angles, where the lower order approximations reproduce the effects of shadowing poorly.

In Figure 7 we present our numerical simulation results for scattering from a 1-d random grating of even symmetry on a perfect conductor. The incident angle is 20° with wavelength at 0.6127 μm. The rms height σ = 1.414 μm, and 1/e correlation length a = 2 μm. A total of 1000 distinct profiles was used in obtaining this plot, and the random segment of the x-axis was divided into 300 equal segments in the numerical solution of the scattering equation. The total incoherent contribution is shown in Figure 7(a); the contribution from the single-scattering processes in Figure 7(b); the contribution from the pure double-scattering processes in Figure 7(c); and the contribution from the single- and double-scattering processes including the interference terms in Figure 7(d).

Figure 8 shows an experimental result of a characterized rough aluminum 1-d random grating with a large slope. The incident angle is 20° with wavelength λ = 0.6328 μm. The 1-d grating was fabricated by exposing photoresist-coated plates with random speckle patterns. A Dektak 3030 profilometer was used to measure the surface profile and estimate the statistics. The rms height σ = 2 μm, and the 1/e correlation length a = 4.4 μm. A very strong peak is observed in the retroreflection direction which indicates a severe enhanced backscattering.

4. ENHANCED TRANSMISSION

It was recently shown that the angular distribution of the intensity of the incoherent component of p-polarized light transmitted through a thin metal film surrounded by vacuum, whose illuminated surface is randomly rough while the back surface is planer, displays a well-defined peak in the antispecular direction in transmission.[7,8] It is believed that the physical origin of this effect is the scattering, by the surface roughness, of the surface polaritons in the film excited, through the roughness, by the incident light. The coherent interference of a doubly scattered light/surface polariton path with its time-reversed partner gives the dominant contribution to this enhanced transmission.

The experimental investigation adopted a He-Ne laser with wavelength λ = 0.6328 μm. Figure 9 shows schematically the structure of the sample employed for the measurements. A plane-parallel plate of BK-7-glass is coated with metal film of thickness d. The bottom surface of the plate is coated with a single layer of Magnesium Fluoride (MgF_2). For the measurement reported here only one sample was investigated: vacuum-deposited silver provided on BK-7 glass substrate. By careful control of vacuum and substrate temperature, the homogeneous thin film coating will have larger crystallites with a rougher surface.

A Dektak 3030 stylus machine was used to contactly measure the surface profile. The stylus radius is about 0.15 microns and the sample interval was 0.025 microns. The profile length ranges from 50 microns to 2000 microns. Each sample was measured 10 times on different areas of the sample surface and this data was used to estimate the different statistical properties required.

The stylus loading is 1 mg. Averaging all 50 micron profiles, the measured rms height of surface roughness of the silver mirror is σ = 118.2 ± 10Å, and in addition to the correlation for the stylus radius the approximate value of 1/e correlation length is 1200Å. Figure 10(a) is a plot of the relative transmittance measurement of p-polarized light through a silver film deposited on a glass substrate whose index of refraction n_d = 1.51. The plane of transmittance coincides with the plane of incidence, and it is the p-polarized component of the transmitted light which has been measured. The angle of incidence $\theta_i = 20°$. A sharp peak is observed at a transmission angle $\theta_t = -20°$, thus attesting to enhanced transmission. The occurrence of an enhanced transmission peak at $\theta_t = -20°$ is due to the fact that the glass substrate in the experiment has finite thickness, and that the two plane surfaces of the glass are parallel. By using Snell's law twice (entrance and exit of the substrate), the peak of the enhanced transmission backs in the $\theta_t = -20°$ direction. Thus it is seen that although the surface of the silver film studied in the present work is a two-dimensional randomly rough surface, rather than the random grating for which the theoretical calculations were based, and the present paper were carried out, enhanced transmission still occurs with the mechanism as described above. Similarly Figure 10(b) is a plot of the measurement of p-polarized light through a silver mirror with incident angle $\theta = 40°$. We see a sharp peak at $\theta = -40°$. Figure 11 shows a comparison of a theoretical calculation and a test of a silver mirror at $\theta = 40°$ where we subtracted the noise from the measured signal.

5. SUMMARY

In summary, four kinds of enhanced scattering from or through metal surfaces has been investigated. For a randomly rough surface with even symmetry there is an interference term that can give rise to enhancements or reductions of the diffuse intensity of the scattered light in the neighborhood of the specular direction. The existence of these coherent effects has been demonstrated. For symmetric surfaces the enhanced mean intensity was measured. For symmetric surfaces with height steps, the predicted reduction in the mean intensity was also found. Both experiments can be explained with Kirchhoff's single scattering approximation.[13,14]

For enhanced backscattering of a large slope aluminum coated surface the in-and-out-of-plane measurements are compared with numerical calculations of a 1-d random metal grating. For enhanced transmission of a smooth silver mirror the experimental results are also compared with the perturbation theory of rough metal film deposited on a dielectric substrate.

It is generally believed that the enhanced backscattering and the enhanced transmission are multiple scattering effects. In the case of a weakly rough metal surface they are due to the multiple scattering of the surface polaritons excited by the incident light, through the surface roughness, by the hills and valleys on the surface, before they are converted back into volume waves in the vacuum, propagating away from the surface. The coherent interference of such a scattering sequence and its time-reversed partner leads to a two-fold enhancement of the intensity of scattering into the anti-specular direction with respect to the intensity of scattering into other directions, when the contribution from single-scattering processes is subtracted off.

6. ACKNOWLEDGMENTS

The authors gratefully acknowledge the research support from the Army Research Office under grants DAAL03-89-C-0036 and DAAL-88-K-0067.

7. REFERENCES

1. A. R. McGurn and A. A. Maradudin, "Localization Effects in the Elastic Scattering of Light from a Randomly Rough Surface", J. Opt. Soc. Am. B4, 910-926 (1987).

2. A. A. Maradudin, A. R. McGurn, R. S. Dummer, Zu-Han Gu, A. Wirgin, and W. Zierau, "Optical Interaction at Rough Surfaces", in Laser Optics of Condensed Matter, Edited by J. L. Birman, H. Z. Cummins, and A. A. Kapyanskii, Penum Publishing Corporation, New York (1988).

3. K. A. O'Donnell and E. R. Mendez, "Experimental Study of Scattering from Characterized Random Surfaces", J. Opt. Soc. Am. A4, 1194 (1987).

4. M. J. Kim, J. C. Dainty, A. T. Friberg, and A. J. Sant, "Experimental Study of Enhanced Backscattering from One- and Two-Dimensional Random Rough Surfaces", J. Opt. Soc. Am. A7, 569 (1990).

5. Zu-Han Gu, R. S. Dummer, A. A. Maradudin, and A. R. McGurn, "Experimental Study of the Opposition Effect in the Scattering of Light from a Randomly Rough Metal Surface", Appl. Opt. 28, 537 (1989).

6. V. Celli, P. Tran, A. A. Maradudin, J. Lu, T. Michel, and Zu-Han Gu, "Waves on Corrugated Surfaces: K-Gaps and Enhanced Backscattering", in <u>Laser Optics Condensed Matter</u>, Vol. 2, Edited by E. Garmire, A. A. Maradudin and K. K. Rebane (Plenum, New York) (1991).

7. A. R. McGurn and A. A. Maradudin, "An Analogue of Enhanced Backscattering in the Transmission of Light Through a Thin Film with a Randomly Rough Surface", Opt. Commun. 72, 279 (1989).

8. Zu-Han Gu, R. S. Dummer, A. A. Maradudin, A. R. McGurn, and E. R. Mendez, "Enhanced Transmission Through Rough Metal Surfaces", accepted for publication in Applied Optics (1991).

9. M. Nieto-Vesperinas and J. M. Soto-Crespo, "Light-Diffracted Intensities from Very Deep Gratings", Phys. Rev. B38, 7250 (1988).

10. M. Nieto-Vesperinas and J. M. Soto-Crespo, "Connection Between Blazes from Gratings and Enhancements from Random Rough Surfaces", Phys. Rev. B39, 8193-8197 (1989).

11. E. R. Mendez, M. A. Ponce, V. Ruiz-Cortes, and Zu-Han Gu, "Coherent Effects in the Scattering of Light from Random Surfaces with Symmetries", Opt. Lett. 16, 123-125 (1991).

12. Zu-Han Gu, A. A. Maradudin, E. R. Mendez, M. A. Ponce, and V. Ruiz-Cortes, "Enhanced Transmission Through Randomly Rough Surfaces", accepted for publication in Waves in Random Media (1991).

13. E. R. Mendez, M. A. Ponce, and V. Ruiz-Cortes, "Light Scattering from One-dimensional Surfaces with an Even Profile", Proc. SPIE 1131, 18-29 (1990).

14. A. A. Maradudin, J. Q. Lu, T. Michel, Zu-Han Gu, J. C. Dainty, A. J. Sant, E. R. Mendez, and M. Nieto-Vesperinas, "Enhanced Backscattering and Transmission of Light from Random Surfaces on Semi-infinite Substrates and Thin Films", accepted for publication in Waves in Random Media (1991).

15. E. R. Mendez, M. A. Ponce, V. Ruiz-Cortes, and Zu-Han Gu, "Photofabrication of One-dimensional Rough Surfaces for Light Scattering Experiments", accepted for publication in Applied Optics (1991).

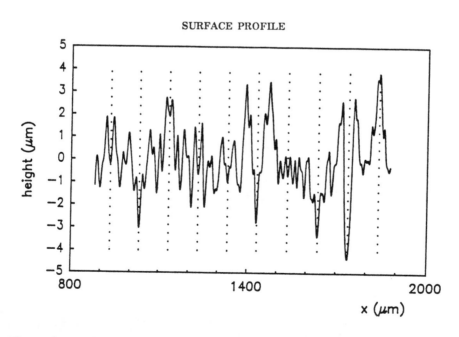

Figure 1.   A Surface Profile of a Gold-Coated 1-D Grating with Centers of Symmetry.

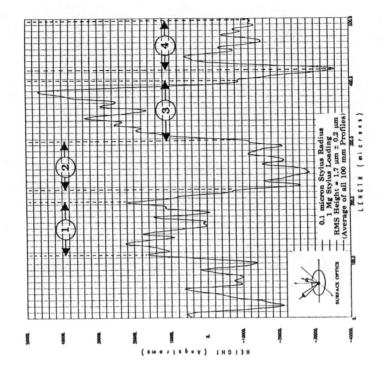

Figure 3. Surface Profile for a Gold-Coated Grating with Even Symmetry and Steps.

Figure 2. A Plot of the Transmission Measurement of the Gold-Coated 1-D Grating with Even Symmetry.

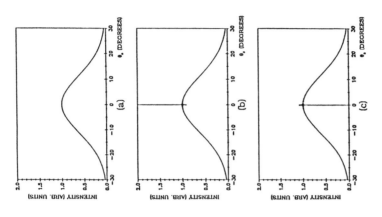

Figure 5. An Analytical Calculation of Mean Intensity as a Function of the Scattering Angle.

(a) A Random Phase Screen.

(b) A Random Phase Screen with a Center of Symmetry.

(c) A Random Phase Screen with a Center of Symmetry and a Relative Phase Shift of π on One Side.

$\lambda = 0.6328$ μm, $\sigma_h = 2.0$ μm, $\alpha = 8$ μm, and $n = 1.51$.

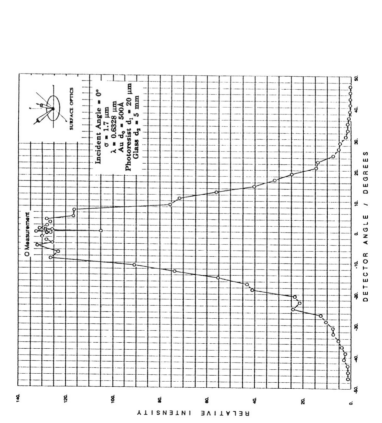

Figure 4. A Plot of the Transmission Measurement of the Gold-Coated 1-D Grating with Even Symmetry and Steps.

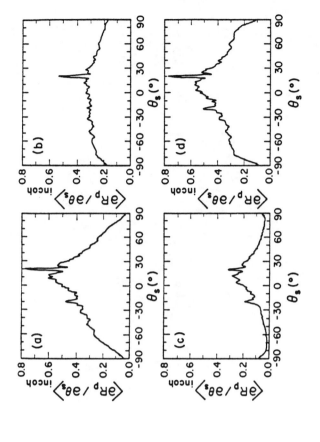

Figure 7. The Incoherent Contribution to the Mean DRC for the Scattering of a P-Polarized Beam of Light from a Random Grating of Even Symmetry on a Perfect Conductor.

(a) The Total Incoherent Contribution to the DRC;

(b) The Contribution from the Single-Scattering Processes.

(c) The Contribution from the Pure Double-Scattering Processes;

(d) The Contribution from the Single- and Double Scattering Processes, including the interference terms.

Figure 6. A Plot of the Reflection Measurement of the Gold-Coated 1-D Grating with Even Symmetry.

ALUMINUM COATED 1-D RANDOM GRATING

$\lambda = 0.6328\ \mu m$
Parallel-Parallel Polarization
$\sigma = 2.0\ \mu m$
$a = 4.4\ \mu m$

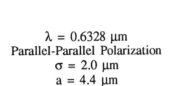

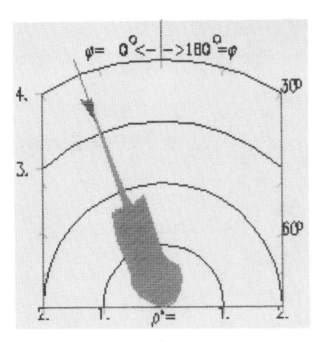

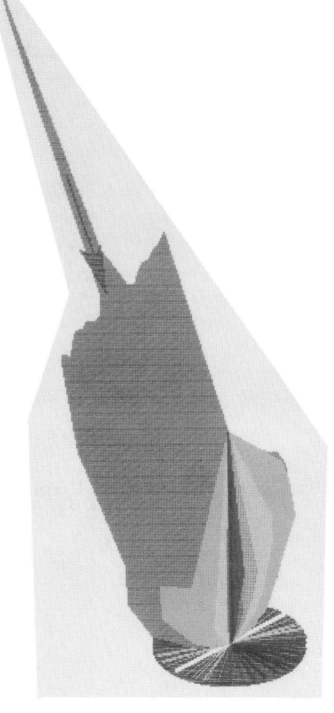

Figure 8.   Measurement of a Characterized Rough Al Coated Random 1-D Grating.

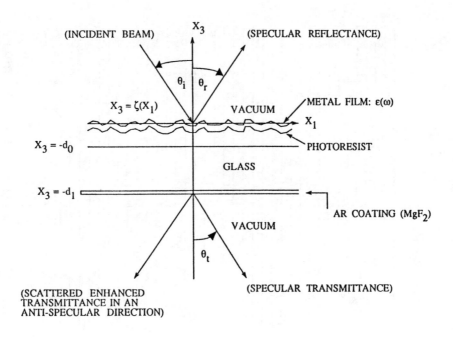

Figure 9. Schematic of Sample for Enhanced Transmission.

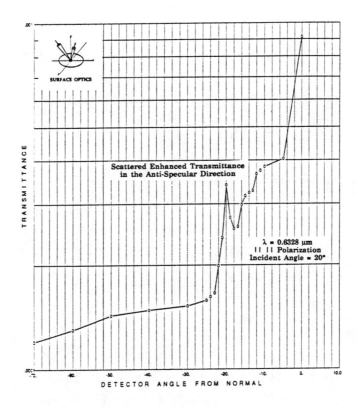

Figure 10 (a). Transmission Measurement of a Silver Mirror at 20° Incident Angle.

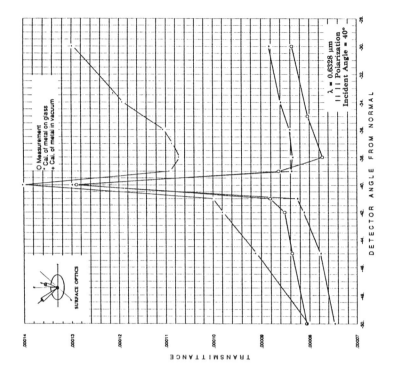

Figure 11. Comparison of Analytic and Test Results for a Silver Mirror.

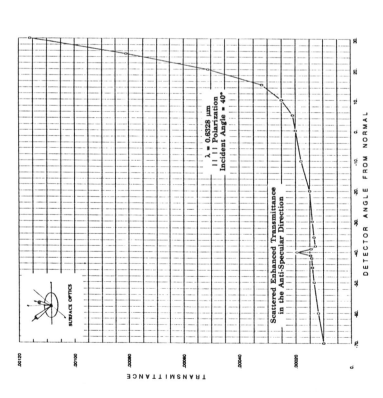

Figure 10 (b). Transmission Measurement of a Silver Mirror at 40° Incident Angle.

The optimal estimation of finish parameters

E. L. Church, USA ARDEC, Picatinny NJ 07806-5000

P. Z. Takacs, Brookhaven National Laboratory, Upton NY 11973-5000

ABSTRACT

This paper discusses basic issues involved in the estimation of surface spectra from laboratory measurements, the development of physically-based spectral models, and the estimation of finish parameters and their associated errors.

1. INTRODUCTION

The finish of optical surfaces is important since imperfect finish leads to scattering and performance degradation. Theory shows that these undesirable effects are functions of the two-dimensional power spectral density of the random phase and amplitude modulation which finish errors impose on the reflected or transmitted wavefronts [1,2,3].

Parametric modelling of such spectra allows the surface to be characterized in terms of specific finish parameters, and, when combined with scattering theory, permits performance specifications to be stated in terms of those parameters.

The classic example of this is the Strehl factor, which relates the reduction of the on-axis image intensity to the mean-square phase fluctuations, which equals the area under the spectrum over the bandwidth of the system [4]. In fact, the entire point-spread function and transfer function can be expressed in terms of finish and system parameters [5].

2. MEASUREMENT OF SURFACE SPECTRA

Most instruments measure finish fluctuations along a linear surface profile rather than over an area, and even area-measuring instruments present data as a series of parallel profiles. It is natural, then, to begin the discussion with the profile power spectrum -- which is the quantity of direct interest for glancing-incidence optics with a slit-detector geometry [5,6] -- and then to relate it to the area spectrum, which is quantity of interest for the more common case of a normal-incidence point-detector geometry [7].

The profile power spectral density is defined by

$$S_1(f_x) = \lim_{L \to \infty} \left\langle \frac{2}{L} \left| \int_L dx \, e^{i2\pi f_x x} z(x) \right|^2 \right\rangle \tag{1}$$

where the subscript 1 denotes that this a one-dimensional or profile power spectrum,

$f_x > 0$ is the surface spatial frequency, L is the trace length, $\langle \ -- \ \rangle$ denotes the ensemble average, and $Z(x)$ is the surface profile. In contrast, the two-dimensional or area power spectrum is:

$$S_2(\vec{f}) = \lim_{A \to \infty} \left\langle \frac{1}{A} \left| \int_A d\vec{r} \ e^{i 2\pi \vec{f} \cdot \vec{r}} Z(\vec{r}) \right|^2 \right\rangle \tag{2}$$

where $\vec{f} = (f_x, f_y)$ and $\vec{r} = (x, y)$ are vectors in the surface plane and A is the surface area.

Despite their similarity these spectra are different. One sign of this is the fact that they have different dimensions: the profile spectrum is length-cubed while the area spectrum is length-fourth. The connection between the two is given by [8]:

$$S_1(f_x) = 4 \int_0^\infty df_x \ S_2(f_x, f_y) \tag{3}$$

An useful consequence of this is the fact that the "areas" under the two spectra are the same. That is, the intrinsic root-mean-square (rms) profile roughness equals the rms area roughness [9], although measured values may differ significantly because of bandwidth effects [10].

Equation 3 gives the profile spectrum in terms of the area spectrum. Unfortunately, this cannot be inverted in general since the integration inherent in the profiling process destroys essential information about the surface spectrum. On the other hand, this information can be restored using a-priori information about the surface. Two canonical cases are machined surfaces measured at a known angle relative to their lay, and isotropically rough surfaces generated, say, by polishing [8].

In the latter case it is easy to see that Eq 3 can be rewritten

$$S_1(f_x) = 4 \int_{f_x}^\infty \frac{f \, df}{\sqrt{f^2 - f_x^2}} \cdot S_2(f) \tag{4}$$

where $f = |\vec{f}|$. This integral equation is in the form of an Abel or half-integral transform, which has a well-known inverse, the inverse-Abel or half-derivative transform:

$$S_2(f) = -\frac{1}{2\pi} \int_f^\infty \frac{df_x}{\sqrt{f_x^2 - f^2}} \cdot \frac{d}{df_x} S_1(f_x) \tag{5}$$

which gives us our desired expression: the area spectrum in terms of the profile spectrum. This in turn can be -- and has been -- used to predict the scattering or BRDF from surfaces in terms of laboratory measurements of their profiles [11,12].

The next problem is how to determine -- or better said -- to ˙estimate˙ the profile spectrum from measured profile data so that we may make use of this mathematical machinery.

3. ESTIMATION OF PROFILE SPECTRA

The definition of the spectrum, Eq 1, is given in terms of an ensemble of infinitely-long continuous profiles, while in the real world we have only a limited number of finite-length profiles measured at discrete sampling points. To determine the spectrum we process these data using an algorithmic procedure called an estimator.

The basic spectral estimator is the windowed periodogram:

$$\hat{S}_1(f_m) = \frac{1}{p} \sum_{j=1}^{p} \frac{2D}{N} \left| \sum_{m=0}^{N-1} e^{i 2\pi m n/N} W(m) z_j(mD) \right|^2 K(m) \qquad (6)$$

where the ˙hat˙ above the S denotes that it is an estimate of the unhatted quantity. Here p denotes the number of independent profiles included in the estimate,

$$f_m = \frac{m}{ND} \qquad , \qquad m = 1 \ (1) \ N/2 \qquad (7)$$

is the spatial frequency, N is the (even) number of profile points, D is the sampling interval, W(n) is a bell-shaped window function, and K(n) is an edge factor [13]. Expressions for W and K are given in the Appendix.

It follows that profile measurements give information about the spectrum only at discrete spatial frequencies and within the limited frequency range, determined by m = 1 and m = N/2:

$$(f_x)_{MIN} = \frac{1}{ND} \qquad , \qquad (f_x)_{MAX} = \frac{1}{2D} \qquad (8)$$

In other words, the longest measurable wavelength is essentially the trace length, L = (N-1)D , and the shortest is twice the sampling interval, 2D -- the reciprocal of the Nyquist frequency. Another feature of the periodogram is that its values show large fluctuations between adjacent frequencies and from profile to profile at a given frequency, which are exactly analogous to speckle in scattering.

Figure 1 is a sketch of a typical periodogram on log-log scales which shows these features -- the finite bandwidth, the discrete frequencies and the fluctuations -- which are artifacts of the measurement-estimation process. The true spectrum, which is also sketched, is expected to be smooth and to cover a much wider range of spatial frequencies than that viewed through the measurement window.

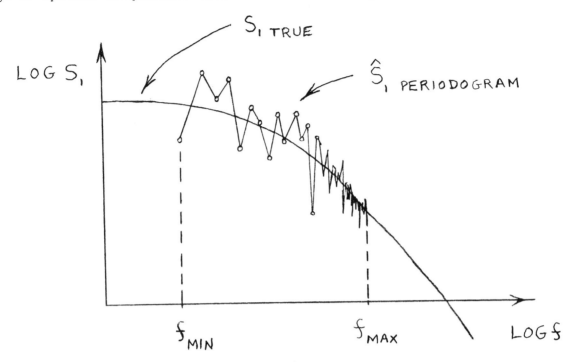

Figure 1. Sketch of the measured periodogram (jagged line) and the true spectrum (solid line) on log-log scales.

4. SPECTRAL MODELS

How can we get rid of these artifacts? The trick is fit the periodogram to a parametric model whose form carries physical information not contained in the original measurements. This has three advantages: it builds in a-priori information in a simple way, it describes the surface in terms of a few "finish" parameters, and it gives an analytic expression which can be used to extrapolate the spectrum outside the measurement bandwidth.

The model we use is the so-called ABC or K-correlation model [11,12,14]:

$$S_1(f_x) = \frac{A}{[1 + (Bf_x)^2]^{c/2}} \qquad (9)$$

where A, B, and C are the adjustable finish parameters. Substituting this into Eq 5 gives the corresponding expression for the two-dimensional spectrum:

$$S_2(f) = \frac{A'}{\left[1 + (Bf)^2\right]^{(c+1)/2}} \qquad (10)$$

where

$$A' = \frac{1}{2\sqrt{\pi}} \cdot \frac{\Gamma\left(\frac{c+1}{2}\right)}{\Gamma\left(\frac{c}{2}\right)} \cdot AB \qquad (11)$$

This model has been chosen in the following way:

1) The spectra of polished surfaces are expected to be smooth functions of frequency. Here we use a three parameter function in contrast to the N/2 parameters of the periodogram.

2) The spectra are expected to have an inverse-power-law or fractal-like character. This is because polishing is not expected to involve a characteristic length scale, and in that case the only mathematically permissible spectral shape is an inverse power law. Eqs 9 and 10 have this character: they behave as $f^{(-C)}$ and $f^{(-C-1)}$ at high frequencies.

3) The fractal form is expected to break down at high frequencies due to the presence of a non-vanishing "inner" length scale, and at low frequencies due to the presence of a finite "outer" length scale. Experiments suggest that the inner scale of polished surfaces may be as small as atomic dimensions [15,16], and that the outer scale may be as large as the part size itself [5].

Some surfaces, however, show spectra with a distinct flattening at low frequencies. That is, they contain an intrinsic outer length scale or "correlation length" which is shorter than the part size [11,12]. This possibility is included in the present model through finite values of the parameter B, where B/(2 pi) is the correlation length.

4) These considerations capture the a-priori information on hand but are not sufficient to determine a unique analytic form for the spectrum. We have based Eq 9 on additional considerations: its simplicity, the fact that its Abel transform can be evaluated analytically, the similarity between the resulting one- and two-dimensional expressions, and the fact that they reduce to familiar forms in special cases.

In particular, Eqs 9 and 10 reduce to white noise when B = 0, pure fractals when B goes to infinity, to the "Lorentzian" forms corresponding to an exponential correlation function when C = 2, and to Gaussians when C goes to infinity.

5) Equations 9 and 10 also correspond to a simple analytic correlation function:

$$C(\tau) = \sqrt{2\pi} \cdot \frac{A}{B} \cdot \frac{2^{-c/2}}{\Gamma(c/2)} \cdot \left(2\pi \frac{\tau}{B}\right)^{\frac{c-1}{2}} \cdot \mathbb{K}_{\frac{c-1}{2}}\left(2\pi \frac{\tau}{B}\right) \qquad (12)$$

where $\mathbb{K}$ is a modified spherical Bessel function of the third kind, which reduces to simple algebraic forms when $C = 2, 4, 6 \ldots$. Also, when $C > 1$ but is otherwise arbitrary, the surface has the finite mean-square roughness (i.e. rms value squared):

$$\sigma^2 = \frac{2\pi}{C-1} \cdot \frac{A'}{B^2} \qquad (13)$$

6) There is the possibility that more complicated spectral shapes can be written as a sum of several terms of the form of Eq 9. Since the one- and two-dimensional spectra are linearly related, the corresponding two-dimensional spectrum would be the sum of terms of the form of Eq 10.

5. ESTIMATION OF SPECTRAL PARAMETERS

Given the model and measured data, how do determine the finish parameters A, B and C and their errors?

Zeroth-order estimates can be found by inspection by simply plotting the spectral data versus frequency on log-log scales, fitting the low-frequency region by a horizontal straight line and the high-frequency region by an inclined straight line. $\hat{A}$ is then the value of S corresponding to the horizontal line, $\hat{B}$ is the reciprocal of the frequency corresponding to the intersection of the two lines, and $\hat{C}$ is the negative of the slope of the inclined line.

A natural analytic procedure would be to fit the ABC curve to the data by least-squares, which could also be used to estimate the errors in the fitted parameters. One complication is that the ABC function is non-linear, so that a numerical non-linear fitting procedure would have to be used. Also, it is not clear whether it the fitting should be done in normal or log space or, in fact, whether least-squares is the best analytic procedure to use.

Ordinary-least-squares fitting is statistically optimal for signals plus white Gaussian noise. Here, however, the fluctuations in the measured periodogram are intrinsic, their distribution is not Gaussian, and their rms values are not constant.

In the limit of large N the individual periodogram values are independent and chi-squared or gamma distributed; that is, their distribution function is (Eq 14):

$$P_k(S_m, \bar{S}_m)\, dS_m = \left(\frac{k}{2\bar{S}_m}\right)^{k/2} \cdot \frac{1}{\Gamma(k/2)} \cdot S_m^{-1+k/2} \cdot e^{-\frac{k}{2}\frac{S_m}{\bar{S}_m}}\, dS_m$$

where S_m is the measured value at $f_x = f_m$, $\bar{S}_m$ is its mean value, and k is the number of degrees of freedom of the measurement:

$$k = 2\,p\,/\,F \tag{15}$$

where p is the number of averaged periodograms in Eq 6 and F is the window factor discussed in the Appendix, which is unity for unwindowed data and is at most ~ 2.

This distribution has the property that the rms value of the scatter of S_m is not constant but proportional its mean value:

$$\mathrm{RMS}(S_m) = \sqrt{\frac{2}{k}} \cdot \bar{S}_m \tag{16}$$

In the case of a single unwindowed profile, k = 2, in which case Eq 14 becomes an exponential or "speckle" distribution, and Eq 16 shows that the rms value of the fluctuations of the measured spectrum about its average is equal the average itself! These properties are very different from those of a Gaussian process and suggest that least-squares may not be the best way of fitting periodogram data.

There is a general procedure for fitting data based on the principle of maximum likelihood. In the case of signals plus white Gaussian noise it leads to ordinary least squares, but to different results for chi-squared signals. Maximum-likelihood estimates not only include information about the known distribution functions but have the advantages of being unbiased and optimal for large N. In other words, they give the true values of the fitting parameters on the average, and the rms errors associated with those values are the lowest-possible: the Cramer-Rao bounds [17].

The first step in deriving these estimates is to form the Likelihood function, which is the product of the probabilities of the individual measurements:

$$\mathrm{Likelihood} = \prod_{m=1}^{M} P_k(S_m, \bar{S}_m) \tag{17}$$

where M is the number of spectral points being fitted (max = N/2), and $\bar{S}_m$ is the model being fitted -- Eq 9 in our case. The maximum-likelihood values of the model parameters are those which maximize Eq 17. In the case of additive white Gaussian noise the maximum-likelihood estimates of the finish parameters are determined by

$$\frac{\partial}{\partial A, B, C} \sum_{m=1}^{M} \left[S_m - \bar{S}_m \right]^2 = 0 \tag{18}$$

which is the expected result. In the case of chi-squared fluctuations they are determined by the different set of equations

$$\frac{\partial}{\partial A, B, C} \sum_{m=1}^{M} \left[\frac{S_m}{\bar{S}_m} - LOG\left(\frac{S_m}{\bar{S}_m}\right) \right] = 0 \tag{19}$$

Note that the chi-squared equations are independent of the order of the distribution, k. This means that averaging a number spectra together does not change the expected values of the finish parameters, although, as discussed later, it does reduce the errors associated with those values and tightens the confidence limits on the predicted spectrum.

Equations 18 and 19 each require the solution of three simultaneous non-linear equations. To explore their differences we consider a simplified form of the ABC model which involves only two parameters -- the fractal model -- which is simpler and of considerable importance in its own right [5,9,18].

6. FRACTAL SURFACES

The profile spectra of fractal-like surfaces can be written

$$\bar{S}_m = K / f_m^L \tag{20}$$

where K and L are the two finish parameters to be determined. As mentioned earlier, this is a special case of the ABC spectrum, Eq 9, rewritten here in simpler notation.

Since this spectrum appears as a straight line in log-log space it is natural to do the least-squares fitting of $\log(\hat{S}_m)$ versus $\log(f_m)$. This leads to the closed-form results:

$$\widehat{LOG} \, K = \frac{1}{DEN} \left[\bar{\sum} P_m^2 \cdot \bar{\sum} Q_m - \bar{\sum} P_m \cdot \bar{\sum} P_m Q_m \right] \tag{21}$$

and

$$\hat{L} = \frac{1}{DEN} \left[\bar{\Sigma} P_m \cdot \bar{\Sigma} Q_m - \bar{\Sigma} P_m Q_m \right] \tag{22}$$

where

$$DEN = \bar{\Sigma} P_m^2 - \left(\bar{\Sigma} P_m \right)^2 \tag{23}$$

and

$$\bar{\Sigma} = \frac{1}{M} \sum_{m=1}^{M} \quad , \quad P_m = LOG \, f_m \quad , \quad Q_m = LOG \, S_m \tag{24}$$

It is easy to see from these results that for k = 2, for example,

$$\langle \widehat{LOG \, K} \rangle = LOG \, K - \gamma \quad , \quad \langle \hat{L} \rangle = L \tag{25}$$

where $\gamma = 0.5772...$ is Euler's constant [18]. That is, although $\hat{L}$ is an unbiased estimate of L, $\hat{K}$ is a multiplicatively-biased estimate of K:

$$\langle \hat{K} \rangle = e^{\langle \widehat{LOG \, K} \rangle} = e^{-\gamma} \cdot K = 0.5615 \cdots K \tag{26}$$

In constrast, the maximum-likelihood estimates based on the chi-squared distribution are solutions of the equations

$$\bar{\Sigma} \, f_m^{\hat{L}} \, S_m \, LOG \, f_m = \bar{\Sigma} \, f_m^{\hat{L}} \, S_m \cdot \bar{\Sigma} \, LOG \, f_m \tag{27}$$

and

$$\hat{K} = \bar{\Sigma} \, f_m^{\hat{L}} \, S_m \tag{28}$$

The first is a non-linear equation for $\hat{L}$, the solution of which is then substituted into the second to give $\hat{K}$. The resulting values of $\hat{K}$ and $\hat{L}$ are both unbiased.

We now consider the errors associated with these estimates. The Cramer-Rao values based on the chi-squared distribution are

$$RMS\left(\frac{\delta\hat{K}}{K}\right) = \left[\frac{NUM}{DEN} \cdot \frac{2}{kM}\right]^{1/2} \tag{29}$$

and

$$RMS(\delta\hat{L}) = \left[\frac{1}{DEN} \cdot \frac{2}{kM}\right]^{1/2} \tag{30}$$

where

$$NUM = \overline{\sum LOG^2 \zeta_M} \tag{31}$$

Equation 29 depends on the number of spectral points being fitted, M , the number of degrees of freedom, k , and the frequency interval, 1/ND. Equation 30, on the other hand, depends only on M and k, and in a familiar way: in the limit of large M and a flat data window it is essentially 1/SQR(pM) where pM is the total number of spectral points used in the estimate.

Table 1 gives values of these bounds evaluated for Wyko-like profiling microscopes with different magnifications:

Magnification	ND (μm)	RMS($\delta\hat{K}$/K) in %	RMS($\delta\hat{L}$) in %
2.5X	5320	15.74	4.53
10 X	1330	9.87	4.53
40 X	332.5	5.10	4.53

Table 1. Cramer-Rao bounds for the fractal parameters K and L based on measurements using Wyko-like profiling microscopes. Values are shown for N = 1024, M = 512 and k = 2.

7. NUMERICAL SIMULATION

We have tested the formalism described above by generating 100 realizations of the process

$$S_m = K/m^L \qquad (32)$$

with an exponential distribution. We then fitted them to Eq 32 using Eqs 21,22 to get least-squares (LS) estimates of the parameters K and L, and to Eqs 27,28 to get the corresponding maximum-likelihood (ML) values.

The results are:

1. $\quad$ K(LS)/K fitted = 0.58 ± 0.18 $\qquad$ expected = 0.56 ± 0.13

$\quad$ K(ML)/K fitted = 1.00 ± 0.23 $\qquad$ expected = 1 ± 0.24

and

$\quad$ L(LS) $\quad$ fitted = $5.6(-3) \pm 5.5(-2)$ $\qquad$ expected = $0 \pm 4.5(-2)$

$\quad$ L(ML) $\quad$ fitted = $5.3(-3) \pm 4.3(-2)$ $\qquad$ expected = $0 \pm 4.5(-2)$

2. For a given realization the observed fitting errors, alpha = $(\hat{K} - K)/K$ and beta = $\hat{L} - L$, are independent of K and L. Also, alpha(LS) does not equal alpha(ML) and beta(LS) does not equal beta(ML). Instead, they differ in an apparently "random" way, with alpha-alpha and beta-beta correlation coefficients of 0.82 in each case.

These results indicate that LS estimates, corrected by Eq 26, are essentially optimal in that they are unbiased and their errors equal the Cramer-Rao bounds. If this also applies to the full ABC form it would be convenient since canned programs are readily available for non-linear least-squares fitting. Inspection suggests that in that case the LS estimate of the parameter A would have the same multiplicative bias factor as K in the fractal case, and that the estimate of C would be unbiased. Simulations are now under way to confirm these speculations and to determine the bias properties of the parameter B .

8. GOODNESS OF FIT

Once we have fitted a curve such as Eq 9 to the measured periodogram, we must determine how good the fit is. A bad fit might suggest that we try a sum of ABC curves instead of one, modify the model, or that some of the data are questionable.

A quick indication of the goodness of fit can be had by seeing where the measured data fall relative to a band drawn about the fitted curve which contains a given percentage of expected points:

$$LOG_{10} \overline{S}_m + A \quad \text{to} \quad LOG_{10} \overline{S}_m - B \tag{33}$$

Table 2 gives values of the displacements A and B for different total percentages divided equally above and below the median. Results are given for for various degrees of freedom, k. As discussed in the Appendix, $k \approx p$ for a Hamming or Blackman data window, and $k \approx 2p$ for a Daniell or Hann window, where p is the number of independent profiles averaged in Eq 6.

Total %	k = 1		k = 2		k = 4		k = 6		k = 8	
	A	B	A	B	A	B	A	B	A	B
0	-.34	.34	-.16	.16	-.08	.08	-.05	.05	-.04	.04
50	.12	.99	.14	.54	.13	.32	.12	.24	.11	.20
80	.43	1.80	.36	.98	.29	.58	.25	.43	.22	.36
90	.58	2.41	.48	1.29	.38	.75	.32	.56	.29	.47
95	.70	3.01	.57	1.60	.45	.92	.38	.69	.34	.56

Table 2. The A and B parameters in Eq 33 that determine the confidence limits for a chi-squared distribution with k degrees of freedom.

9. APPENDIX: THE PERIODOGRAM ESTIMATE

The periodogram estimate of the profile power spectral density is given by Eq 6, which we repeat here for completeness:

$$\hat{S}_1(f_m) = \frac{1}{p} \sum_{j=1}^{p} \frac{2D}{N} \left| \sum_{m=0}^{N-1} e^{i 2\pi m m / N} W(m) Z_j(mD) \right|^2 K(m) \tag{34}$$

p is the number of averaged spectra, D is the sample spacing, N is the number of measured points, W is a data window and K is the edge factor: $K(n)=1$ except $K(0) = K(N/2) = 1/2$.

This factor ensures that the profile variance evaluated in coordinate space for each realization, Z,

$$\hat{\sigma}_z^2 = \frac{1}{N} \sum_{m=0}^{N-1} Z(mD)^2 \tag{35}$$

is identical with the corresponding quantity

$$\hat{\sigma}_S^2 = \frac{1}{ND} \sum_{m=0}^{N/2} \hat{S}_1(S_m) \tag{36}$$

evaluated in frequency space, for W(n) = 1. This is a mathematical identity which is useful for checking the programming of Eq 34.

If W is not unity the values of Eqs 35 and 36 need not be equal for a given profile, but their ensemble averages will be equal providing the window function is normalized as

$$\frac{1}{N} \sum_{m=0}^{N-1} W(m)^2 = 1 \tag{37}$$

Typical data windows, each normalized according to Eq 37, are given below [13]. We usually use the Hamming or Blackman windows.

Daniell: $W(m) = 1$ (38)
(unwindowed)

Hann: $W(m) = \sqrt{2/1728}\,[24 - 24\cos\theta_m] \tag{39}$
(raised cosine)

Hamming: $W(m) = \sqrt{2/1987}\,[27 - 23\cos\theta_m] \tag{40}$

Blackman: $W(m) = \sqrt{2/1523}\,[21 - 25\cos\theta_M + 4\cos 2\theta_M] \tag{41}$

where

$$\theta_m = 2\pi \, m / N \qquad (43)$$

The purpose of the data window is to taper the sharp ends of the data set to reduce the strong edge ringing which they would otherwise generate. Equation 37 guarantees that this windowing does not affect the mean value of the spectra of the tapered data, but it does change the shape of their distribution function.

Although the mean spectrum is the same, the mean-square fluctuations about that average are increased by the factor

$$F = \frac{1}{N} \sum_{m=0}^{N-1} W(m)^4 \qquad (45)$$

which has the values of 1, 1.056, 1.817 and 2.348 for the four windows given above. The usual assumption is that the new distribution function is also chi-squared but with its order modified by this factor according to Eq 15.

The Hamming and Blackman windows suppress ringing more effectively than the Daniell or Hann windows, but they need twice as many profiles to give the same accuracy in the finish coefficients and the fitted spectrum.

There is an immense literature on spectral estimators and their statistical properties. For example, refs [20-23].

10. REFERENCES AND FOOTNOTES

1. "Finish" errors are not limited to topographic height fluctuations, but can include contributions from variations in the optical constants of isotropic materials [2] and domain-orientation effects in birefringent materials [3]. In the that case Z is an equivalent (complex) height which may include contributions from a variety of sources.
2. E. L. Church and P. Z. Takacs, "Subsurface and volume scattering from smooth surfaces", Proc. SPIE 1165 31-41 (1989).
3. E. L. Church, P. Z. Takacs and J. M. Stover, "Scattering by anisotropic grains in beryllium mirrors", Proc SPIE 1331 205-220 (1990).
4. M. Born and E. Wolf, "Principles of Optics", Pergamon Press, Oxford, 1970; pp 462-464.
5. E. L. Church and P. Z. Takacs, "The prediction of the performance of glancing-incidence mirrors from laboratory measurements" Proc SPIE 1160 46-55 (1988). "The metrology and specification of x-ray mirror surfaces", Proc SPIE 1319 518-519 (1990). "The specification of glancing-incidence mirrors in terms of system performance", Proceedings of the Annual meeting of the Society for Precision Engineering, ASPE 2 103-105 (1990). See also 7. below.
6. E. L. Church and P. Z. Takacs, "The interpretation of glancing-incidence scattering measurements", Proc SPIE 640 126-133 (1986).

7. E. L. Church and P. Z. Takacs, "A unified figure-finish relationship for mirror testing and specification". Verbal presentation at the annual meeting of the

SPIE, San Diego, July 1990. Manuscript in preparation.

8. E. L. Church and P. Z. Takacs, "Effects of the optical transfer function on surface finish measurements", Proc SPIE **1164** 46-59 (1989).

9. E. L. Church, H. A. Jenkinson and J. M. Zavada, "Relationship between surface scattering and microtopographic features", Opt. Engr. **18** 125-136 (1979).

10. E. L. Church, G. M. Sanger and P. Z. Takacs, "The comparison between WYKO and TIS measurements of surface finish", Proc SPIE **749** 65-73 (1987).

11. E. L. Church, P. Z. Takacs and T. A. Leonard, "The prediction of BRDF and TIS from surface profile measurements", Proc SPIE **1165** 136-150 (1989).

12. W. K. Wong, D. Wang, R. T. Benoit and P. Barthol, "Comparison of low-scatter-mirror PSD derived from multiple-wavelength BRDFs and WYCO profilometer data", paper 1530-10 presented at the Annual Meeting of the SPIE, San Diego, July 1991.

13. E. L. Church and P. Z. Takacs, "Instrumental effects in surface finish measurements", Proc SPIE **1009** 46-55 (1988).

14. E. L. Church and P. Z. Takacs, "Statistical and signal-processing concepts in surface metrology", Proc SPIE **645** 107-115 (1986).

15. E. Spiller, R. A. McCorkle, J. S. Wilczynski, L. Golub, G. Nystrom, P. Z. Takas and C. Welch, Opt. Engr. **30** 1109-1115 (1991).

16. D. Windt, "X-ray reflectometry of multilayer mirrors at AT&T", paper 1547-19 presented at the Annual Meeting of the SPIE, San Diego, July 1991.

17. See, for example, S. Brandt, "Statistical and Computational Methods in Data Analysis", Elsevier-North-Holland, New York, 1970; chapter 7.

18. E. L. Church, "Fractal surface finish", Appl. Opt. **27** 1518-1526 (1988).

19. Equation 23 is a particular case (k = 2) of

$$\langle \widehat{LOG K} \rangle = LOG K + \Psi(k/2) - LOG(k/2) \tag{46}$$

where Ψ is the psi or digamma function.

20. G. M. Jenkins and D. G. Watts, "Spectral Analysis and its Applications", Holden-Day, San Francisco, 1968.

21. P. Bloomfield, "Fourier Analysis of Time Series: an Introduction", Wiley-Interscience, New York, 1976.

22. M. B. Priestley, "Spectral Analysis and Time Series. Volume 1: Univariate Series", Academic Press, London, 1981.

23. D. R. Brillinger, "Time Series; Data Analysis and Theory", Holden-Day, San Francisco, 1981.

A comparison of low-scatter-mirror PSD derived from multiple wavelength BRDF's and WYKO profilometer data

Wallace K. Wong, Dexter Wang, Robert T. Benoit
SSG, Inc., Waltham, Massachusetts 02154

Peter Barthol
University of Wuppertal, Federal Republic of Germany

ABSTRACT

An electroless nickel plated over aluminum mirror was tested for BRDF' and surface profiles at two stages. Firstly, after a standard polish for optical figure and secondly, after a "super finishing process" which is designed to minimize optical scatter. BRDF measurements corresponding to spatial frequency range between .0016 μm^{-1} to 1 μm^{-1} were obtained using 0.633 μm, 1.06 μm, 3.39 μm and 10.6 μm lasers. The conversion formula used to derive PSD_{2D} from BRDF data is based on the Rayleigh-Rice vector theory. Measurements of the same spots by an optical profilometer (WYKO TOPO-2D) with several objectives were used to cover similar spatial frequency limits. The surface finish statistics, extracted from the profile data, was processed to produce PSD_{1D}. Then, composite PSD_{1D}s were fitted to an analytic function for PSD_{1D} using the K-correlation model approach. The derived PSD_{2D} from profile data was readily determined from the A,B,C coefficients associated with the K-correlation model.

The derived PSD_{2D}s from BRDF and profile measurements were compared to quantify the difference in surface finish statistics between the standard polished mirror and the same mirror after "super finishing". Good correlation is found for the standard polished surface. This corroborates the "topographic scatter" behavior for the clean, standard polished nickel surface which promise useful interpolation of BRDF values to wavelengths at which direct BRDF data is unavailable. Correlation for the "super finished" surface is worse, probably due to unplanned surface contamination between tests.

2. INTRODUCTION

Recently, an experiment was designed to investigate the surface finish statistics of a low scatter mirror during its fabrication process. The test mirror is part of a set of flight mirrors designed to be used in a cryogenically cooled IR sensor named *CRISTA* (*CR*yogenic *I*nfrared *S*pectrometers and *T*elescopes for the *A*tmosphere). Low scattering finishing are required to produce high off axis rejection telescope performance. CRISTA, integrated in the free-flyer ASTRO-SPAS, will be launched in 1993 by the space shuttle for a 9 day limb measurement mission. The CRISTA mirrors were built by SSG, Inc. under contract to University of Wuppertal at Germany.

These low scatter mirrors have BRDF (10.6 μm, 1°) specification of $\leq$ 2.0e-4 per sr which must be complied after gold overcoating. A flat mirror with integral mounting feet was selected for further investigation because it exhibited surprisingly good IR BRDF performance with average visible BRDF performance prior to the "Super Finishing" (S/F) process. An experiment designed to quantify the surface statistics changes before and after the S/F process was conceived and carried out. The chronology of the BRDF and profile measurements is listed in Section 3. Then it is followed by the PRE-S/F surface data and analysis in Section 4. The equations used to convert BRDF to PSD_{2D} are based on the Rayleigh-Rice vector theory.[1,2,3] The procedures to convert surface profile data to PSD_{2D} are more involved.[4,5,6,7,8,9,10] The K-correlation model is required as an intermediate step to represent the PSD_{1D}, so that the PSD_{2D} may be calculated for isotropic surfaces. In Section 5, the POST-S/F data "after S/F" (i.e., POST-S/F) are presented. In Section 6, lessons learned and preliminary conclusions are stated.

0-8194-0658-9/91/$4.00

3. TEST SURFACE & TEST CHRONOLOGY

Description. The optical surfaces used for experimentation is a flat mirror with clear aperture measuring 50 mm x 75 mm. This mirror, designated as M-2b s/n 2, was fabricated from a 6061-T6 aluminum block, plated with .005" of nickel and to be finish coated with a layer of gold, has integral mounting feet. This mirror was originally processed at Capricorn, a division of SSG, between December 1989 and January 1991. The normal mirror manufacturing steps are (a) substrate fabrication, (b) pre-plating grinding, (c) nickel plating, (d) post-plating thermal cycling, (e) optical figuring, (f) super-finishing, (g) BRDF testing, (h) gold overcoating and (i) final BRDF testing.

As part of the in process BRDF evaluation, it was discovered that this mirror in a standard polished state had exhibited BRDF (10.6 μm, 1°) much lower than expected (~ 3e-4 per sr) while the BRDF (.633 μm, .1°) appears to be nominally high (~ 2e-2 per sr). At this point in time, an extraordinary set of tests were planned on this mirror through the various stages of its manufacturing process.

Chronicle of Mirror Tests. There were three evaluation tests before this mirror was "super finished" (i.e., PRE-S/F) nickel on aluminum surface.

Test 1. PRE-S/F BRDF @ SSG @ 10.6 μm and .633 μm
Test 2. PRE-S/F BRDF @ TMA @ 10.6 μm, 3.39 μm, 1.06 μm and .633 μm and
Test 3. PRE-S/F linear profile measurements at WYKO.

Next, the mirror was processed for "super-finishing". During this process, unacceptable amount of pits became evident. The surface was re-polished to removed the pits and then processed for the "super-finishing". The mirror was successfully super-finished. The following evaluation tests were associated with the "POST-SUPER-FINISHED (i.e., POST-S/F) surface.

Test 4. POST-S/F BRDF @ SSG @ 10.6 μm and .633 μm
Test 5. POST-S/F-#1 BRDF @ TMA @ 10.6 μm, 3.39 μm, 1.06 μm and .633 μm
Test 6. POST-S/F-#2 BRDF @ TMA again
Test 7. POST-S/F linear profile measurements at WYKO

Handling. The mirror was freshly washed prior to the 10.6 μm BRDF measurement (Test 1). The 0.633 μm measurement (Test 1) was performed immediately after the mirror was washed. The mirror was packed and sealed in a clean aluminum container under a laminar flow clean bench. The double bagged container was used for round trip shipments between SSG and TMA. The TMA BRDF tests (Test 2) were performed as received. The mirror was returned to SSG and inspected before hand carried to WYKO (Test 3). The mirror was freshly washed prior to the SSG POST-S/F test (Test 4). The mirror was shipped to TMA and tested without washing (Test 5). The mirror was returned to SSG, washed and covered with a strippable polymer (Opti-Clean) and then sealed and shipped to TMA for re-testing. The protective polymer was removed just prior to testing at TMA (Test 6). The mirror was returned to SSG, inspected and sent to WYKO for POST-S/F profile tests (Test 7).

Test Spots. Three test locations were specified on the mirror surface as Spot A, Spot B and Spot C so that the same locations and orientation are used in different tests. All BRDF tests are in plane measurements at 5 deg. angle of incidence with a nominal 5 mm spot size. The sample length of the WYKO TOPO-2D measurements varies between .33 mm to 8.87 mm depending on the objective used.

At the writing of this paper, this mirror has been sent to be gold overcoated. The BRDF data on the POST-S/F surface after gold coating is unavailable.

4. PRE-S/F DERIVED PSD DATA ANALYSIS

__Test 1__. The 10.6 μm BRDF and .633 μm BRDF data (Figure 1) illustrates the Spot A surface scatter characteristics which is representative of the other spots. The BRDF (10.6 μm; 1°) was ~ 2-3e-4 per sr. The BRDF (.633 μm, 1°) was 2-3e-2 per sr. The 1°, 2° and 3° data points are instrument profile.

__Test 2__. The TMA BRDF data were extracted at intervals of 1° and plotted in the same scale at the SSG data (see Figure 2). The 4 test wavelengths were the laser lines at 0.633, 1.06, 3.39, and 10.6 μm. Comparison of the SSG to TMA BRDF data (see Figure 3) illustrates slightly higher BRDF values when measured at TMA. It is assumed that there are no systematic calibration bias between the test facilities.

4.1 Conversion from BRDF to PSD-2D

To convert BRDF(β-β_o) to PSD$_{2D}$, the following equation is used so that PSD$_{2D}$ has unit of (μm^2 μm^2). The BRDF test wavelength is in μm and the spatial frequency is in unit of μm^{-1}.

$$PSD_{2D}(f) = \frac{\lambda^4 BRDF(\beta - \beta_o)}{16\pi^2 \cos\theta_i \cos\theta_s Q} \tag{1}$$

where
θ_i = angle of incidence
θ_s = angle of scatter
Q = $\sqrt{(R_i R_s)}$
R_i = Sample specular reflectance at θ_i
R_s = Sample specular reflectance at θ_s
f = Spatial frequency vector in μm^{-1}; ($f = (\beta-\beta_o)/\lambda$)
β-β_o = $\sin(\theta_s)$ - $\sin(\theta_i)$

The specular reflectance values for electroless nickel are 88%, 81.5%, 82% and 76% at 10.6, 3.39, 1.06 and .633 μm respectively. The derived PSD-$_{2D}$ from the PRE-S/F BRDF data from SSG and TMA are shown in Figure 4 and Figure 5. An attempt has been made to represent the composite PSDs in the K-correlation model which is described below. The K-model parameters {A,B,C} can define both PSD$_{1D}$ and PSD$_{2D}$ for all spatial frequencies. The parametric numerical values fitted to the derived PSD trend are noted on the PSD$_{1D}$ figure. Comparison of the TMA and SSG derived PSD's on the same graph (see Figure 6) illustrated that the TMA data is generally higher by a factor of 5 because the fitted ABC's are differing only in A by a factor of 5. It is concluded that the mirror BRDF's may have changed between these tests (probably due to contamination) or there is a systematic bias between the facilities.

__Test 3__. The mirror was hand carried to WYKO by SSG personnel for TOPO-2D and TOPO-3D measurements. TOPO-2D linear profile data were recorded (26 October 1990) on the same areas that were BRDF tested. Since the TOPO-2D samples a line segment only, both X-oriented (in-plane BRDF) and Y-oriented (out-of-plane BRDF) scans were collected. This was repeated on each spot with 1.5x, 10x, and 40x magnification objectives so that the spatial frequency regions corresponds to that of the BRDF tests. Figure 7 shows typical TOPO-2D surface profiles recorded with different objectives. Figure 8 illustrates the corresponding TOPO-3D profile with the 10X objective. The TOPO-3D data was not used in the current analysis.

4.2 Conversion from linear profile data to PSD$_{2D}$

The procedures to generate PSD$_{2D}$ from a typical WYKO TOPO-2D data file has been developed by Church et al.[4,5,6,7,8,9,10] and adopted in this analysis.

Step 1. Extract the raw detector counts from (WYKO) floppy disk.
Step 2. Convert counts to surface heights, Z(j)
Step 3. Remove piston, tilt and curvature from surface heights
Step 4. Calculate PSD_{1D} (i.e., 1 dimensional periodogram)
Step 5. Restored PSD_{1D} with inverse filters
Step 6. Use restored data range only if S/N > 1.
Step 7. Repeat steps 1-6 for other objectives
Step 8. Fit K-correlation model to composite restored PSD_{1D} data
Step 9. Generate PSD_{2D} from K-correlation model

The result after Step 7 above for Spot A is shown in Figure 9 along with the K-correlation model using A = .0002, B = 50 and C = 2.0. The K-correlation model generates $S_{1D}(f)$ which is a fitted PSD_{1D} for any spatial frequency by the following analytic expression with only 3 parameters, namely A, B and C. Then the equivalent two dimensional analytic model (fitted PSD_{2D}) may be determined as S_{2D} which can be compared directly with the BRDF derived PSD_{2D}.

$$S_{1D}(f) = \frac{A}{[1+(Bf)^2]^{c/2}} \qquad (2)$$

$$S_{2D}(f) = \frac{A'}{(1+[Bf]^2)^{(c+1)/2}} \qquad (3)$$

where

$$A' = \frac{1}{2\sqrt{\pi}} \frac{\Gamma(\frac{c+1}{2})}{\Gamma(\frac{c}{2})} AB \qquad (4)$$

The comparison of the PRE-S/F derived PSD's from BRDF and profile data is shown in Figure 10.

5.0 POST-S/F DERIVED PSD DATA ANALYSIS

This section discusses Test 4, Test 5, Test 6 and Test 7 corresponding to the POST "super finished" (POST-S/F) mirror surface conditions.

5.1 Post-S/F BRDF Data

Test 4. The POST-S/F SSG BRDF data (Figure 11) illustrated that the optical scatter can be characterized as BRDF (10.6 μm, 1°) ~ 2e-4 per sr and BRDF (.633 μm, 1°) ~ 5-8e-3 per sr.

Test 5. The POST-S/F TMA BRDF data set #1 (Figure 12) did not scale with respect to wavelength as predicted by equation (1). Furthermore, there was significant disagreement relative to the SSG's POST-S/F data. Mirror contamination between SSG and TMA tests was speculated and confirmed in a second BRDF test at TMA after the following precautionary measures.

Test 6. Prior to the retest at TMA, the mirror was cleaned at SSG, then coated with "Opti-Clean" before shipping the mirror to TMA by UPS. "Opti-Clean" is a strippable polymer which had been used satisfactorily by SSG in situation when a wet cleaning was deemed unfeasible or undesirable. Instructions were supplied along with the mirror to TMA personnel regarding removal of the Opti-Clean protective film. This new data set, denoted as TMA POST-S/F-#2 data showed lower BRDF than TMA POST-S/F #1 data (Figure 13). The direct comparison between SSG and TMA BRDF has shown good agreement of levels at .633 μm but still a general bias (higher for TMA) at 10.6 μm. Furthermore, the BRDF data for TMA has shown significant variant from a smooth slope. We currently attribute it to contamination problem inherent in shipping mirror between facilities. Similar difficulties had been noted in the industry wide CO_2 round robin[11].

The BRDF derived PSD's are shown in Figure 14 and Figure 15 with the respective K-correlation parameters noted on the plots. The direct comparison of the .633 μm and the 10.6 μm data and the fitted K-parametric curves for both TMA #2 and SSG BRDF data illustrated the TMA #2 data is higher by a factor of 3 for all spatial frequencies (see Figure 16).

5.2 Post-S/F Profile Data

Test 7. The WYKO TOPO-2D measurement were done on 20 May 1991. Due to the unavailability of the 1.5X objective at the time, the TOPO-2D measurements were done with 2.5X, 10X, and 40X. These linear profiles were processed as shown above and a typical composite derived PSD_{1D} is shown in Figure 17. The data fitted K-correlation parameters of A = .00024, B = 100, C = 3.3.

Along with the TOPO-2D measurement, some TOPO-3D data was collected also. A typical TOPO-3D area profile is reproduced for reference (Figure 18).

5.3 Derived PSD_{2D} Comparison

The POST-S/F PSD_{2D}s are compared in Figure 19. The overall agreement is estimated to be within a factor of 2 which is excellent considering the variability across a practical mirror is often a factor of 2 to 5.

6.0 SUMMARY AND CONCLUSIONS

This comparison case study had illustrated several points:

(1)   The comparison of the PRE-S/F and POST-S/F PSD_{2D} using the K-correlation parametric forms (Figure 20) demonstrated the "super finishing" process apparently had modified the surface PSD mainly in the spatial frequency range corresponding to small angle scatter in the short visible wavelength region. The spatial frequency region determining the small angle scatter for LWIR (i.e., 10.6 μm) wavelengths was not significantly changed by the "super finishing" process.

(2)   The correlation between BRDF derived PSD_{2D} and profile derived PSD_{2D} is reasonable and encouraging. The K-correlation model parameters A,B,C provides a very simple and useful representation for the surface finish assessment through (linear) profile measurements.

(3)   The WYKO profilometer data showed much higher degree of fluctuation at all spatial frequencies. This is due to the fact that the periodogram is derived from a very small area which provides no ensemble averaging effects. The fitting of the K-correlation model to the derived PSD_{1D} was very subjective in this investigation. For future works, a more rigorous (i.e., method of least squares[13]) algorithm should be applied to the data expressed in the log (PSD_{2D}) vs. log (Spatial Frequency) space. Also, a larger set of profile data should be assessed in aggregate for better statistic significance.

(4)   The direct BRDF measurements in multiple spots using multiple lasers had been found to be a critical diagnostic tool of unintended contamination during this investigation. Contamination is a relative issue

which became more important as the mirror finish quality improved even further. The fact that the "topographic" behavior would be violated for a "contaminated" surface may be developed into a practical diagnostic criterion for quantification of actual surface contamination conditions.

(5)   A test case had been established on the degree of contamination when mirrors were shipped between facilities for BRDF measurements. An improved handling procedure was demonstrated with Optic-Clean as a protective coating. What if anything is left behind due to this coating is still unknown. The best proven technique is still a proper mirror wet cleaning procedure by a well trained person under a clean room environment.

7.0 ACKNOWLEDGEMENT

The authors would like to thank John Stover for pointing out an earlier error in using the 1-dimensional surface statistic conversion equations for analysis of an isotropic surface. Also, much appreciation goes to Eugene Church for his encouraging and helpful tips in applying the algorithms used in the conversion of the WYKO linear profile data to PSD_{2D}.

8.0 REFERENCES

1.   Schiff, T. F., Stover, J. C., "Surface Statistics Determined from IR Scatter,". Proc. SPIE, 1165-06, (1989) pp. 52-61.
2.   Stover, J. C., Optical Scattering: Measurement and Analysis, McGraw-Hill, New York, 1990.
3.   Stover, J. C., Serati, S. A., Gillespie, C. H., "Calculation of Surface Statistics from Light Scatter", OPTICAL ENGINEERING, Vol. 23 No. 4 July/August 1984, pp. 406-412.
4.   Church, E. L., Takacs, P. Z., "Subsurface and Volume Scattering From Smooth Surfaces", Proc. SPIE, 1165-04, (1989) pp. 31-41.
5.   Church, E. L., Takacs, P. Z., Leonard, T. A. "The Prediction of BRDF's from Surface Profile Measurements", Proc. SPIE, 1165-10, (1989).
6.   Church, E. L., Takacs, P. Z., "Effects of the Optical Transfer Function in Surface Profile Measurements", Proc. SPIE 1164-09 (1989) pp 203-211.
7.   Church, E. L., Takacs, P. Z., "Instrumental Effects in Surface Finish Measurement", Proc. SPIE 1009 (1988) pp. 46-55.
8.   Church, E. L., Vorburger, T. V., Wyant, J. C., "Direct Comparison of Mechanical and Optical Measurements of the finish of Precision Machined Optical Surfaces", OPTICAL ENGINEERING Vol. 24 No. 3 May/June 1985 pp. 388-395.
9.   Church. Eugene L., "Fractal Surface Finish", APPLIED OPTICS Vol. 27, No. 8, 15 April 1988, pp. 1518-1526.
10.   Private communication (Church, E. L. July 91).
11.   Leonard, T. A., Pantoliano, J. R., "Results of a CO_2 BRDF Round Robin", Proc. SPIE, 1165-12. (1989) Pg 444-449.
12.   Vernold, C. L., "Application and Verification of Wavelength Scaling for Near Specular Scatter Predictions," Proc. SPIE, 1165-03, (1989) pp. 18-30.
13.   DeRusso, P. M., Roy, R. J., Close, C. M.; State Variables for Engineers, John Wiley and Sons, Inc., New York, 1965.

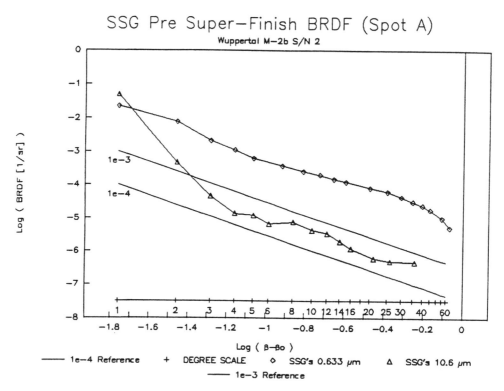

Figure 1. SSG PRE-S/F BRDF

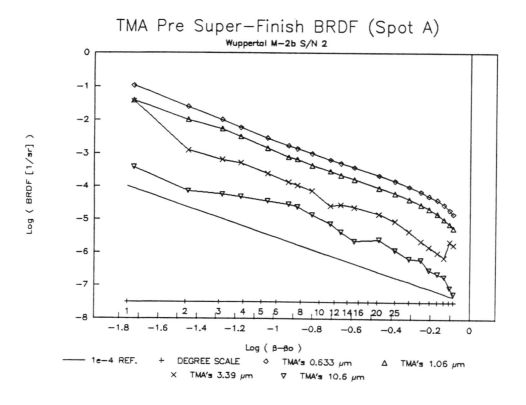

Figure 2. TMA PRE-S/F BRDF

SSG—TMA Pre S/F BRDF Comparison Spot A

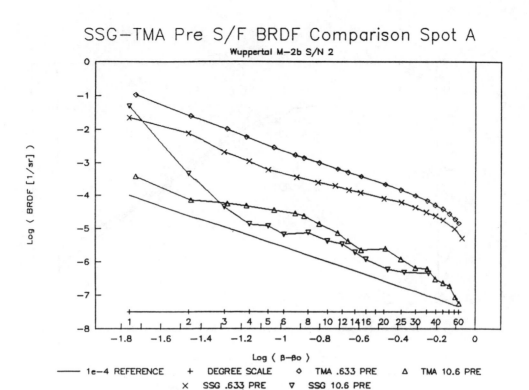

Figure 3. SSG-TMA PRE-S/F BRDF

K—MODEL: SSG PRE—SUPER—FINISH

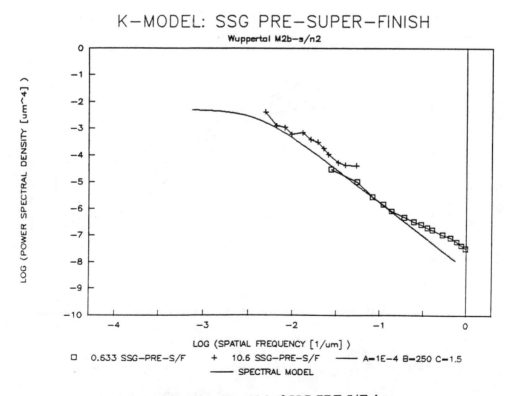

Figure 4. K-model of SSG PRE-S/F data

K-MODEL: TMA PRE-SUPER-FINISH
Wuppertal M2b-s/n2

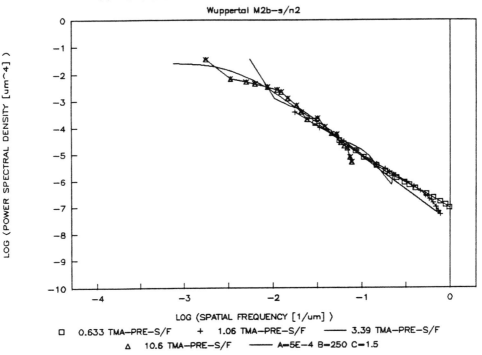

□	0.633 TMA-PRE-S/F	+	1.06 TMA-PRE-S/F	——	3.39 TMA-PRE-S/F
△	10.6 TMA-PRE-S/F	——	A=5E-4 B=250 C=1.5		

Figure 5. K-model of TMA PRE-S/F data

K-MODEL: SSG VS TMA PRE-SUPER-FINISH
Wuppertal M2b-s/n2

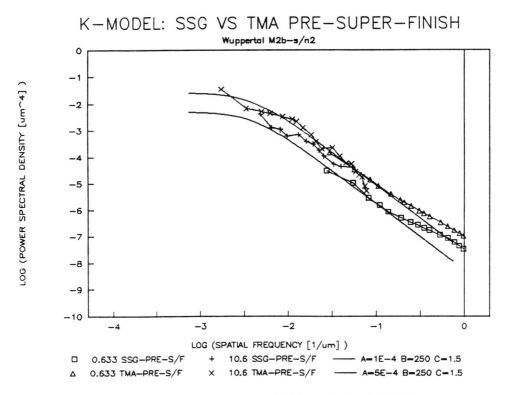

□	0.633 SSG-PRE-S/F	+	10.6 SSG-PRE-S/F	——	A=1E-4 B=250 C=1.5
△	0.633 TMA-PRE-S/F	×	10.6 TMA-PRE-S/F	——	A=5E-4 B=250 C=1.5

Figure 6. K-model of SSG-TMA PRE-S/F data

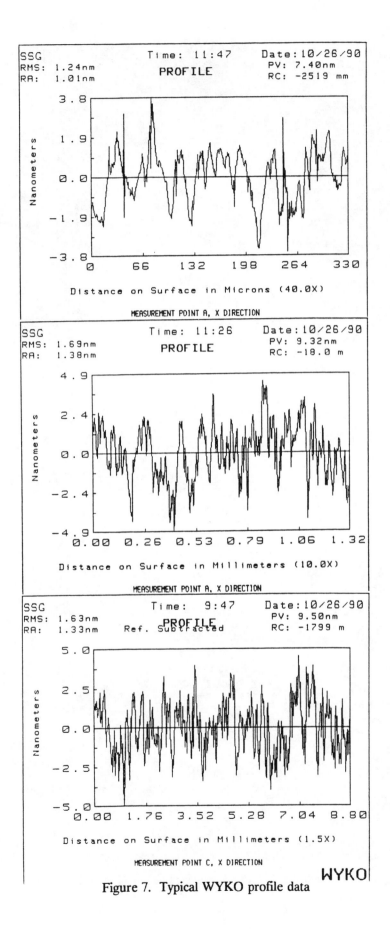

Figure 7. Typical WYKO profile data

Figure 8. TOPO-3D surface height map (PRE-S/F)

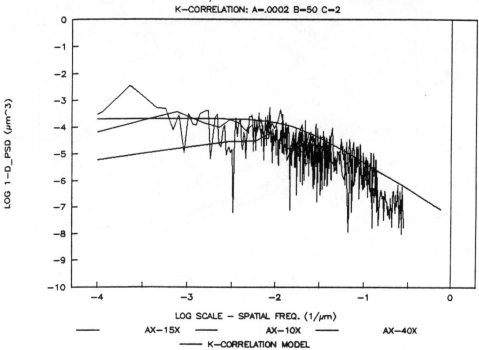

WYKO PRE−S/F SPOT A X−DIRECTION

K−CORRELATION: A=.0002 B=50 C=2

AX−15X   AX−10X   AX−40X

K−CORRELATION MODEL

Figure 9. WYKO PRE-S/F derived PSD$_{1D}$

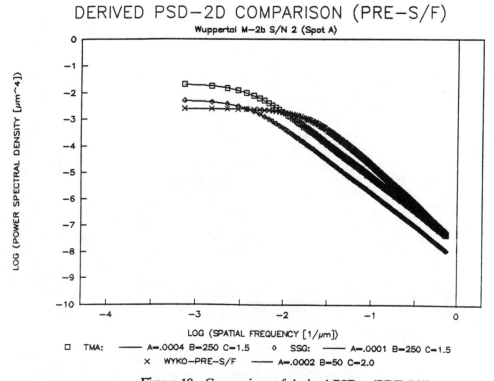

DERIVED PSD−2D COMPARISON (PRE−S/F)

Wuppertal M−2b S/N 2 (Spot A)

□ TMA: A=.0004 B=250 C=1.5   ◇ SSG: A=.0001 B=250 C=1.5

× WYKO−PRE-S/F   A=.0002 B=50 C=2.0

Figure 10. Comparison of derived PSD$_{2D}$ (PRE-S/F)

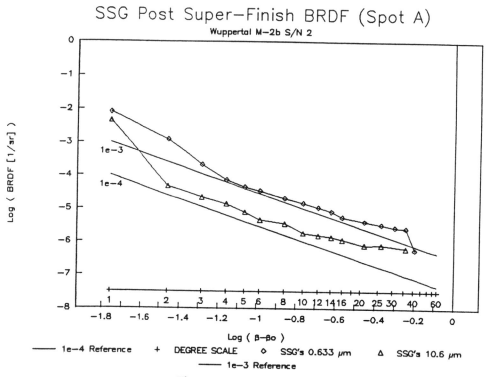

Figure 11. SSG POST-S/F BRDF

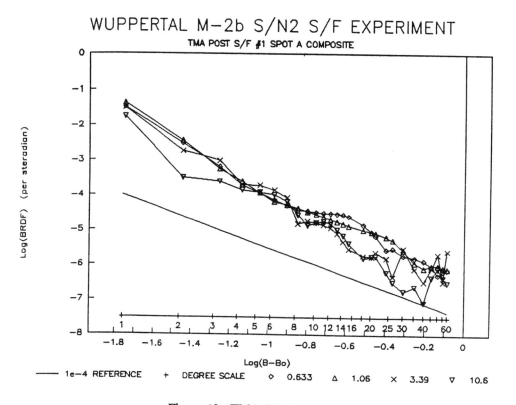

Figure 12. TMA POST-S/F-#1 BRDF

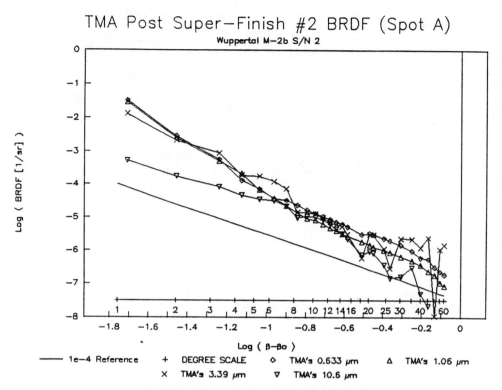

Figure 13. TMA POST-S/F-#2 BRDF

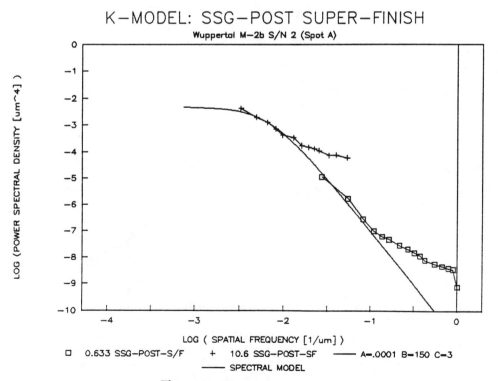

Figure 14. K-model of SSG POST-S/F data

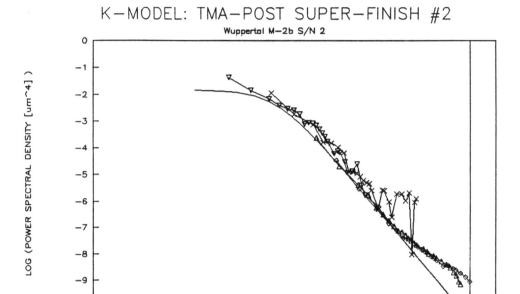

Figure 15. K-model OF TMA POST-S/F-#2 data

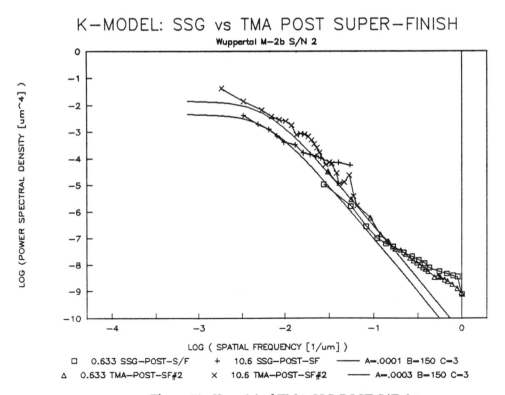

Figure 16. K-model of TMA-SSG POST-S/F data

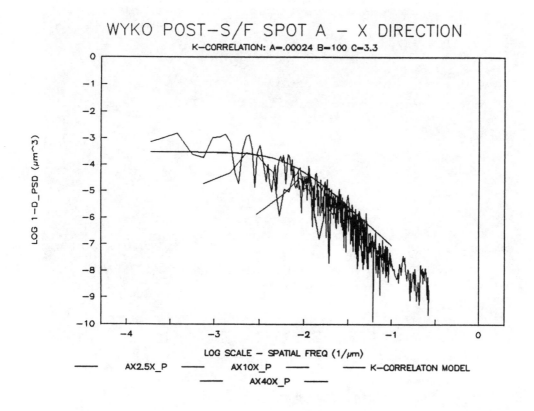

WYKO POST-S/F SPOT A – X DIRECTION

K-CORRELATION: A=.00024 B=100 C=3.3

AX2.5X_P   AX10X_P   K-CORRELATON MODEL

AX40X_P

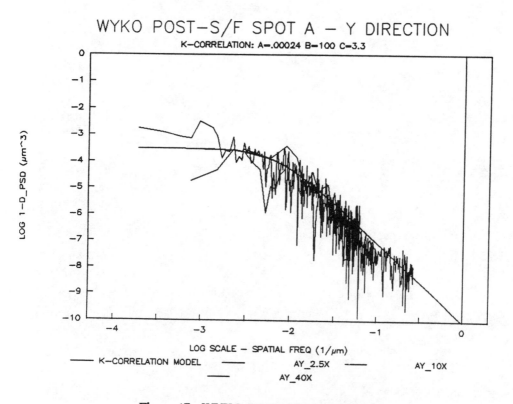

WYKO POST-S/F SPOT A – Y DIRECTION

K-CORRELATION: A=.00024 B=100 C=3.3

K-CORRELATION MODEL   AY_2.5X   AY_10X

AY_40X

Figure 17. WYKO POST-S/F derived PSD$_{1D}$

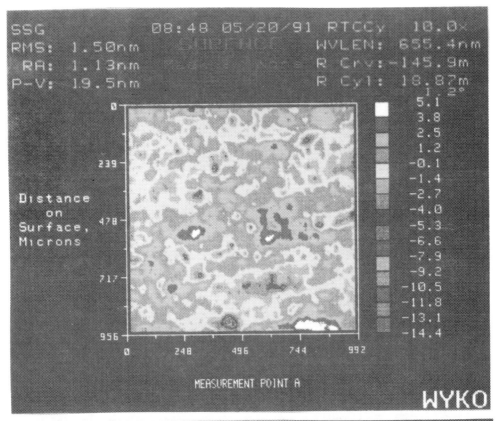

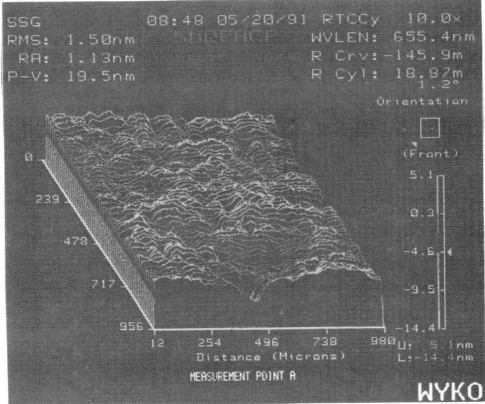

Figure 18. TOPO-3D surface height map (POST-S/F)

DERIVED PSD–2D COMPARISON (POST–S/F)

Wuppertal M–2b S/N 2 (Spot A)

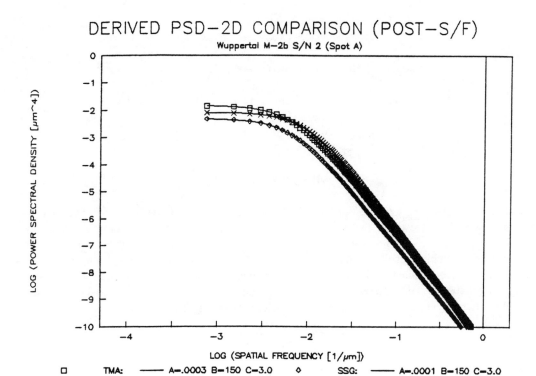

□ TMA: ——— A=.0003 B=150 C=3.0   ◇   SSG: ——— A=.0001 B=150 C=3.0

× WYKO: ——— A=.00024 B=100 C=3.3

Figure 19. Comparison of derived PSD$_{2D}$ (POST-S/F)

DERIVED PSD–2D COMPARISON SUMMARY

Wuppertal M–2b S/N 2

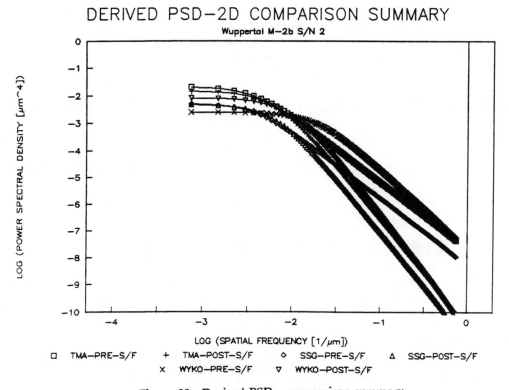

□ TMA–PRE–S/F   + TMA–POST–S/F   ◇ SSG–PRE–S/F   △ SSG–POST–S/F

× WYKO–PRE–S/F   ▽ WYKO–POST–S/F

Figure 20. Derived PSD$_{2D}$ comparison summary

Scatter and roughness measurements on
optical surfaces exposed to space

Dirk-Roger Schmitt, Helmut Swoboda*), Helmut Rosteck

Deutsche Forschungsanstalt für Luft- und Raumfahrt e. V. (DLR)
Institut für Flugführung
Flughafen, W-3300 Braunschweig, Germany

ABSTRACT

Optical systems which are used in the low orbits of the US Space Shuttle can be destroyed within some hours due to degrading by atomic oxygen. To investigate these effects the Surface Effects Sample Monitor SESAM was developed. SESAM will be mounted on the ASTRO-SPAS satellite and will carry different sets of optical samples. Different samples will be exposed during the launching, orbit and landing phases. Before and after each exposition phase the surfaces are hermetically sealed. The scattering characteristic of the samples are measured with a modified TIS apparatus, which allows the investigation of bare transparent substrates. Further information on the surface characteristics are collected using Nomarski differential interference contrast microscopy to obtain highly resolving 2D-images.

1. INTRODUCTION

Nowadays a lot of optical systems as telecopes (optical and x-ray), scanning systems, and earth observing cameras are launched into space. To achieve an improvement in resolution the requirements on the optical quality of all components are increasing. Especially the roughness of the optical interfaces plays an important role for the reduction of stray light and scattering.

Unfortunately most systems only can be launched into the rather low orbits of the Space Shuttle. In these low orbits an appreciate concentration of atomic oxygen exists which can destroy any optical surface. In previous Shuttle flights these surface degradation effects have been clearly observed. Especially a glow on different types of materials was seen[1,2]. But it was found out, that materials which are unstable in the low earth orbit environment were weak glow producers; stable materials produced more intense glow[3].

From our knowledge the details of the degradation are not yet fully investigated or understood. Only few informations exist on the damage threshold of different optical surfaces. Especially a degradation of these materials will roughen the front surfaces of the optical systems and increase the amount of scattering.

To get more information on the damage threshold with respect to scattering, we developed the **Surface Effects Sample Monitor SESAM**. This experiment will give us the possibility to expose different sets of optical materials to space conditions and to examine the samples with respect to their changes in scattering. The results obtained by this experiment are of great importance for the design of future optical systems for space applications.

2. CARRIER FOR THE EXPERIMENTAL UNIT

The experiment will be mounted (amongst many others) on the ASTRO-SPAS satellite, where SPAS means Shuttle Pallet Satellite. The systems serves in general as a vehicle for a cooperative German/US science programme with 4 missions. The German part of the programme is being managed by Deutsche Argentur für Raumfahrtangelegenheiten (DARA), and the satellite is being developed and manufactured by Messerschmitt-Bölkow-Blohm (MBB).

The system consists of a reusable Space Shuttle didicated satellite for some days of operation in the vicinity of the Shuttle Orbiter. The satellite will be deployed and some days later retrieved by the orbiter. So the system is an ideal carrier to expose

---

*) TD-TB

0-8194-0658-9/91/$4.00

samples with well characterized surfaces for some days to space conditions. After the end of the mission the samples can be investigated with respect to surface roughness and scattering.

During the landing phase the cargo bay of the shuttle will be vented. Unfortunately a lot of contaminating burned materials which detach from the heat shieldings of the spacecraft, will enter into the cargo bay and will overlay all non-protected surfaces. For this reason it is important to seal the samples before the landing phase.

3. SESAM EXPERIMENTAL UNIT

The concept of the SESAM experiment deals with the different effects degrading optical surfaces which are being launched into space. To get best information on these effects, three sets of samples are exposed to different conditions during the whole space shuttle mission.

Phase 1

The first set will be exposed during the shuttle launching phase. This special part of the experiment is carried out to collect information on the contamination with dust particles. This is of great importance for future telescope systems which shall be launched by the shuttle. In most cases the surfaces are not protected, however, they are sensitive on contamination with dust. For only less information on the clean conditions in the cargo bay during launching exists, this part of the experiment will improve the knowledge how to keep optical surfaces clean during launching. After that phase these samples are sealed.

Phase 2

In the second phase of the experiment another set of samples is ready for the exposition to space conditions. After the satellite has been developed from the orbiter the samples will be exposed to obtain information on the degradation due to atomic oxygen. Before the retrieval the samples will be sealed again.

Phase 3

A third set of samples will be exposed during the landing phase to collect information on the contaminations by the burned material entering the cargo bay after venting.

A good vacuum sealing of the samples of phases one and two is fundamental to get them into the laboratory within any other contamination.

The SESAM experimental box is shown in Fig. 1.

The unit consists of a closed box on which a cap is mounted. The cap has a set of holes with 50 mm in diameter. Through these holes the test surfaces are exposed. The samples (25 mm diameter) are mounted on a movable support behind the cap. The movement of the support provides, that only some samples, corresponding to the different phases of the experiment, are exposed through the holes to the environmental conditions. The principle will become clear regarding Fig. 2.

Twenty samples are mounted on the support. During phase 1, the shuttle launching phase, four samples are behind the holes and are exposed to the environment. The other samples are shielded behind the cap (Fig. 2a). At the end of that phase the support is shifted one step that now the eleven samples of phase 2 are exposed (Fig. 2b). Before the retrieval of the system the support is shifted a second time to cover the samples of phase 1 and 2 and to expose the five samples of phase 3. This system provides an easy mechanism to expose the selected samples to well defined environment conditions and to cover the samples after each phase. Needless to say that only the movement of the samples behind the cover is not sufficient to protect them against the contamined air entering the cargo bay during the shuttle landing phase. So additional mechanisms are installed to keep each sample under high vacuum conditions using 0-ring seals.

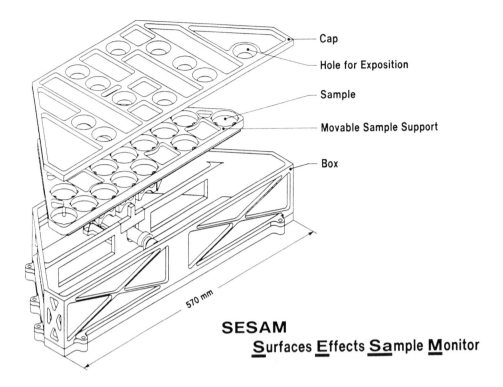

SESAM
Surfaces **E**ffects **Sa**mple **M**onitor

Fig. 1. SESAM experimental unit.

4. MEASUREMENT EQUIPMENT

To achieve information on the surface degrading effects, the total integrated scattering (TIS) of the samples is being measured after the misson. For this purposes a scatter apparatus using an integrating Ulbricht sphere is being used. The system is shown in Fig. 3.

The light of a helium-neon laser is being s-polarized and chopped. Some of the light intensity passes through mirror 1 to reach the reference detector to give the signal I_{ref}. The reflected intensity is deflected again at mirror 2, which has a radius of 2 m. The beam is incident at $\pi/6$ (30°) on the sample that is mounted at the Ulbricht sphere. The specularly reflected light leaves the sphere and is absorbed in a light sink. The sample is mounted to a diaphragm that has a hole of 2 mm in diameter to block unwanted light scatter by the rear side of the transparent substrate. Only some retroscattered light from the rear surface can travel back again into the sphere. If the rear side of the sample is polished with the same quality as the front side, this light intensity is at least a magnitude lower than the scattered light from the front side. Also the light that is reflected at the rear side of the substrate is blocked by the diaphragm. To avoid scattering of this reflected light beam on the diaphragm, a sink made of a thin film-coated silicon wafer with ultralow scattering and reflection is mounted at the incident point on the diaphragm.

The sphere has a radius of 0.125 m. Inside it is coated with specially conditioned $BaSO_4$ material, which is mechanically stable and hard; it has a reflectance of 0.975. The scattered light intensity I_s is measured using a shuttered detector. Both signals, I_{ref} and I_s, are processed in lock-in amplifiers. The electronically calculated ratio I_s/I_{ref} is calibrated to 0.975 using a $BaSO_4$ reference surface that is made of the same material used for the sphere. Then the ratio is equal to the scattering:

$$S = \frac{I_s}{I_{ref}} \tag{1}$$

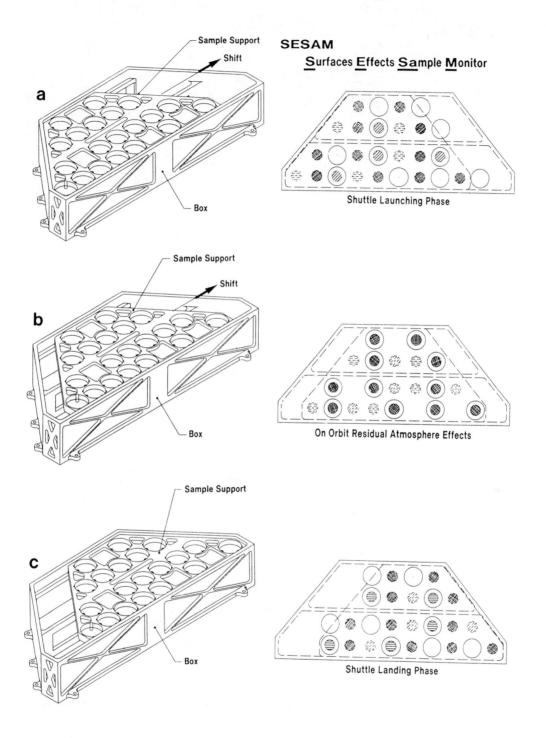

SESAM

Surfaces Effects Sample Monitor

Shuttle Launching Phase

On Orbit Residual Atmosphere Effects

Shuttle Landing Phase

Fig. 2. Principle of sample exposition and sealing at the different phases.
Left side: View on top of the box through the cap.
Right side: Box without cap showing the movable sample support.

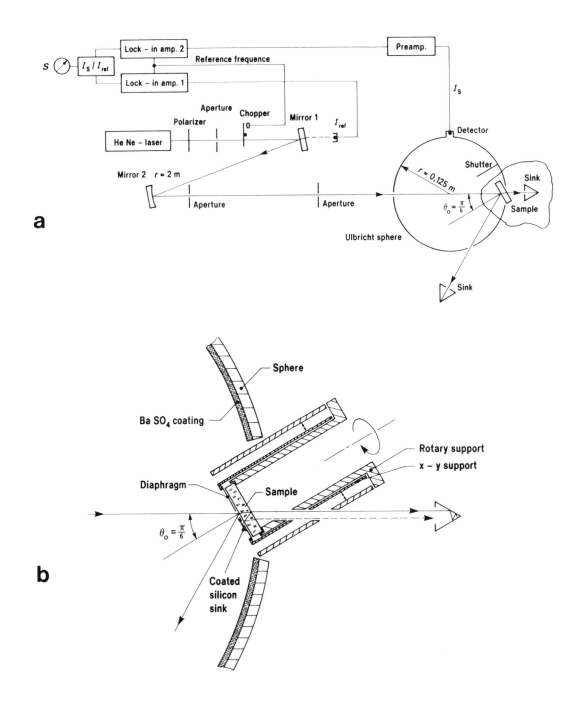

Fig. 3. Instrument for measuring S_{TIS}: (a) schematic diagram; (b) section of the sphere showing the beam path at the sample support.

the total integrated scattering becomes

$$S_{TIS} = \frac{S}{\rho_o} \qquad (2)$$

where ρ_o is the reflectance of the sample.

The DLR TIS equipment is installed in a superclean room of class 100. The scatterometer has an electronic noise of $\Delta S = 0.2 \cdot 10^{-6}$, which is the limit of the resolution. The optical background level due to diffraction and maybe due to air scattering inside the sphere is a constant of $S_o = 1.0 \cdot 10^{-6}$. The collected spatial wavelengths are from 0.6 μm to 16.5 μm. The rms roughness δ can be calculated from the S_{TIS} value if $\delta << \lambda$[5]:

$$\delta = \frac{\lambda}{4\pi \cos\theta_o} \sqrt{S_{TIS}} \qquad (3)$$

where θ_o is the angle of incidence.

With this system the TIS values of bare transparent substrates can be determined.

Another quite easy method to characterize rough surfaces is high resolution Nomarski differential interference contrast microscopy (DIC). With an optimized system rms roughness values down to 0.01 nm can be resolved [4,6].

An example is shown in Fig. 4.

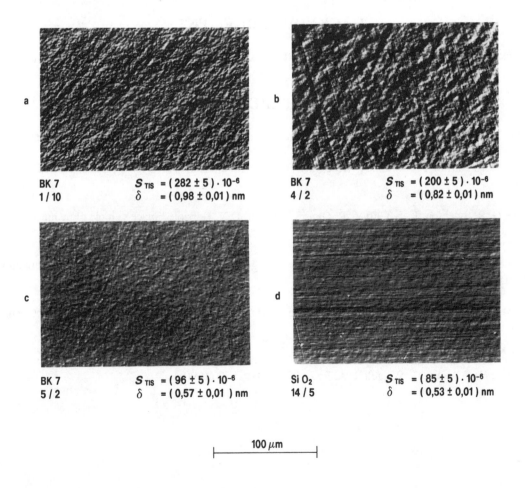

BK 7 1 / 10	$S_{TIS} = (282 \pm 5) \cdot 10^{-6}$ $\delta = (0,98 \pm 0,01)$ nm	BK 7 4 / 2	$S_{TIS} = (200 \pm 5) \cdot 10^{-6}$ $\delta = (0,82 \pm 0,01)$ nm
BK 7 5 / 2	$S_{TIS} = (96 \pm 5) \cdot 10^{-6}$ $\delta = (0,57 \pm 0,01)$ nm	Si O_2 14 / 5	$S_{TIS} = (85 \pm 5) \cdot 10^{-6}$ $\delta = (0,53 \pm 0,01)$ nm

100 μm

Fig. 4. Photomicrographs of different polished surfaces. The rms roughness δ has been calculated from the total integrated scattering S_{TIS}.

The photomicrographs were taken of different optical surfaces using DIC. With this system a rather high contrast could be achieved. The total integrated scattering S_{TIS} was measured using the instrument described above. Then the rms roughness δ was calculated using equation (3). It clearly can be seen that the roughness data obtained by the scattering method can be correlated to the qualitative interpretation of the photomicrographs.

5. MISSION SCENARIO FOR THE SESAM EXPERIMENT

The ASTRO-SPAS system as carrier for SESAM will be launched on 1 Dec 1992. Inclination will be $28.5° \pm 0.1°$ with an orbital height of 296 km $\pm$ 50 km.

A further mission is planned for 1993. Two identical SESAM units will be mounted to the satellite system, so that for each mission 40 places for samples are available.

6. CONCLUSIONS

We presented an extraordinary application of scatter measurements. The SESAM optical experiment has been developed to investigate the changes in scattering behaviour of optical samples exposed to space conditions, including launching and landing phases. Care must be taken to seal the samples after each exposition phase. Sample investigation will be carried out using a total integrated scatter apparatus, modified for the application on transparent substrates, and a Nomarski microscope system.

7. ACKNOWLEDGEMENTS

We want to thank R. Wattenbach of DARA and the ASTRO-SPAS Team of MBB for their good cooperation in the SESAM project.

8. REFERENCES

1. P. M. Banks, P. R. Williamson, W. J. Raitt, "Space Shuttle glow observations", Geophys. Res. Letters **10**, 118 (1983).

2. S. B. Mende, O. K. Garriot, P. M. Banks, "Observations of optical emissions on STS-4", Geophys. Res. Letters **10**, 122 (1983).

3. S. B. Mende, G. R. Swenson, K. S. Clifton, R. Gause, L. Leger, O.K. Garriot, "Space vehicle glow measurements on STS 41-D", J. Spacecraft **23**, 189 (1986).

4. D.-R. Schmitt, "Characterization of mirror surfaces for laser-gyro applications", in: Surface Measurement and Characterization, J.M. Bennett, ed., Proc. SPIE **1009**, 155 (1989).

5. H. E. Bennett, J. O. Porteus, "Relation between surface roughness and specular reflectance at normal incidence", J. Opt. Soc. Am. **51**, 123 (1961).

6. D.-R. Schmitt, "Characterization of supersmooth surfaces with roughnesses below 0.1 nm", in: Optics in Complex Systems, F. Lanzl, H.-J. Preuss, G. Weigelt, eds., Proc. SPIE **1319**, 572 (1990).

Surface roughness measurements of spherical components

Chen Yi-Sheng   Wang Wen-Gui

Shanghai Institute of Optics and Fine Mechanics, Academia Sinica
P.O. Box 800-211, Shanghai, 201800, China

ABSTRACT

In this paper, we deal with the laser measurement of surface roughness from spherical optical components (both convex and concave samples) by means of a self-made total integrated scattering apparatus and set out some experimental results. The radius of curvature of the spherical surface can be as much as 50 mm.

1. INTRODUCTION

No matter whether the surface properties of optical surfaces or other precision mechanical components are concerned, surface roughness is an significant parameter which has direct relation with the performance and life-spans of products. As to transparent optical components, the scattering caused by surface roughness is more important than scattering caused by inhomogeneities in the bulk material. And as for metal surfaces and optical coating surfaces with high reflectance, scattering is more a surface effect than any thing else. Therefore, measuring and assessing surface roughness, which have arisen much interest, become an item of exploitation in technique development. Some mechanical profilometers that are used for direct measurement of surface profile such as Telestep, Telesurf and some non-contact methods based on the light scattering measurement technique have emerged accordingly.

The root mean square (rms) roughness consisted of surface microirregularities is an effective method of evaluating the surface qualities. It is more direct to measure the rms microroughness with total integrated scattering (TIS). This method, extended from university and government laboratories into industry, has got a wide application and formed the ASTM standard (American Society for Testing and Materials) in measuring the statistical average values for optical surface roughness. The total integrated scattering apparatus[1,2] in our laboratory has accumulated a lot of data and experience in measuring planar optical components. As far as optical manufacturing industry is concerned, however, the roughness measurement of spherical surfaces is as important. It is beneficial to notice and analyse the scattering loss and the microstructure of these surfaces. The stylus on conventional mechanical profilers can follow spherical surface to do the measuring, but will bring such damage and unreliability to extremely high precision surfaces, surfaces of soft materials and semiconductor wafers that the measuring may become unable to be carried on. As a result, a non-contact method to measure the roughness of spherical surfaces is often necessary. This paper deals with the roughness measurements of spherical surfaces by means of TIS technique.

2. PRINCIPLES OF MEASUREMENT

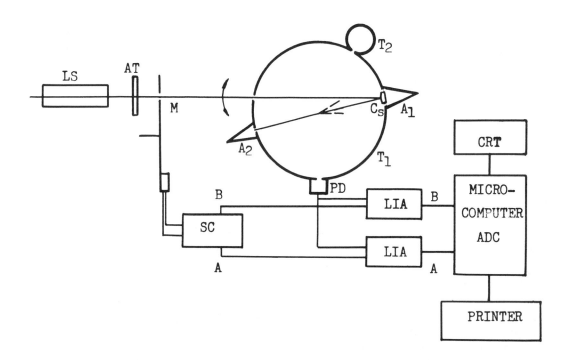

LS: He-Ne laser,   M: Light modulator (chopper)
AT: Attenuator, T_1: Measuring sphere
T_2: Reference sphere, A_1: Transmitting light trap
A_2: Reflecting light trap,   PD: Photoelectric detector
C_s: Sample,   SC: Synchronizer
LIA: Lock-in amplifier

Figure 1.   Schematic diagram of the measuring principle

We have made a total integrated scattering apparatus as shown in Figure 1. The incident light I (see also Figure 2.) enters the measuring integrated sphere and hits the planar sample to be measured and its reflected light I' in the direction stipulated by the law of light reflection will gather at the entrance (shown in Figure 1) of absorbing light trap A_2 on integrating sphere. The principle for designing the aperture of the light trap entrance is : This light beam and all the light in dimensions corresponding to the diffraction limit are absorbed by the light trap so that the scattered light caused by surface roughness is collected by an integrating sphere and then detected with a photoelectric detector (PD). The total integrated scattering (TIS) and the surface roughness σ (the rms height of surface microirregularities) from the sample surface can be obtained from the following equations

$$TIS = \frac{1}{1-f}\left(1 - f\frac{C}{A}\right)\frac{B}{C} \qquad (1)$$

$$\sigma = \frac{\lambda}{4\pi}\sqrt{TIS} \qquad (2)$$

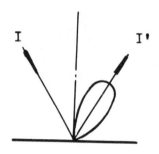

Fig. 2

In (1), (2) equations

f: the opening rate of the integrating sphere.

A: the measured value of the incident light entering the referential sphere T_2

B: the measured value of scattered light from the sample when the incident light shoots the sample C_s

C: the measured value of the incident light hitting the wall of the measuring sphere T_1

λ: the wavelength of the incident light.

but as for a spherical surface, it is a different case. Take the spherical sample as shown in Fig. 3, for example. The incident light beam has always its fixed size. When it is irradiated on the surface of the component, the normal in various positions in the irradiating area of the light beam have different directions. Thus, after the light beam goes through reflection on the sample, the parallel light beam of incidence IK turns into divergent reflection beam I'K', which can't be absorbed completely by the light trap. In the meantime, it mixes the reflective light and the scattering light. Therefore, the ability of apparatus in providing surface information is damaged. The more the divergence, the more the harm. The extent of divergence in light beam has something to do with the radius r of the light beam incident upon the sample surface and the radius of curvature of the sample surface R. On hitting the sample, the light forms an illuminated spot, and when its angle of spread towards the curvature center of the sample is 2θ, there is an approximate as shown below:

$$R \sin\theta = r \qquad (3)$$

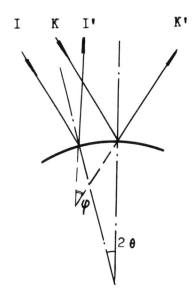

Fig. 3

When the angle $\varTheta$, in other words the ratio r/R is reduced to a certain extent, the surface of spherical sample approximate a plane. Experiments show that when

$$\text{Sin } \varTheta = r/R \leq 0.01 \,,$$

we can measure these spherical samples.

If the aperture of the reflective absorbing light trap on the integrating sphere is D, the light trap will absorb completely the reflected beam with a divergent angle φ as well as the light within its corresponding diffraction limit, thus forming the following approximate relation as shown in Fig. 4

$$\text{L Sin } \varphi = D \mp 2r - d \qquad (4)$$

In this equation, L represents the distance between the sample and the reflect-ive absorbing light trap; d is the addition to the illuminated spot diameter brought about by diffraction; 2r is the diameter of the incident light beam; $\varphi = 4\varTheta$. As to the mathematical sign before the diameter 2r of the illuminat-ed spot, the positive sign is valid for a concave sample while the negative sign for a convex sample.

Equation (4) can decide the relation between D and L . The larger the aper-ture of the absorbing light trap is, the more the loss of scattered light inten-sity is. This leads to measuring error in the amount of scattered light. When the scattered light closes to the specular reflection, the energy is assembled;

when it drifts off the specular reflection beam, the energy declines quickly, The smoother the surface, the quicker the decline. As a result, the value D should be within proper range. In other words, it should be decided according to individual situations. From equation (3), we learn that angle θ is decided by the ratio r/R . And once angle θ is fixed, angle φ is defined readily.

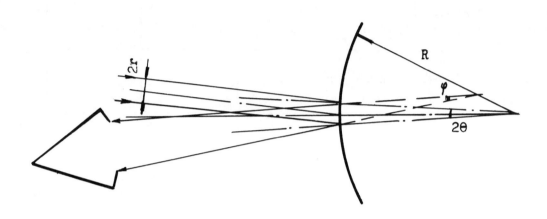

Fig. 4

In order to reduce the value D, we must reduce the value L, or shift the reflective absorbing light trap towards the sample until it is on the position of the dotted line as shown in Fig. 1. Adding accessories properly and moving the reflective absorbing light trap towards the sample as near as possible will enlarge the range of measurement. In our experiments, we have accomplished the measurement of spherical samples (both concave and convex) with the radius of curvature R = 55.46 mm.

3. MEASUREMENT RESULTS

Owing to restrictions on the apparatus structure, the reduction of the value L is very limited, The minium L is 60 mm.

The surface roughnesses measured for high reflective coating samples are given in Tables 1 and 2.

Table 1

The measuring value (unit Å) of surface roughness from high reflective coating
samples:
light beam diameter $2r = \phi 3$; the distance between sample and reflective absorbing light trap L = 60; the aperture of the reflective absorbing light trap $D = \phi 6$; Sin θ = 0.009.

Radius of curvature of samples	1000	586.9	474.2	300	165.12
Average value(Å)	20.9	21.2	22.7	22.8	24.1
Probability error (Å)	± 0.061	± 0.059	± 0.1	± 0.1	± 0.062

Table 2

The measuring value (unit Å) of surface roughness from high reflective coating
samples:
light beam diameter $2r = \phi 1$; the distance between sample and reflective absorbing light trap L = 60; the aperture of the reflective absorbing light trap $D = \phi 4$; Sin θ = 0.009.

Radius of curvature of samples	104.23	80	65.5	55.46	
Average value (Å)	21.0	19.3	23.5	24.0	
Probability error (Å)	± 0.051	± 0.079	± 0.082	± 0.06	

4. CONCLUDING REMARKS

In optical manufacturing industry, the surface roughness measurements from spherical components is significant. Stover [3] from the view of angle scattering technique, analyses the relation between the radius of curvature of curved surfaces and the surface power spectral density (PSD) function, and points out the effect of the radius of surface curvature on achieving useful data.

When the radius r of the illuminated spot on the sample is reduced to 0.5 mm and the radius of surface curvature on the sample is no less than 50 mm, reasonable results are obtainable. We have experimented on the total integrated scattering of spherical samples and got some surface roughness data, which coincide with the above results.

The results presented in this paper show that such an additional mechanism extends the application of TIS apparatus to the measurement of spherical surfaces.

5. REFERENCES

1. YiSheng Chen and WenGui Wang, " The measurement of the absolute spectral scattering and total integrated scattering ", ICO-13 Conference Digest, 546 (1984) Sapporo, Japan
2. YiSheng Chen and WenGui Wang, " Laser measurements of roughness for supersmooth surfaces ", Applied Lasers, 9 (5), 211(1989)
3. J.C. Stover, " Surface roughness measurements of curved surfaces by light scatter ", Opt. Engr., 21 (6), 987(1982)

Frequency spectrum analysis and assessment of optical surface flaws

Wenliang Gao

Department of Professional Technology Education
Hangzhou Technical College,Hangzhou, PR CHINA 310012

Xiao Zhang, Guoguang Yang

Dept. of Optical Engineering
Zhejiang University, Hangzhou, PR CHINA 310027

ABSTRACT

The assessment of optical surface flaws requires a standard which objectively reflects their influences upon an optical system and which is also widely accepted. Can the area of flaws, which has been used in most national standards, reflect objectively the performances of optical surface flaws? In this paper, calculation and analysis of the frequency spectrum distributions of typical surface flaws (including scratches and digs) in an optical system have been performed. The relationship between the spectrum intensity distribution and flaw area, depth, shape has been given. The corresponding experimental results of several 20 um wide scratches with different depth or shape of cross section have been obtained. From the results obtained above, a novel method which is called "Spectrum Energy Function Assessment" has been put forward. This method is based on the performance of surface flaws (ie. spectrum energy) instead of the area of flaws which is widely used in traditional assessment method. The comparison between the two methods has also been described. The results show that the performances of two surface flaws, which are thought to be of the same quality based on the traditional method, are greatly different in the same optical system obtained from the experimental results.

1. INTRODUCTION

Optical surface imperfections can be generally classified into two main types:[1] localized flaws and distributed imperfections. Distributed imperfections are usually represented by roughness, to which much attention has been given, and for which many corresponding testing or measuring methods have been developed;[2,3] localised flaws, including scratches, digs,cracks, etc. For a polished surface, the common flaws are scratches and digs, which are important for surface quality. The effects of surface flaws on a optical system exists in two aspects: 1. system performance effects, eg. brightness loss,

0-8194-0658-9/91/$4.00

glare phenomena, etc. "noise"of a system. surface of spetacles. performance effects. These effects can be looked at as the 2. cosmetic effects, eg. glare from the Here we discuss the first aspect, ie.

In an incoherent system under common illumination, flaws induce glare light which makes background increasing on the output plane, (this background from flaws is badly distributed) and decreases the contrast of the output. In an image system, additional fringe will appear on the expected image resulted from surface flaws.

In a laser system, because of its high energy and good coherence, the disturbance of surface flaws is more severe besides the effects stated above.

On the one hand, flaws will produce partial diffraction patterns, and different diffraction patterns which are from different surface flaws will interfere, not only with flaws on the same surface but also flaws on different surfaces in a system because the laser coherence length is usually long enough.

On the other hand, surface flaws absorb more energy than non-flaw regions, and this heat absorption can cause locked-in stress which destroy the surface and the whole optical system.

In order to assess optical surface quality, objective standard is required. The effects stated above should be considered. For functional applications of optical surfaces, objective standard should be a performance standard. There are many standards which are now widely used.[4] A brief discussion is as follows:

(1). British Scratches Standard (BS4301).[5] This standard is a visibility standard. Its assessment and traceability are based on the visibility of flaws. Reference flaws with different widths and depths should be manufactured and there is no master reference standard.[6]

(2). US MIL standard (MIL-O-13830A).[7] This standard is based on visibility through comparision of the specimen and glass artifacts (ie. standard scratches). There is master standard which is used in calibrating the second and submaster standards.

(3). German Standard (DIN3140 PART 7).[8] The German standard, which is greatly different from BS4130 and MIL-O-13830A, is based on flaw area and represented by AxS, where A is the number of approved flaws, S is the flaw grade number from flaw area. This standard neglects the other factors of flaws.

(4). Chinese National Standard --- Surface Flaws of Optical Element(Code No.87470354).[9] This standard is similar to the

German Standard (only considering the area of flaws).

Can the visibility or area exactly reflect the flaws behavior and easily compare the different flaws and standards? A detailed discussion of flaw characteristics is required.

2. FREQUENCY SPECTRUM ANAYLSIS

As shown in Fig.1, the incident light is a parallel laser beam. It goes vertically onto a specimen surface which is placed on the front focal plane of a Fourier lens. On the back focal plane is the Fourier Transform of the illuminated surface region of the specimen.

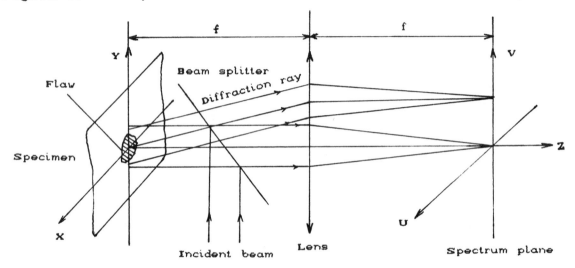

Fig.1 Theory of assessment

If the surface reflectivity function of the illuminated surface region of the specimen is $R(x,y)$, and the FT on the back focal plane is $F(u,v)$, then the intensity distribution on the plane is $I(u,v) = |F(u,v)|^2$,

where:
$$F(u,v) = \int_{-\infty}^{+\infty} \int_{-\infty}^{+\infty} I_{in} \cdot R(x,y) e^{-2\pi i(ux+vy)} dxdy \qquad (1)$$

Since the size of the illuminated surface region is much larger than the size of flaws, in order to simplify the calculation, we do not consider the effect of the limitation of the size of the incident beam.

The Frequency Spectrum Analysis of scratches and digs are shown as follows:

2.1. Scratches.

In real conditions, there are many kinds of scratches on surface with different depth, width, shape etc. We calculate two kinds of scratches : one is a scratch with a rectanglar cross section, the other is a scratch with a isosceles triangle cross section , as shown in Fig.2a and Fig.2b.

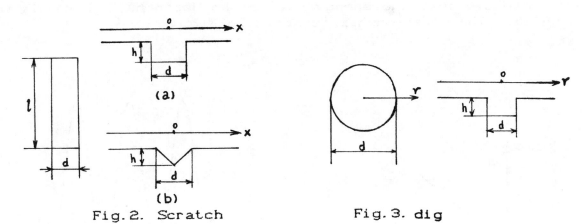

Fig.2. Scratch   Fig.3. dig

In Fig.2a, assume the amplitude reflectivity of the surface is η_o, the wavelength of the incident beam is λ and its amplitude is unit incident. Then the surface reflectivity function is

$$R(x,y)=\eta_o\left[1-\left(1-e^{j\frac{2\pi}{\lambda}2h}\right)rect\left(\frac{x}{d}\right)rect\left(\frac{y}{l}\right)\right] \qquad (2)$$

so,

$$F(u,v)=\eta_o\left[\delta(u,v)-\left(1-e^{j\frac{2\pi}{\lambda}2h}\right)dl\cdot sinc\,(du)sinc\,(lv)\right] \qquad (3)$$

The intensity is

$$I(u,v)=\eta_o^2\left[\delta^2(u,v)-4dl\cdot sin^2\left(\frac{2\pi}{\lambda}h\right)\delta(u,v)+4d^2l^2sin^2\left(\frac{2\pi}{\lambda}h\right)\cdot\right.$$

$$\left. sinc^2(du)\cdot sinc^2(lv)\right] \qquad (4)$$

The direct reflecting light is focused on the back focal point, which is a pulse function(δ) in the above equation. We do not consider the energy of the direct reflecting light. The remainder of the Frequency Spectrum Intensity distribution is

$$I'(u,v)=4\eta_o^2 d^2 l^2 sin^2\left(\frac{2\pi}{\lambda}h\right)sinc^2(du)sinc^2(lv) \qquad (5)$$

So the light energy is

$$E= \int\limits_{-\infty}^{+\infty} \int\limits_{-\infty}^{+\infty} I'(u,v)\,dudv$$

$$=4\eta_o^2 \cdot dl \cdot \sin^2\left(\frac{2\pi}{\lambda}h\right) \tag{6}$$

We can see that the energy relates to depth(h), width(d) and length(l). The relationship is shown in Fig.4.

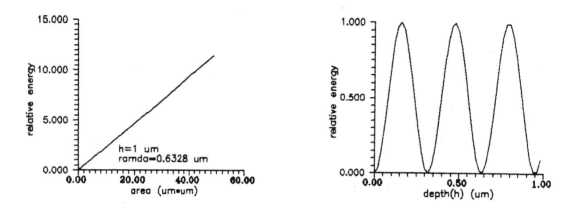

Fig.4. Relationship

In Fig.2b

$$R(x,y)=\eta_o\left[1-\left(1-e^{\,j\frac{2\pi}{\lambda}(\frac{d}{2}-|x|)\frac{4h}{d}}\right)\text{rect}\left(\frac{x}{d}\right)\text{rect}\left(\frac{y}{1}\right)\right] \tag{7}$$

The Fourier transform is

$$F(u,v)=\eta_o\left[\delta(u,v)-dl\cdot\text{sinc}(du)\text{sinc}(lv)+\frac{dl}{2}e^{\,j(\frac{\pi h}{\lambda}+\frac{\pi d}{2}u)}\text{sinc}(lv)\cdot\right.$$

$$\left.\text{sinc}(\frac{d}{2}u-\frac{h}{\lambda})+\frac{dl}{2}e^{\,j(\frac{\pi h}{\lambda}-\frac{\pi d}{2}u)}\text{sinc}(lv)\text{sinc}(\frac{d}{2}u+\frac{h}{\lambda})\right] \tag{8}$$

The intensity distribution is

$$I(u,v)=|F(u,v)|^2 \tag{9}$$

as we did above, the result of the intensity distribution, omitted pulse function, is:

$$I'(u,v)=\eta_o^2 d^2 1^2 \text{sinc}^2(lv)\left[\ \text{sinc}^2(du)+\frac{1}{4}\ \text{sinc}^2\left[\frac{d}{2}(u-\frac{t}{\lambda})\right]\right.$$

$$+\frac{1}{4}\ \text{sinc}^2\left[\frac{d}{2}\left(u+\frac{t}{\lambda}\right)\right]-\text{sinc}(du)\cos\left[\frac{\pi d}{2}\left(\frac{2t}{\lambda}+u\right)\right]\text{sinc}\left[\frac{d}{2}\left(u-\frac{t}{\lambda}\right)\right]$$

$$-\text{sinc}(du)\cos\left[\frac{\pi d}{2}\left(\frac{2t}{\lambda}+u\right)\right]\text{sinc}\left[\frac{d}{2}\left(u+\frac{t}{\lambda}\right)\right]$$

$$+\frac{1}{2}\cos(\pi du)\text{sinc}\left[\frac{d}{2}\left(u-\frac{t}{\lambda}\right)\right]\text{sinc}\left[\frac{d}{2}\left(u+\frac{t}{\lambda}\right)\right]\Bigg\}\tag{10}$$

where

$$t=\frac{4h}{d}\tag{11}$$

The light energy is

$$E=\int_{-\infty}^{+\infty}\int_{-\infty}^{+\infty}I'(u,v)dudv$$

$$=\eta_o^2 l\left[\ 2d-\frac{4\lambda}{\pi t}\cos\left(\frac{\pi dt}{\lambda}\right)\sin\left(\frac{\pi dt}{2\lambda}\right)\right]$$

$$=\eta_o^2\cdot dl\cdot\left[\ 2-\frac{\lambda}{\pi h}\cos\left(\frac{4\pi}{\lambda}h\right)\sin\left(\frac{2\pi}{\lambda}h\right)\right]\tag{12}$$

Also we can see that the energy relates to width(d), depth(h) as well as the shape of cross section. The relationship is shown in Fig.5.

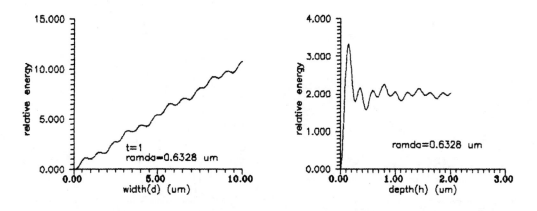

Fig.5. Relationship

2.2. Digs.

In real surface conditions, the edge shapes of digs are irregular, but they are appoximatly circular. In order to simplify the calculation, we considered the dig as a regular circle.

In Fig.3, all assumptions at scratches calculation remained, but we use polar coordinates.

$$R(r) = \eta_o \left[1 - \left(1 - e^{j\frac{2\pi}{\lambda}2h} \right) circ\left(\frac{r}{d} \right) \right] \tag{13}$$

$$I(u,v) = |F(u,v)|^2 = \eta_o^2 \left[\delta^2(\rho) - 4\sin^2(\frac{2\pi}{\lambda}h)\delta(\rho) \right.$$
$$\left. + 4\sin^2(\frac{2\pi}{\lambda}h)\frac{J_1^2(2\pi d\rho)}{\rho^2} \cdot d^2 \right] \tag{14}$$

$$I'(u,v) = 4\eta_o^2 d^2 \sin^2(\frac{2\pi}{\lambda}h)\frac{J_1^2(2\pi d\rho)}{\rho^2} \tag{15}$$

The energy is

$$E = \eta_o^2 N \cdot \pi(\frac{d}{2})^2 \cdot \sin^2(\frac{2\pi}{\lambda}h) \tag{16}$$

where

$$N = 32\int_0^{+\infty} \frac{J_1^2(x)}{x} dx \tag{17}$$

As stated above, the result also relates to depth, diameter(d) and cross section shape. Fig.6 shows the relationship.

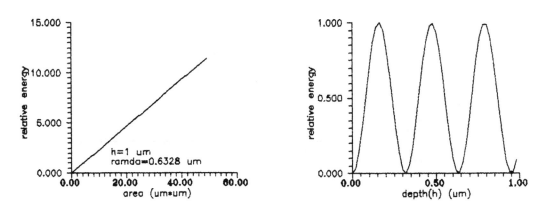

Fig.6. Relationship

We can calculate the isosceles triangle cross section digs and the same result can be obtained.

So, in a optical system, the performance of flaws relates to not only area, but also depth, shape, cross section, etc.

3. EXPERIMENTAL RESULTS AND ASSESSMENT

As we stated in section 2, the main standards for scratches and digs are based on visibility or area. From the analysis of frequency spectrum, it is clear that the performance of scratches or digs is related to not only area, but also other factors, such as depth, shape, etc. So flaw area itself can not completely reflect the performance of the flaws. Visibility, to a certain extent,[10] can reflect the performance of flaws, but existing visibility inspection schemes are, by their nature, subjective. These schemes require scratch artifacts, which make the credibility of submasters or second standards decrease.

In order to objectively assess the flaws, the authors have defined an assessment function which is called the "Spectrum Energy Assessment Function (SEAF)", as shown in Fig.1.

$$SEAF = \frac{\iint_{\sigma_{all} - \sigma_o} |F(g(x,y))|^2 \, dudv}{\iint_{\sigma_{xy}} I_{in} \, dxdy} \qquad (18)$$

where

σ_{all} : the whole spectrum plane (u,v plane).

σ_o : the theoritical focal point, Airy disc in real.

σ_{xy} : the region of detected surface which is covered by the illuminating beam.

$g(x,y)$: the reflected wavefront at the surface of specimen.
$$g(x,y) = A \cdot R(x,y)$$

A : amplitude of incident plane wavefront.

$R(x,y)$: surface reflectivity function of the illuminated surface region.

I_{in} : the intensity of incident beam.

F : Fourier Transform

From the defination of the SEAF, the SEAF includes area, depth, shape etc, it also implies the visibility of flaws, but no reference scratches or digs are required when spcimens are inspected. This makes flaw inspection objective.

A experimental setup has been developed to inspect flaws by using the SEAF.[11] The setup is shown in Fig.7.

A laser beam goes through a lean-placed mirror with a slot in the center and onto the surface of the specimen. The result is obtained through a PMT which receives the spectrum energy. A micropro-cessbased system controls scanning of the specimen and processing of the signal.

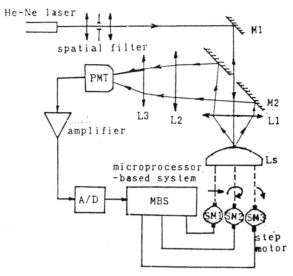

Fig. 7. Experimental setup

Several scratches have been inspected, the experimental result is shown in Fig. 8. Because we let the Iin equal to 1 in the experiment, the denominator of the SEAF is a constant.

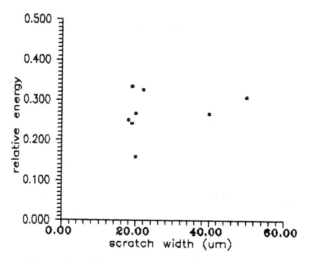

Fig. 8. Experimental result

In the above Figure, we can see that scratches of approximately equal area (width) have greatly different spectrum energies, this indicates they will have greatly different performances in an optical system.

Additionally, an assessment method using the SEAF does not require reference scratches or digs. Precise reference flaws not only are difficult to obtain , but also are not convenient for comparison of different flaws and different standards. Correspondingly , we can also define the transmissive SEAF.

Of course, the SEAF does not include every aspect of flaw performance. Some effects of even the same flaw are different from system to system .

4. DISCUSSION

The real condition of any optical surface is rough, so, for a practical inspection system, on the back focal plane is the flaw frequency spectrum plus speckle from the roughness of the illuminated region of the specimen. When the surface of a specimen is very rough, the background becomes large, and the SNR decreases.

But for a fine polished surface, statistically, the surface scatter resulting from roughness is only the background to surface flaw detection. Because the surface condition is approximately the same from region to region, the background is only a constant parameter. But what we are interested in is the variation. The definition of the SEAF has omitted the approximately constant background. The background can be substracted by signal processing.

Because of the instability of a laser beam, the background varies during the inspection of the SEAF. In order to decrease this effect, detection of the incident laser beam should be introduced, eg. adding a PMT to detect the reference light. In our experimental system, due to the limited scale of the mirror M2, the high frequency factors could not be detected. But these factors are small factors relating to frequency energy, therefore, the error is small. In a developed system, a slot in the lens should be replaced with a hole.

5. ACKNOWLEDGEMENT

The authors would like to thank associate professor Cheng Shang-Yi's discussion on the control section of the experimental setup. Thanks go also to Dr. J.C Stover's discussion at the final stage of the writing of this paper.

6. REFERENECES

1. S.Martin, "Effect of defects and surface finish on the performance of optical systems." SPIE Vol.525, pp82-93, 1985

2. D.Gilsinn,etc. "Optical roughness measurments of industrial surfaces" SPIE Vol.665, pp8-16, 1986

3. J.C Wyant "Optical profilers for surface roughness" SPIE Vol.525, pp174-180, 1985

4. Lionel.R.Baker "Scratch and dig measurement -- a way ahead"

SPIE Vol.680, pp88-94, 1986

5. A.J.Cormier, "Assessment of current scratch standards" SPIE Vol. 805, pp152-159, 1987

6. J.Slater,D.Cox "Scratches: At what price quality" SPIE Vol.654, pp68-75, 1986

7. M.Young "Scratch-and-dig standard revisited" Applied Optics Vol.25, No.12, pp1922-1929, June 1986

8. DIN 3140: Part 7,1978, "Dimensions and tolerence data for optical components, surface defects."

9. "Surface Flaws of Optical Elements," National Standard of PR CHINA. GB(CODE NO. 87470354)

10. M.Young, "The Scratch Standard is only a cosmetic standard" SPIE Vol.1164, pp185-190, 1989

11. G.Yang, W.Gao, S.Cheng, "Automatic Inspection Technique For Optical Surface Flaws." SPIE Vol.1332-08, 1990

OPTICAL SCATTER:
APPLICATIONS, MEASUREMENT, AND THEORY

Volume 1530

SESSION 3

Scatter from Be Mirrors

Chair
Donald J. Janeczko
Martin Marietta Electronic Systems

Solution for anomalous scattering of bare HIP Be and CVD SiC mirrors

Cynthia L. Vernold

Hughes Danbury Optical Systems, Inc.
100 Wooster Heights Road, M/S 826
Danbury, Connecticut 06810

ABSTRACT

Infrared (IR) Bi-directional Reflectance Distribution Function (BRDF) curves from bare super-polished hot isostatic pressed (HIP) beryllium (Be) and chemical vapor deposited (CVD) silicon carbide (SiC) mirrors are much higher and have drastically different slopes than those predicted from either visible (VIS) BRDF or surface profilometry data. The end result is that current state-of-the-art HIP Be and CVD SiC will not meet some low scatter requirements. This paper presents data showing that this anomalous IR scattering effect can be easily "covered-up" by coating the bare super-polished substrates.

1. INTRODUCTION

IR scatter from super-polished HIP Be and CVD SiC mirrors has consistently failed to obey topographic surface scatter models. [1,2] When BRDF data is plotted in power spectral density (PSD) versus frequency space, this anomalous behavior becomes obvious.

If a sample is smooth with respect to the test wavelength, clean, and front surface reflecting, then it will have "well-behaved" scattering characteristics. Based on both Fourier linear systems theory[3] and a more vigorous Rayleigh-Rice vector perturbation theory,[4] a well-behaved surface can be represented by a topographic PSD function. Since this PSD function is a result of errors that exist on the surface that are only topographic in nature, the magnitude of the PSD will not change as the test wavelength or the measurement technique is varied.

By evaluating HIP Be and CVD SiC with respect to being "well-behaved" surfaces, one can begin to understand the source of the anomalous scattering. Since bare HIP Be and CVD SiC samples have been shown to be well polished with low surface roughness, they qualify as being smooth. In addition, if care is taken while performing BRDF measurements, the samples can also be kept clean. This indicates that both the smoothness and cleanliness criteria are satisfied for representing the scatter from these surfaces using a topographic PSD function. Thus, only the front surface reflecting criterion is not satisfied.

Given that these sample types do not appear to be front surface reflectors, the anomalous scattering problem can be solved by not allowing the incident light from penetrating into the substrate material. One approach to accomplishing this is to coat the bare polished substrate with a thin opaque coating or a thick polished coating.

This paper describes the two most commonly used techniques for evaluating polished surfaces:

- Multi-wavelength scatterometry (BRDF)
- Optical phase measuring interferometry (surface roughness)

BRDF data indicating anmalous behavior, is shown for substrates of HIP Be and CVD SiC. Surface roughness is evaluated before and after coating to show that the coating process does not change the measured surface roughness of the substrates. Additionally, BRDF data at 0.6328 µm, 3.39 µm, and 10.6 µm is plotted in PSD space to illustrate the impact of coating an anomalously scattering sample. In conclusion, this paper shows that coating bare super-polished HIP Be and CVD SiC eliminates their anomalous scattering characteristics.

0-8194-0658-9/91/$4.00

2. TECHNIQUES USED FOR SURFACE CHARACTERIZATION

A Bi-directional Scatter Distribution Function (BSDF) scatterometer system is used to measure the angular dependence and magnitude of scattered radiance with respect to the incident irradiance. It is an optical non-contact method of measuring the scattered light distribution that results from a rough surface. Figure 1 is an optical schematic of the Hughes Danbury Optical System's (HDOS) BSDF scatterometer.

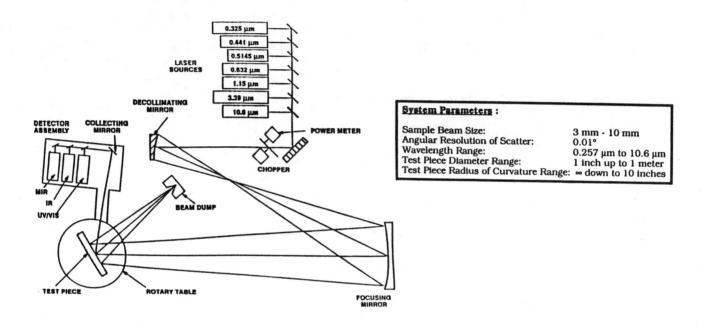

System Parameters:	
Sample Beam Size:	3 mm - 10 mm
Angular Resolution of Scatter:	0.01°
Wavelength Range:	0.257 µm to 10.6 µm
Test Piece Diameter Range:	1 inch up to 1 meter
Test Piece Radius of Curvature Range:	∞ down to 10 inches

Figure 1. Optical schematic of HDOS BSDF scatterometer

The micro-phase measuring interferometer (µ-PMI) is an optical non-contact method for measuring surface micro-roughness. A surface profile is measured relative to an internal reference surface, where the phase change of the reflected light is interpreted as a height difference on the surface. Figure 2 illustrates the way surface microtopography is imaged onto the pixels in a µ-PMI detector system.

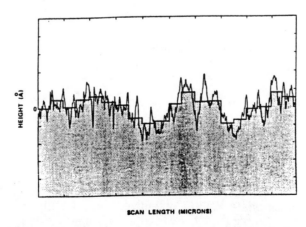

Figure 2. Illustration of how surface roughness is imaged onto linear detector array

BSDF instrumentation is used to evaluate the scattering characteristics of a surface while µ-PMI optical profilometry is used to evaluate the roughness or topography of a surface. In comparing measurement results from both instruments, it is imperative to know the spatial period bandwidth that each instrument is sensitive to.

The spatial period bandwidth sensitivity of the BSDF instrument can be varied using a combination of the test wavelength and incident angle. The WYKO TOPO-2D, the µ-PMI used for this paper, has an operating wavelength of 0.633 µm. However, the bandwidth sensitivity of the WYKO µ-PMI is not varied by changing the source wavelength, but rather by changing the objective head magnification. This changes the WYKO field-of-view (FOV) which means that the evaluation area on the surface changes. Figure 3 compares the spatial bandwidths associated with the different WYKO objectives with the spatial bandwidths associated with the BSDF system for different wavelengths and an incident angle of one degree.

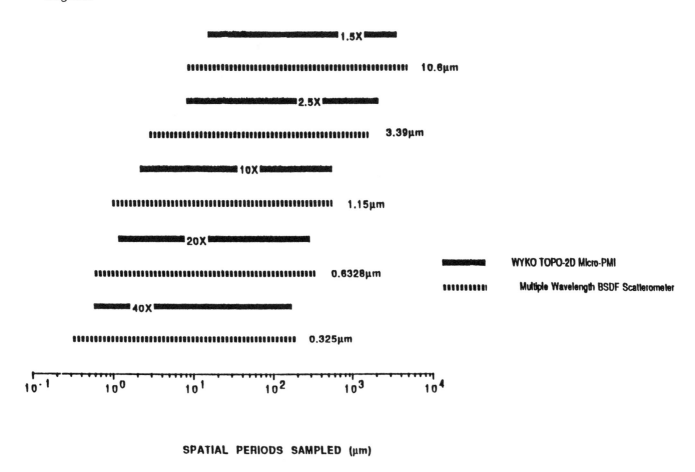

SPATIAL PERIODS SAMPLED (µm)

Figure 3. Comparison of spatial period bandwidth sensitivities of the WYKO µ-PMI and the BSDF scatterometer

For instance, if one were interested in how a surface would scatter at 10.6 µm, then the most appropriate WYKO objective to use to determine the root-mean-square (rms) roughness of the surface would be the 1.5X or the 2.5X. On the other hand, if the surface under test was to be used in an ultraviolet (UV) system, then the most appropriate WYKO objective would be the 40X. Although the spatial bandwidths in these two examples do not exactly overlap, they are very close, and thus the rms derived roughness from both instruments will be similar if the surface under test is smooth, clean, and front surface reflecting.

Making assumptions that the scattering behavior of a surface is isotropic allows the use of BRDF data to calculate a BRDF derived rms roughness value for a surface. This value can then be compared to the rms roughness value derived from optical profilometry. Mapping BRDF data and surface profile data into a PSD versus frequency format is another method used to compare data from both measurement techniques. Even though both comparison techniques have limitations associated with them, their use is very advantageous when trying to discern the sources of scatter from a surface.

Surface roughness determined from an optical profiler, and visible scatter (BRDF) measurements do not accurately predict the level of IR scatter from both bare HIP Be and CVD SiC. Surface roughness data indicates that even though both types of samples exhibit anomalous scattering behavior, they have extremely smooth surface topography. That is to say that their surface finish parameters are similar to state-of-the-art glass surfaces. Their rms surface roughness values are below 10 Å over a spatial bandwidth of approximately 0.5 mm down to about 1 μm.

3. WAVELENGTH SCALING RESULTS

If a surface qualifies as smooth, clean, and front surface reflecting, or in other words, purely a topographic scatterer, then the BRDF data can be used to calculate the surface power spectral density (PSD) function. Figure 4 illustrates how to convert "BRDF" versus "Angle From Specular" data into "PSD" versus "Frequency" data.

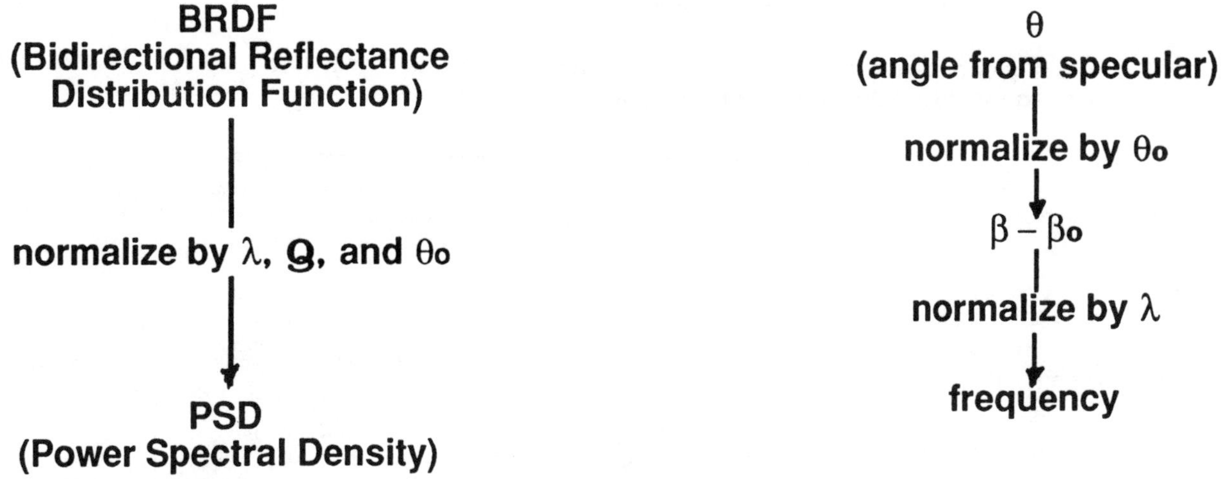

BRDF
(Bidirectional Reflectance Distribution Function)

$\downarrow$

normalize by λ, Ω, and θ_0

$\downarrow$

PSD
(Power Spectral Density)

θ
(angle from specular)

$\downarrow$

normalize by θ_0

$\beta - \beta_0$

$\downarrow$

normalize by λ

$\downarrow$

frequency

$$PSD_{2D} = \frac{BRDF \ \lambda^4}{16 \ \pi^2 \cos\theta_i \ \cos\theta_s \ \Omega}$$

$$frequency = \frac{\beta - \beta_0}{\lambda} = \frac{\sin(\theta_0 + \theta) - \sin\theta_0}{\lambda}$$

Figure 4. Flow diagram illustrating the conversion of BRDF data into PSD data

Representing BRDF data in PSD space uses a wavelength scaling law based on Fourier linear systems theory. This theory can be useful in determining if a surface has "well-behaved" scattering characteristics. If a surface is well-behaved, it will:

- Exhibit linear-shift-invariant behavior. [5]
- Wavelength scale. [1,6]

Linear shift-invariant behavior, in this context, means that the scattering function of a surface does not change as the incident angle of the test wavelength is varied. If a sample "wavelength scales", its scattering function,or its BRDF, at one wavelength can be ascertained from scatter data taken at another test wavelength.

This wavelength scaling law and linear shift invariance allows one to predict scattering behavior at various wavelengths and incident angles by measuring scatter at only one wavelength and incident angle. The elegance of this is that a surface's scattering function is expressed in terms of its surface PSD, which is invariant regardless of measurement technique or measurement parameters. Once BRDF data is plotted in PSD space, all the BRDF curves that represent the scattering characteristics of a surface at different wavelengths and incident angles must fall on top of one another if the sample is said to wavelength scale.

Some optical materials do not exhibit this wavelength scaling over a broad wavelength region. Two of those samples are super-polished bare HIP Be and CVD SiC. It is important to note that this non-wavelength scaling is apparent only when these surfaces are well polished such that their topographic roughness is not the dominant scattering mechanism.

Figure 5 is plot of bare HIP Be which shows four BRDF curves from the UV through the VIS that have been converted to "PSD" versus "Frequency" space.

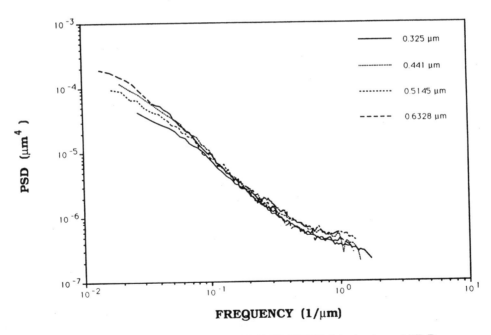

Figure 5. PSD vs. frequency of UV/VIS BRDF data for bare HIP Be

This data indicates that bare HIP Be wavelength scales from the UV through the VIS. Thus, one can measure the scattering characteristics of this surface at only one of these wavelengths and accurately predict how the surface will scatter at another wavelength. In addition, optical μ-PMI data can be used to accurately predict the scattering characteristics of this sample since the scatter is primarily due to topographic errors.

Figure 6 indicates the same wavelength scaling results for CVD SiC as for the bare HIP Be.

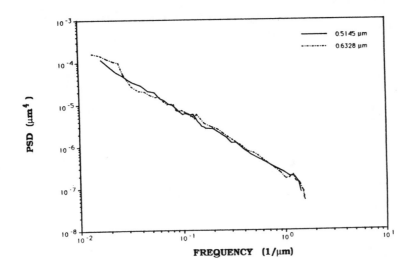

Figure 6. PSD vs. frequency of VIS BRDF data for bare CVD SiC

Figure 7 is an example of a nickel sample that not only wavelength scales from the UV through the visible, but also through the mid-infrared (MIR).

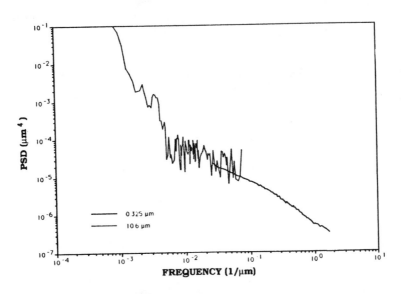

Figure 7. PSD vs. frequency of UV/MIR BRDF data for bare nickel

Note that the PSD data derived from MIR BRDF extends the PSD derived from UV BRDF into a lower frequency band. Scatter in the MIR wavelength regime is due to larger spatial period errors on the surface than scatter in the UV/VIS wavelength regime. One can measure the BRDF of this sample at a UV wavelength and accurately predict its scattering characteristics at a MIR wavelength. This also indicates that surface topography is the dominant scattering mechanism, and optical μ-PMI measurements can be used to accurately predict the scattering characteristics of this surface over a large spatial period range.

Figure 8 is a plot of the PSD derived from the UV scatter measurements on the bare HIP Be sample and the PSD derived from the MIR scatter measurements.

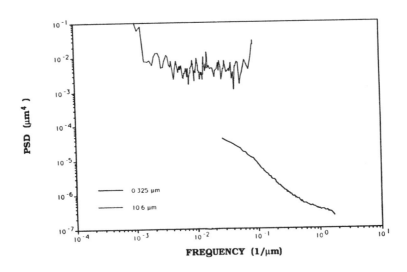

Figure 8. PSD vs. frequency of UV/MIR BRDF data for bare HIP Be

The curve that corresponds to the MIR BRDF data is much higher than we expected based on the UV measurements. It also has a drastically different slope, almost zero, in contrast to the slope of the UV measurements which is approximately -2. Thus, topographic wavelength scaling does not apply for super-polished bare HIP Be. One can not use UV or VIS scatter measurements, or optical μ-PMI measurements to predict how this type of sample will scatter in the MIR.

Figure 9 shows similar non wavelength scaling results for the CVD SiC sample.

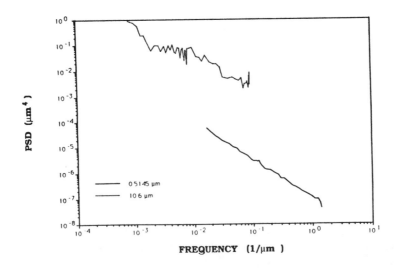

Figure 9. PSD vs. frequency of VIS/MIR BRDF data for bare CVD SiC

Again, this sample does not wavelength scale through the MIR. Thus, neither VIS scatter measurements nor μ-PMI data can be used to accurately predict how this surface will scatter in the MIR.

Some questions that immediately come to mind:

- What is the source of this increase in scatter?
- Is there any wavelength dependence of this anomalous scattering behavior?
- Does this non-wavelength scalability become more extreme as the IR wavelength increases?

Figure 10 is another plot of PSD versus frequency for a bare HIP Be sample, but this time the data includes the MIR wavelength 3.39 μm.

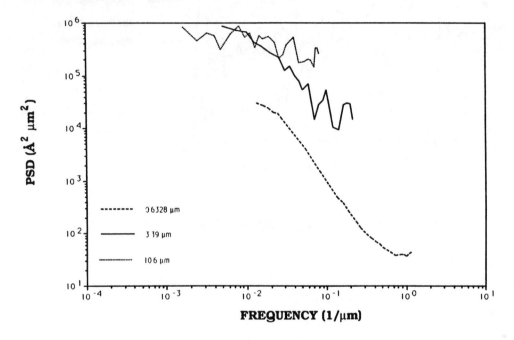

Figure 10. PSD vs. frequency of UV/MIR BRDF data for bare HIP Be

It is obvious from Figure 10 that the anomalous scattering is indeed wavelength dependent. It is also apparent that the magnitude of this effect increases as the test wavelength gets larger. The source of this anomalous scatter is still under investigation in the scatter community, but in the meantime a solution is needed.

5. ELIMINATION OF ANOMALOUS SCATTERING EFFECT BY COATING

Since the surface roughness of bare HIP beryllium over a spatial bandwidth of 1.3 μm to 0.66 mm is approximately 8 Å rms and CVD silicon carbide is approximately 5 Å rms, there is not much the optician can do to improve upon the polishing technique. This surface topography is comparable to state-of-the-art glass surfaces. So, the anomalous scattering problem does not appear to be related to the topographic defects present in the surface.

5.1 Coating with amorphous aluminum

To verify that this effect is non-topographic, half of a bare HIP beryllium sample and half of a bare CVD silicon carbide sample were coated with a thin, opaque (1000 Å) layer of aluminum (Al). This thin coating layer served to preserve the topographic characteristics of the surfaces while acting as a barrier to the test wavelength. The coating did not allowing the test wavelength to "see" the bare substrate. Figure 13 and

14 are the "BRDF" versus "Angle From Specular" results for the bare and Al coated HIP Be sample and the bare and Al coated CVD SiC sample respectively.

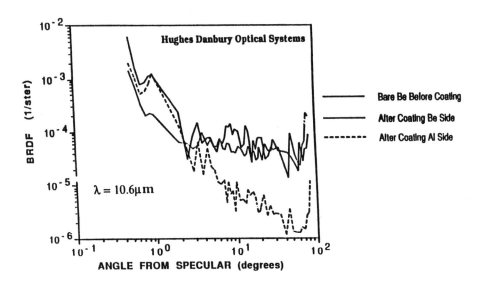

Figure 13. BRDF vs. angle from specular for bare HIP Be (λ = 10.6 µm)

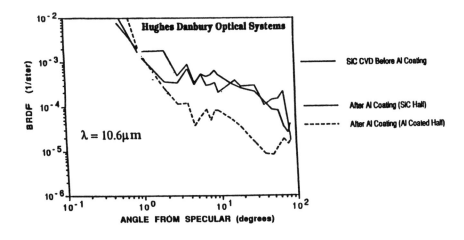

Figure 14. BRDF vs. angle from specular for bare CVD SiC (λ = 10.6 µm)

Once coated, these two samples no longer exhibit anomalous scattering behavior at the test wavelength of 10.6 µm. Both samples now scatter topographically as predicted by µ-PMI data and VIS BRDF data. This data illustrates that one solution to the anomalous scattering phenomena is to coat the substrates with a thin layer of aluminum.

5.2 Coating with amorphous-like beryllium

When using a Nomarski microscope, the polarization of the light used to view a surface can be varied. When viewing a bare Be surface using this method, the different Be grains that make up the surface are distinguished by varying shades of grey across the surface. This indicates that the reflectivity of the bare beryllium surface is polarization dependent, and varies as a function of the grain orientation. In addition, this points to a possible source of the anomalous scattering: the grain structure present in both bare HIP Be and CVD SiC.

The individual Be grains from the samples used in this paper are approximately 10 - 15 μm in size. Figure 11 is a Nomarski photomicrograph of a super-polished bare HIP Be surface showing this polarized reflectivity effect due to the Be grains. The photo was taken using a magnification of 200X.

Figure 11. Nomarski photomicrograph of bare super-polished HIP Be

This substrate's surface roughness was evaluated using a WYKO μ-PMI. The grain structure that is visible in the Nomarski photomicrograph is not apparent to any degree in the surface profile as shown in Figure 12. The rms surface roughness of this super-polished bare HIP Be substrate is approximately 7 Å over a spatial period bandwidth of 1.3 μm to 0.66 mm.

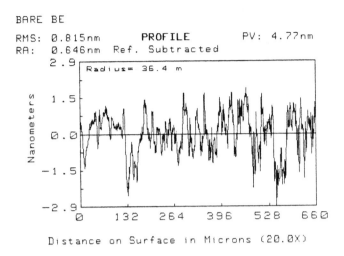

BARE BE
RMS: 0.815nm   PROFILE   PV: 4.77nm
RA: 0.646nm Ref. Subtracted

Figure 12. WYKO μ-PMI surface profile for bare HIP Be substrate

Previously in this paper, a thin aluminum coating was used to drastically alter the scattering characteristics of a bare CVD SiC or HIP Be substrate. It appears to have covered-up the polarization-sensitive grain structure of the substrates without changing their inherent topographic surface roughness. However, from a thermal matching viewpoint, the best coating match for a substrate made from beryllium is beryllium rather than aluminum. Ideally, this coating would have extremely small or nonexistent grain structure associated with it. The HDOS Be coating capability produces amorphous-like Be coatings as thin as 1000 Å and as thick as 10 μm. The thin coatings are as smooth as the substrate that they are applied to, while the thicker Be coatings are polishable to approximately 6 Å rms. Figure 13 is the WYKO μ-PMI surface profile of super-polished thick Be coating that was deposited on a bare HIP Be substrate.

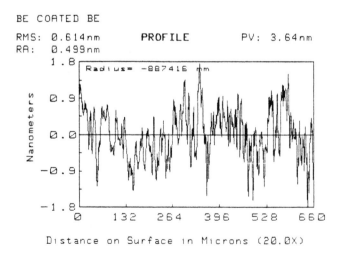

BE COATED BE
RMS: 0.614nm   PROFILE   PV: 3.64nm
RA: 0.499nm

Figure 13. WYKO μ-PMI surface profile for super-polished thick Be coating on bare HIP Be substrate

The microstructure of HDOS beryllium coatings is amorphous-like. Viewing the surface in a Nomarski microscope no grain structure is evident. Figure 14 is a Nomarski photomicrograph of the super-polished thick Be coating that was deposited on a bare HIP Be substrate.

Figure 14. Nomarski photomicrograph of super-polished thick Be coating on bare HIP Be substrate

PSD results show that once a bare Be substrate is coated with a polished Be coating it no longer exhibits anomalous scattering. Be coated Be samples wavelength scale from the VIS through the MIR. Figure 15 shows the BRDF derived PSD functions for a Be Coated Be sample.

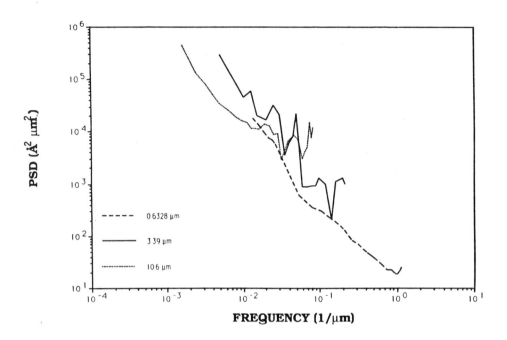

Figure 15. PSD versus frequency of Be coated Be substrate indicating wavelength scaling

6. CONCLUSIONS

Bare HIP Be and CVD SiC samples with well polished, low surface roughness surfaces were characterized. IR scatter from these super-polished HIP Be and CVD SiC mirrors consistently fails to obey topographic surface scatter models. These sample types do not appear to be front surface reflectors, and appear to have polarization dependent reflectivity characteristics. It was thus concluded that the anomalous scattering problem can be solved by not allowing the incident light to contact the substrate material.

Data was presented showing that the anomalous IR scattering effect seen on these mirrors can be "covered-up" by coating the bare super-polished substrates. HDOS's approach to accomplishing this is to coat the bare polished Be substrate with a thin opaque Al or Be coating or a thick polishable Be coating.

7. ACKNOWLEDGMENTS

The author would like to thank Robert Radomski for his participation in the collection of data for this paper, and also Sandeep Dave and Warren Wilczewski for their work in production of the beryllium coatings. The author would also like to express gratitude to Robert Harned who was very supportive in my "quest-to-find-the-answer".

8. REFERENCES

1. C. L. Vernold, Application and verification of wavelength scaling for near specular scatter predictions," Proc. SPIE 1165,18 (1989).
2. J. C. Stover, M. L. Bernt, D. E. McGary, and J. Rifkin, "An investigation of anomalous scatter from beryllium mirrors," Proc. SPIE 1165, 100 (1989).
3. J. E. Harvey, "Surface scatter phenomena: A linear shift-invariant process," Proc. SPIE 1165, 87 (1989)
4. E. L. Church, H. A. Jenkinson, and J. M. Zavada, "Relationship between surface scattering and microtopographic features," Opt. Eng. **18**, 125 (1979).
5. J. E. Harvey, "Light-Scattering Characteristics of Optical Surfaces", Ph.D. Dissertation, University of Arizona, (1976).
6. J. C. Stover, J. Rifkin, D. R. Cheever, K. H. Kirchner, T. F. Schiff, "Comparison of wavelength scaling data to experiment," Proc. SPIE 967, 44 (1988).

"Effective" surface PSD for bare hot isostatic pressed (HIP) beryllium mirrors

Cynthia L. Vernold

Hughes Danbury Optical Systems, Inc.
100 Wooster Heights Road, M/S 826
Danbury, Connecticut 06810

James E. Harvey

Center for Research in Electro-Optics and Lasers (CREOL)
The University of Central Florida
12424 Research Parkway
Orlando, Florida 32826

ABSTRACT

Our understanding of the relationship between optical surface topography and scattering behavior has improved in recent years to the point where the scattering characteristics of optical systems are routinely controlled by placing specifications upon the root-mean-square (rms) surface roughness. Bare polished hot isostatic pressed (HIP) beryllium (Be) has consistently failed to obey the established topographic surface scatter models. Surface scatter in the form of the Bidirectional Reflectance Distribution Function (BRDF) from bare beryllium mirrors using several test wavelengths will be presented in this paper. This experimental data indicates that non-topographic scattering effects are significant for bare beryllium mirrors. An empirical "effective" surface power spectral density (PSD) function will be developed that can be used to predict the scattering behavior of bare polished HIP Be from topographic rms surface roughness data.

1. HISTORICAL BACKGROUND

- Stover (1975), Church (1979), and Elson (1979), among others, have contributed significantly to our understanding of the theoretical relationship between optical surface finish and angular scattering behavior.

- Harvey (1976) illustrated that empirical BRDF data of many optical surfaces obeys an inverse power law. He also demonstrated that wavelength scaling can be predicted for well-behaved surfaces.

- Bennett, Glenn, Takas, and Wyant, among others, have made continuous improvements in surface topography measurement capabilities.

- Janeczko (1989), proposed a surface power spectrum standard for optical surface roughness specifications.

- Vernold (1989) and Stover (1989) reported on the non-wavelength scaling of the scattering behavior of bare polished HIP beryllium and chemical vapor deposited (CVD) silicon carbide.

2. BACKGROUND ON NON-WAVELENGTH SCALABILITY OF BARE BERYLLIUM

When topographic errors are the dominant scattering mechanism on a surface, the scattering function produced by that surface, its BRDF, can be used to calculate a topographic surface PSD function. Representing BRDF data in PSD space uses a wavelength scaling law based on scalar diffraction

0-8194-0658-9/91/$4.00

phenomena and Fourier linear systems theory. This theory predicts that the scattering function from a well-behaved optical surface will be invariant with respect to changes in the incident angle.[1] This "shift invariance" with respect to incident angle leads to the derivation of a "surface transfer function", which in turn relates scattering properties to surface topography with a simple wavelength scaling law based on an "angle spread function".[2] If departures from shift invariance are severe or if polarization effects are significant, then a more vigorous vector treatment is required.[3-5] Equation 1 illustrates the formula used to convert BRDF data into PSD data:[6,7]

$$PSD_{2D} = \frac{BRDF \; \lambda^4}{16 \; \pi^2 \cos\theta_i \cos\theta_s \; Q} \qquad [\mu m^4]$$

(1)

The test wavelength is represented by λ. The incident angle of the test wavelength is represented by θ_i, while the scattered angle of the light is θ_s. If polarization effects are not significant, then the polarization factor, Q, is the surface reflectivity.

This wavelength scaling law allows one to measure the scattering characteristics of a surface at only one wavelength and be able to accurately predict how the surface will scatter at another wavelength. This allows a surface's scattering function to be expressed in terms of its surface PSD, which is invariant regardless of measurement technique. Thus, once BRDF data is plotted in topographic PSD space, all the BRDF curves that represent the scattering characteristics of a surface at different wavelengths and incident angles must fall on top of one another if the sample is scattering solely due to topographic defects.

Figure 1 is a plot of scatter data taken from a bare HIP Be mirror that has grain sizes on the order of 10 - 15 μm. The curves represent BRDF data from the ultraviolet (UV) through the visible (VIS) that has been converted to PSD data.

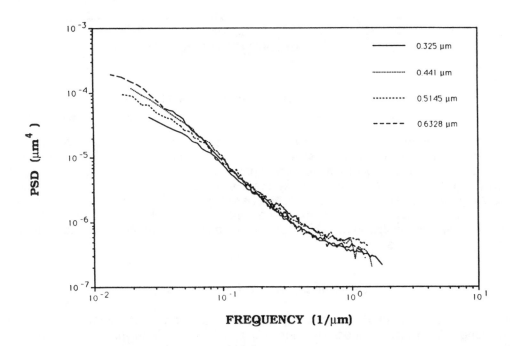

Figure 1. PSD vs. frequency of UV/VIS BRDF data for bare HIP Be

The fact that the PSD curves predicted from BRDF data at different wavelengths agree so well indicates that we can use this simple topographic model to predict scattering behavior at one wavelength from measurements made at another wavelength. This agreement also indicates that topographic scatter is the dominant scatter mechanism, and we can use this theory with confidence in specifying surface finish tolerances to satisfy scattered light requirements.

This simple topographic theory is not applicable when using BRDF data in the IR from bare HIP Be mirrors[6,8]. Figure 2 is a plot of the PSD derived from UV BRDF measurements on bare HIP Be and the PSD's derived from two MIR BRDF measurements at 3.39 μm and 10.6 μm.

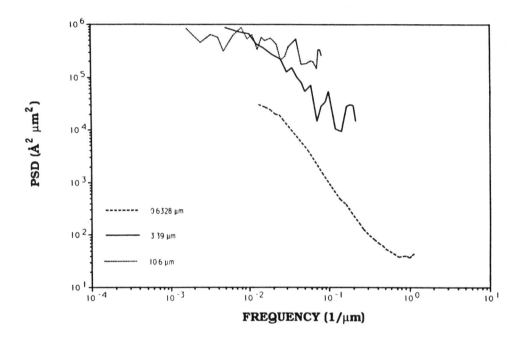

Figure 2. PSD vs. frequency of UV/MIR BRDF data for bare HIP Be

It is obvious from Figure 2 that the anomalous scattering is wavelength dependent, with the magnitude of this effect increasing as the test wavelength is increased. The curves that correspond to the MIR BRDF data are much higher than we expect based on the UV measurements. They also have different slopes than the UV measurement which has a slope of approximately -2. Thus, topographic wavelength scaling does not apply for super-polished bare HIP Be. One can not use UV or VIS scatter measurements, or surface finish measurements in conjunction with current scattering models to predict how this type of sample will scatter in the IR. Thus, in the case of IR scatter, new criteria are needed to specify surface finish tolerances.

3. EMPIRICAL FIT TO BRDF DATA

Theoretical investigations of various non-topographic scattering mechanisms are being made to determine the origin of this anomalous behavior in bare HIP Be. Meanwhile, an empirical fit to the smoothed PSD data from the BRDF measurements at 0.6328 μm, 3.39 μm, and 10.6 μm has been made as shown in Figure 3.

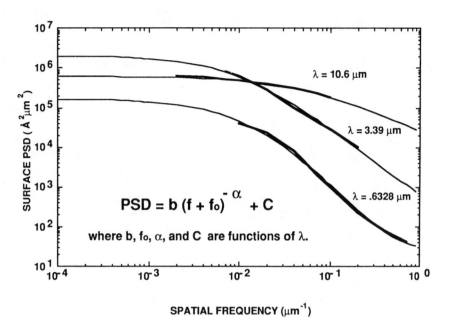

Figure 3. Empirical fit to BRDF derived PSD data

The result of this empirical fit allows us to formulate a wavelength-dependent "effective" surface PSD for this type of bare polished HIP beryllium. This "effective" surface PSD can, in turn, be used when deriving optical surface finish specifications. Surface finish is a parameter that can easily be measured using an optical or mechanical profiler.

4. THE EFFECTIVE SURFACE PSD OF BARE BERYLLIUM MIRRORS

The mathematical fit that describes the the effective surface PSD function is illustrated in Equation 2 and Figure 4:

$$PSD(f)_{eff} = b(\lambda) \, [f+f_o(\lambda)]^{-\alpha\lambda} + C(\lambda) \quad [\mu m^4] \qquad (2)$$

where:

$b(\lambda) = -0.5624\lambda^2 + 0.8485\lambda - 0.1402$
$f_o(\lambda) = 0.0589\lambda^2 - 0.3764\lambda + 1.6166$
$\alpha(\lambda) = 0.0107\lambda^2 - 0.2600\lambda + 2.5600$

$C(\lambda) = 0.00004\lambda$

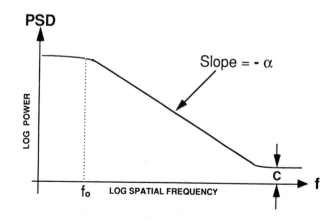

Figure 4. Effective surface PSD functional fit

A constant lambertian scatter term,C, proportional to λ was added to improve the fit at large spatial frequencies. The magnitude of the PSD curve is represented by b(λ).

5. SPECIFYING OPTICAL SURFACE TOLERANCES FOR BARE BERYLLIUM

Not everyone in the fabrication community currently uses BRDF instrumentation to verify the scatter performance of an optical surface. Even though the use of BRDF scatterometers is fast becoming popular, optical and mechanical profilometry methods are still used more frequently. For well-behaved optical surfaces, the relationship between rms surface roughness values, from profilometry, and BRDF values, from scatterometery, is well understood and is readily applied. However, when using purely topographic modeling, profilometry alone is **not** sufficient to predict the scattering characteristics of samples that are not well-behaved. Until every fabrication house has easy access to a BRDF scatterometer, there exists a need for new models that do relate profilometry to BRDF accurately for samples that are not well-behaved. There is a need for a simple way to verify scatter specification adherence.

Given the inverse power law form of the empirical equation for the "effective" PSD of beryllium, we can integrate over the spatial frequency band-limits of the instrument to be used for measuring surface roughness in the fabrication shop. Using this "effective" PSD for super-polished bare HIP Be mirrors, profilometry data in the form of rms surface roughness can easily be related to a scatter specification. This process can be used iteratively to derive the surface roughness tolerance necessary to meet a given scatter requirement based on Equation 3:

$$\int_{a}^{\,f_1}_{f_2} PSD(f) \;=\; \sigma^2 \;\; \alpha \;\; b(\lambda)$$

(3)

This equation states that rms surface roughness, σ, can be related to the PSD of a surface by integrating that PSD and taking the square root of the integrated value. Figure 5 is an illustration of this equation.

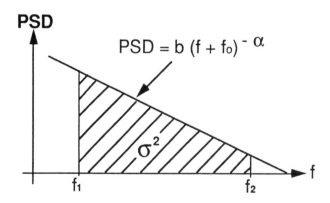

Figure 5. Illustration of the rms surface roughness relationship with surface PSD

The resulting closed form solution indicates that $b(\lambda)$ is proportional to σ^2. Since we have already shown that $b(\lambda)$ is proportional to the PSD function, it is now easy to see how we can relate the effective surface PSD to surface roughness values derived from profilometry.

For example, suppose a given application has the requirement that the BRDF curve for a particular wavelength, (λ_o), can nowhere lie above a given line or curve. We can now vary the value of $b(\lambda)$ in the "effective" surface PSD equation for bare HIP beryllium until the predicted BRDF curve lies just below the requirement. Let the value of the requirement be called $b(\lambda_o)_{spec}$. The values of $b(\lambda_o)_{ref}$ and σ_{ref} are already known from the empirical fit performed previously on the PSD data from the reference HIP Be surface.

The appropriate specified tolerance upon surface roughness is thus given by Equation 4:

$$\sigma_{spec} = \sigma_{ref} \sqrt{\frac{b(\lambda_o)_{spec}}{b(\lambda_o)_{ref}}} \tag{4}$$

A super-polished bare HIP beryllium mirror can now be tested during the polishing process until the bandlimited microroughness, as measured by optical or mechanical profilometry, σ_{spec}, is reduced to the specified value.

6. CONCLUSIONS

Experimental data illustrates the non-wavelength scaleability of bare super-polished HIP beryllium. This non-wavelength scalability varies as a function of the test wavelength, and the effect increase as the wavelength increases. An empirical fit of the experimental measured data results in an "effective" surface PSD for bare HIP beryllium mirrors. The "effective" surface PSD can be used to relate profile measurements to the scattering characteristics of bare super-polished HIP Be mirrors, and can then be used as a tool to define surface roughness tolerances based on scatter requirements. This in turn relieves the fabrication shop from the requirement of using BRDF measurements to predict final scatter specification adherence.

7. REFERENCES

1. J. E. Harvey, "Light-Scattering Characteristics of Optical Surfaces", Ph.D. Dissertation, University of Arizona, (1976).
2. J. E. Harvey, "Surface scatter phenomena: A linear shift-invariant process," Proc. SPIE 1165, 87 (1989).
3. J. M. Elson and J. M. Bennett, "Vector scattering Theory", Opt. Eng. **18**, 116-124 (1979).
4. E. L. Church, H. A. Jenkinson, and J. M. Zavada, "Relationship between surface scattering and microtopographic features," Opt. Eng. **18**, 125 (1979).
5. Y. Wang and W. L. Wolfe, "Scattering from microrough surfaces: Comparison of theory and experiment", J. Opt. Soc. Am. 73, 1596-1602 (1983).
6. C. L. Vernold, Application and verification of wavelength scaling for near specular scatter predictions," Proc. SPIE 1165,18 (1989).
7. J. C. Stover, J. Rifkin, D. R. Cheever, K. H. Kirchner, T. F. Schiff, "Comparison of wavelength scaling data to experiment," Proc. SPIE 967, 44 (1988).
8. J. C. Stover, M. L. Bernt, D. E. McGary, and J. Rifkin, "An investigation of anomalous scatter from beryllium mirrors," Proc. SPIE 1165, 100 (1989).

Cryo-scatter measurements of beryllium

Barret Lippey
Wilfried Krone-Schmidt

Hughes Aircraft Company
2000 E. El Segundo Blvd., El Segundo, California 90245

ABSTRACT

Bi-directional Reflection Distribution Function measurements were performed as a function of cryogenic temperature for various substrates. Substrates investigated include HIPed and sputtered beryllium produced from different powders and by various manufacturing and polishing processes. In some samples investigated, the BRDF at 10.6 microns increased by a factor of 2 to 5 during cooling from 300 to 30 Kelvin. On repeated temperature cycling the change in BRDF appeared to be totally elastic. The cryo-scatter effect does not occur for all types of beryllium.

2. INTRODUCTION

This experiment was undertaken to determine the effect of cryogenic cooling on the BRDF optical scatter of beryllium mirrors. Space-looking long-wave-infrared sensors require low-scatter, cryogenic operation to achieve low background and high sensitivity. Beryllium mirrors are primary candidates for these systems because of beryllium's high stiffness to weight ratio, high thermal conductivity, and high nuclear hardness.

During cryo-scatter testing of a large variety of substrate materials and coatings over the past two years, it was observed that the scatter remained constant during cool-down or warm-up. Thus it was surprising to find that while testing a low-scatter beryllium mirror at 30 K, the scatter was found to have increased at low scatter angles. After thorough inspection of the chamber and procedure, all possible reasons for this change in scatter were eliminated except for changes in the beryllium mirror itself.

Additional experiments were conducted to quantify changes in the scatter of various types of beryllium mirror materials at low temperatures. The experimental strategy was to measure the 10.6 micron scatter of a number of hot-isostatically-pressed (HIPed) beryllium samples of various types both while the temperature was being lowered, again after the temperature had stabilized at a low value, and again while the temperature was raised in a controlled manner.

3. EXPERIMENTAL METHOD

3.1 Design

The measurement of beryllium cryo-scatter required that the cryo-scatter measurement facilities be reliable and accurately calibrated. The temperature of the samples needed to be accurately measured, and cryofilm contamination had to be eliminated as a possible source of scatter increase. The temperature of the sample was the independent variable in all of the cryo-scatter experiments. The temperature was measured at the copper sample holder on which the samples were mounted. The samples were pressed against the sample mounting plate and indium foil was used between the samples and the plate to ensure good thermal contact. The temperature of the sample was varied between 30 K and room temperature. The temperature sensors were calibrated with liquid nitrogen at 77 K and the accuracy was better than plus or minus 1 K.

10.6 micron optical scatter as a function of scatter angle was the dependent variable. The scatter was measured at 10 discrete angles from 2 to 50 degrees away from the specular reflection. The scatter of a flame-sprayed-aluminum reference sample was measured before and after each group of scatter measurements to calibrate the scatterometer. A low instrument scatter background and low scatter due to self-contamination was verified by measuring the apparent scatter from a low scatter aluminum-coated fused silica sample. The instrument background was shown to be far less

150 / SPIE Vol. 1530 Optical Scatter: Applications, Measurement, and Theory (1991)

than the scatter of any of the beryllium samples.

When the samples were cooled, contraction of the sample and the mount resulted in a minor shift of the sample mounting position. To determine the amount of shift, a HeNe laser (coaligned with the scatterometer 10.6 laser) was reflected from the sample and its angle of reflection was monitored. This reflection angle drifted by, at most, a few tenths of a degree during cooling. No significant change in the scatter curves was caused by this small angular shift.

Quartz crystal monitors (QCMs) were used to measure the amount of molecular contamination that condensed onto the samples. The deposition rate was measured to be less than 1 Angstrom per hour in VICS1 and less than 5 Angstroms per hour in VICS2. Based on extensive previous measurements of scatter as a function of cryofilm thickness, we know that the cryofilms present during the cryo-scatter measurements did not contribute to the cryo-scatter increase seen for beryllium. The fused silica samples provided another check that the cryo-scatter increase was not caused by cryofilm contamination. Care was taken to reduce particulate contamination to minimum possible levels, and they also contributed negligibly to the scatter.

3.2 Mirror samples

Three super-polished (2 inch diameter) fused silica samples were used for background scatter checks. Each had SiO_x on aluminum coating and were designated part numbers 6, 15 and 18. Beryllium mirror samples were obtained from various suppliers. Most of the samples were chosen to be O-50 type beryllium due to high homogeneity and superior polishability. The following beryllium mirror samples (between 1 inch and 2 inches in diameter) were tested for cryo-scatter:

Material Type	Part Number	Comments
O-50	H04071	Incompletely Polished
O-50	30	Thermal Cycled to 200°C
O-50	35	Thermal Cycled to 200°C
O-50	44	Thermal Cycled to 200°C
O-50	3-10	Thermal Cycled to 200°C
O-50	892C	Low HIPing Temperature
O-50	892D	Low HIPing Temperature
O-50	B3	High HIPing Temperature
O-50	B4	High HIPing Temperature
O-50	892A	
O-50	892B	
O-50	H0565 B1	
O-50	H0565 B2	
Spherical Powder	S4C1	
Be sputtered on Be	U-46B	
I-70	733	

3.3 Apparatus

The following commercially available apparatus was used:

Description	Supplier	Model Number
QCM	QCM Research	Mark-16
Mass Spectrometer	VG Quadrupoles	Micromass PC
Temperature Sensor	Lakeshore	DRC-82C and DT 470

The following specially designed apparatus was used:

1. Vacuum-In-Situ-Cryo-Scatterometer 1 (VICS1). One sample can be measured per vacuum pump-down. The sample is cooled with a closed-cycle liquid helium cooler. See Figure 1.

2. Vacuum-In-Situ-Cryo-Scatterometer 2 (VICS2). Seven samples can be measured per pump-down. The samples are cooled with an open-cycle liquid helium transfer line. See Figure 2.

The accuracy and repeatability of the VICS1 and VICS2 measurements were determined by comparing the respective vacuum chamber data with data taken at the Hughes Scatter Measurement Facility. The Hughes Scatter Measurement Facility has been proven reliable and accurate during previous Rome Air Development Center round-robin scatter exercises. Excellent agreement between all three scatter stations was achieved for flame-sprayed aluminum samples (used as lambertian references), low-scatter beryllium mirrors, and for very low-scatter aluminum-coated fused silica mirrors (used as background checks). The estimated repeatability of the cryo-scatter BRDF is plus or minus 20%. Plus or minus 20% repeatability is typical for BRDF scatter measurements performed at a variety of well-known scatter measurement facilities.

3.4 Cryo-scatter measurement procedure in VICS1

Since the BRDF is the ratio of the signal from the test sample to the signal from a Lambertian reference, a reference sample measurement was performed prior to and after every pump-down. The reference, a diffuse flame sprayed aluminum disk with a reflectance of 87%, was mounted in the same location as the beryllium witness sample. The chamber was closed up, but not pumped down during the reference readings. Reference measurements taken at ambient and under vacuum were always within 20% of the same values. When in-situ diffuse reference measurements were being made, the reference sample was rotated about its horizonal axis at 250 rpm for speckle suppression.

After the scatterometer system was completely checked out and the sample under test was measured and compared to the results the Hughes Scatter Facility had obtained, the chamber was closed off, and the pump down was performed. After pump down, another (vacuum) BRDF scan was taken. Before proceeding, the BRDF scan in the Hughes Scatter Measurement Facility, the ambient scan, and the vacuum scan must agree to within plus or minus 20%.

The total pressure in the VICS1 vacuum chamber was approximately 1×10^{-8} Torr. Mass spectrometer measurements indicated that this total pressure was caused almost entirely by nitrogen gas. The only gases present in measurable amounts that would condense on the sample at 30 K are water and carbon dioxide. The partial pressure of water was measured at approximately 1×10^{-9} Torr and the partial pressure of carbon dioxide and other organics was measured at less than 1×10^{-11} Torr. The temperature stability of the samples in VICS1 (measured at the sample mount) was plus or minus 0.1 K.

3.5 Cryo-scatter measurement procedure in VICS2

The samples in VICS2 were measured in five experimental groups over a period of three months. To determine repeatability, multiple scatter measurements of the same samples were made in some of the groups, and some samples were measured in more than one group. The first group consisted of samples 18, H04071, S4C1, and 30. The second group consisted of samples S4C1, 35, 44, and 3-10. The third group consisted of samples 15, 892D, B4, and U-46B. The fourth group consisted of samples 15, 892C, and B3. The fifth group consisted of samples 15, 892A, 892B, H0565 B1, and H0565 B2. Some of the samples were cooled while the scatter was being monitored and some of the samples were stabilized at each temperature before measuring the scatter.

Each group of scatter measurements were performed by doing the following steps in the VICS2 chamber:

1. Measure the lambertian reference sample (average of lowest and highest glints when rotated about the center of the sample holder).

2. Measure the scatter of each sample in the group. Align each sample into the proper rotary position by using a sighting telescope from outside the vacuum chamber.

3. Cool all test samples in the group and the diffuse reference sample.

4. Measure the scatter of each sample in the group. Align each sample into the proper rotary position by using a sighting telescope from outside the vacuum chamber.

5. Check the tilt alignment of each sample by checking the reflection angle of a HeNe laser coaligned with the 10.6 micron laser. (No significant tilt errors were found at room temperature, 81 K, or 30 K.)

6. Check that the measured scatter from the lambertian reference sample had not changed.

7. Warm the test samples back to room temperature.

The total pressure in the VICS2 vacuum chamber was approximately 1×10^{-6} Torr. This total pressure was caused almost entirely by nitrogen gas. The only gases present in measurable amounts that would condense on the sample at 30 K are water and carbon dioxide. The partial pressure of water was measured at approximately 1×10^{-10} Torr and the partial pressure of carbon dioxide and other organics was measured at less than 1×10^{-11} Torr. The temperature stability of the samples in VICS2 was approximately plus or minus 1 K.

4. RESULTS

4.1 Summary

Cooling caused the 10.6 micron scatter of the polished beryllium material to increase by up to a factor of five at 2 degrees from the specular beam. Repeated measurements and cross checks of the data and equipment show the cryo-scatter effect to be repeatable, and not due to sample contamination or measurement errors. Fused silica mirrors and some beryllium material types did not exhibit an increase in cryo-scatter at low temperatures.

4.2 Cryo-scatter data

Typical cryo-scatter curves are shown in Figures 3 through 8. Figure 3 shows the measurement of the very-low-scatter fused silica samples in VICS1. Figure 4 shows a typical measurement of another very-low-scatter fused silica sample in VICS2 (note that the BRDF scale is different for this figure). The instrument signatures are below these levels. Figure 5 shows typical O-50 beryllium scatter during cooling. Figure 6 shows a relatively low scatter increase of O-50 beryllium at stabilized temperatures. Figure 7 shows a relatively high scatter increase of O-50 beryllium at stabilized temperatures. Figure 8 shows typical spherical powder beryllium scatter. Figure 9 shows I-70 scatter. Figure 10 shows the scatter of polished, sputtered beryllium on a beryllium substrate.

5. DISCUSSION

By inspecting the scatter curves of O-50 beryllium, it can be seen that the scatter increases primarily between the temperatures of 250 K and 100 K. A graph of cryo-scatter as a function of temperature and angle is shown in Figure 11. The scatter increase occurs primarily at small scatter angles. This causes the curves of BRDF versus scatter angle to appear to hinge about a single scatter angle of approximately 10 degrees.

The spherical powder samples show only a factor of two increase in scatter when the temperature is reduced. This is less than the cryo-scatter increase measured for the O-50 samples.

The H04071 sample shows a high scatter before cooling and a similarly high scatter at low temperatures. This indicates that the amount of scatter increase due to cooling may not be measurable for high scatter beryllium surfaces. Rather than characterizing beryllium by a factor by which the beryllium scatter increases, it may be more useful to characterize beryllium by the absolute scatter level to which it rises.

O-50 beryllium samples processed with low and high HIPing temperatures showed increases in scatter due to cooling. The O-50 material samples that were thermal cycled to 200°C during polishing also showed increases in scatter due to cooling.

Both the Be on Be and I-70 samples did not show any cryo-scatter increase. Further experiments must be performed to confirm these results.

6 BRIEF THEORETICAL CONSIDERATIONS

The beryllium mirrors used in this study were hot-isostatically pressed (HIPed) from beryllium powder and were polycrystalline. Since beryllium has a hexagonal-close-packed crystal structure, the c-axis expansion coefficient is different than the a-axis expansion coefficient. Temperature change, therefore, can produce stress and strain at the surface because the grains are oriented randomly. Thus, differing grain contraction can cause a roughening of the surface.

In addition to the above topographic effect, beryllium is also known to have non-topographic scatter effects at 10.6 microns that cause "anomalous" scatter behavior.[1] The stress from temperature change might increase subsurface strain that affects scatter. Since the extinction coefficient for beryllium at 10.6 microns is much lower than for other common mirror materials such as aluminum and gold,[2] the 10.6 micron radiation penetrates about 10 times farther into the surface. The transmittance of beryllium metal is down by a factor of approximately 10^{-4} at a depth of 0.3 microns. Current experiments are attempting to determine whether the mechanism of cryo-scatter increase is primarily a topographic effect or a non-topographic effect.

As the temperature drops, the expansion coefficient of beryllium approaches zero.[3] Below about 100 Kelvin, the expansion coefficient is relatively small. This could explain why the scatter stops increasing as the temperature drops below about 80 Kelvin.

From the dependence of the scatter increase on scatter angle, the spatial frequency of the increased surface roughness (or subsurface defects) can be roughly calculated.[4] Using the grating equation, the increased scatter at a scatter angle of 2 degrees corresponds to a spatial frequency of 300 microns. Even larger spatial frequencies may be present (1 degree corresponds to 600 microns), but scatter at angles below 2 degrees is not measurable in the current experimental apparatus. The spatial frequency calculated above is much larger than the grain size of the beryllium (approximately 15 to 20 microns). This is to be expected as the elastic deformation due to differing grain contraction would spread out over several grains.

7. ACKNOWLEDGEMENTS

This research was performed at and funded by Hughes Aircraft Company under IR&D. The authors wish to thank Werner Brandt, Karl Gast, Darrell Gleichauf, Ernest Gossett, David Michel, James Sickert, Alvin Trafton, and Roger Withrington for their assistance.

8. REFERENCES

1. E. L. Church, "Scattering by Anisotropic Grains in Beryllium Mirrors," Stray Radiation in Optical Systems, SPIE, Vol. 1331, pp. 12-17, 1990.

2. D. E. Gray, American Institute of Physics Handbook, Section 6, McGraw-Hill, New York, 1972.

3. Y. S. Touloukian, Thermophysical Properties of Matter, p. 25, IFI/Plenum Data, New York, 1975.

4. E. L. Church, "Relationship between Surface Scattering and Microtopographic Features," Optical Engineering, Vol. 18, No. 2, pp. 125-136, March-April, 1979.

Figure 1
Vacuum In-Situ Cryo Scatterometer 1
(VICS-1)

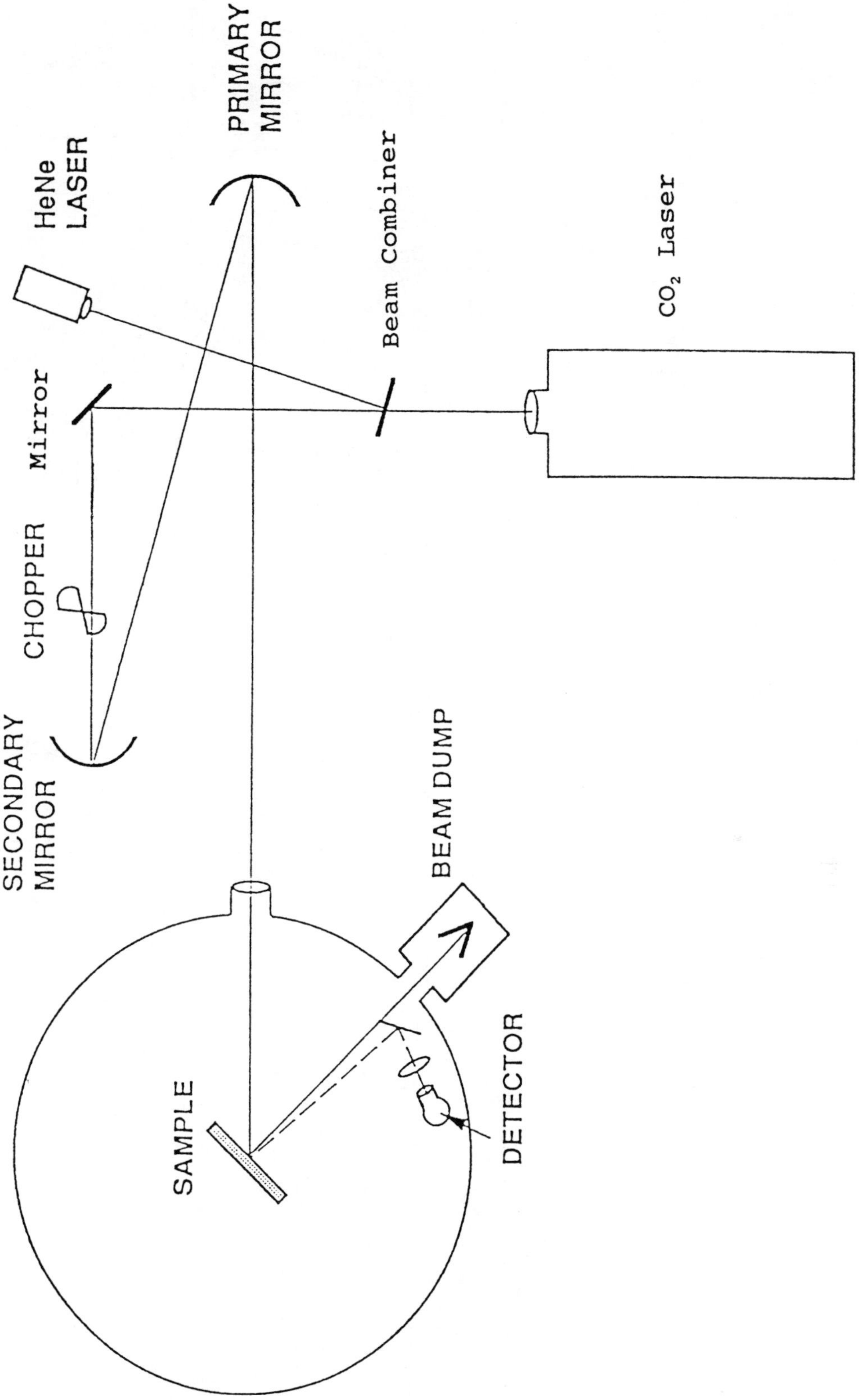

Figure 2
Vacuum In-Situ Cryo Scatterometer 2
(VICS-2)

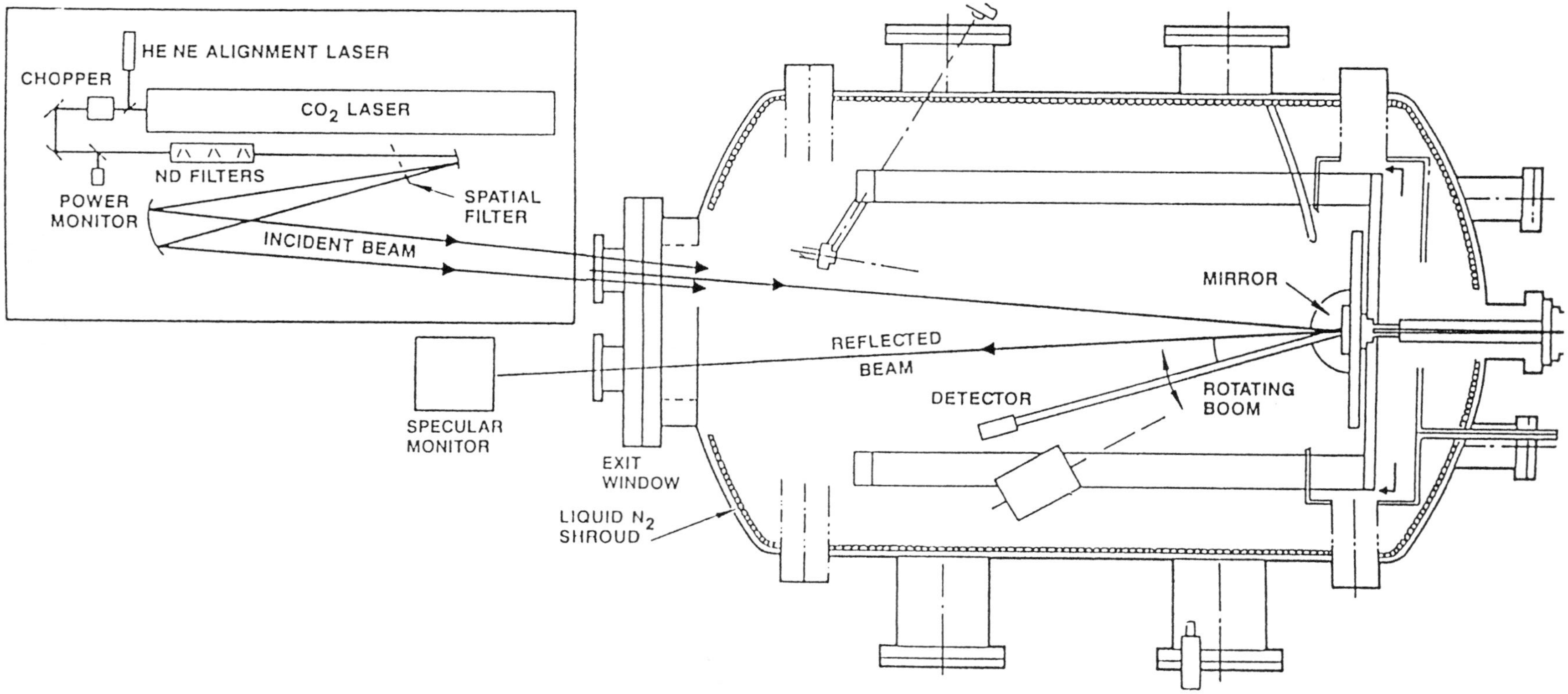

Figure 3
Cryo-Scatter of Fused Silica
Sample #6 Measured in VICS-1

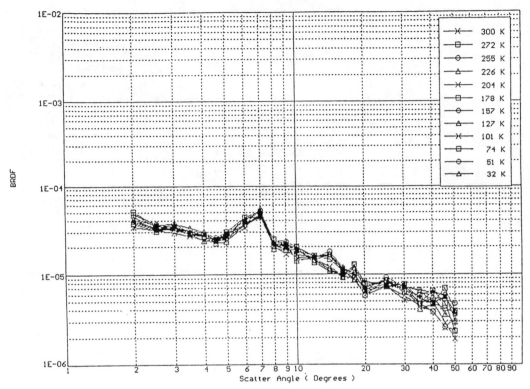

Figure 3
Cryo-Scatter of Fused Silica
Sample #6 Measured in VICS-1

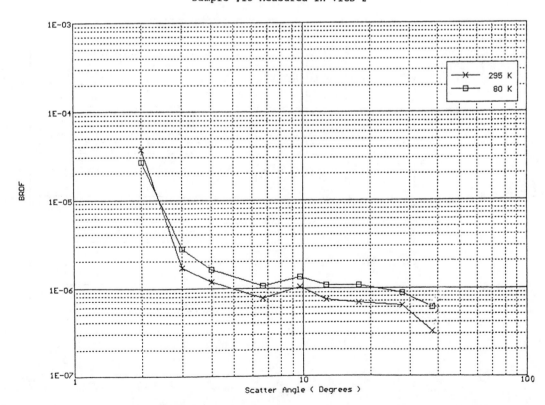

Figure 4
Cryo-Scatter of Fused Silica
Sample #15 Measured in VICS-2

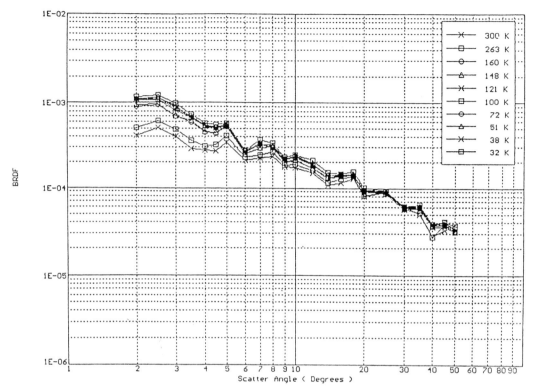

Figure 5
Cryo-Scatter of 0-50 Beryllium
Sample #30 Measured in VICS-1

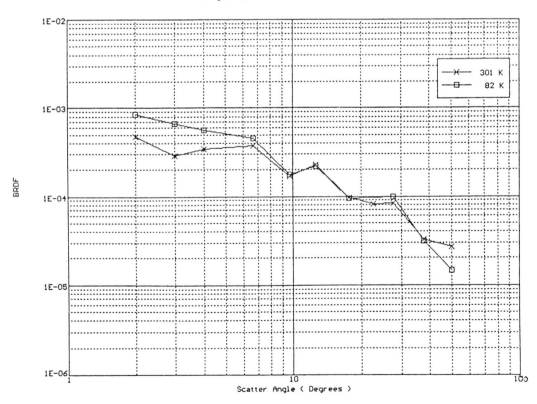

Figure 6
Cryo-Scatter of 0-50 Beryllium
Sample #44 Measured in VICS-2

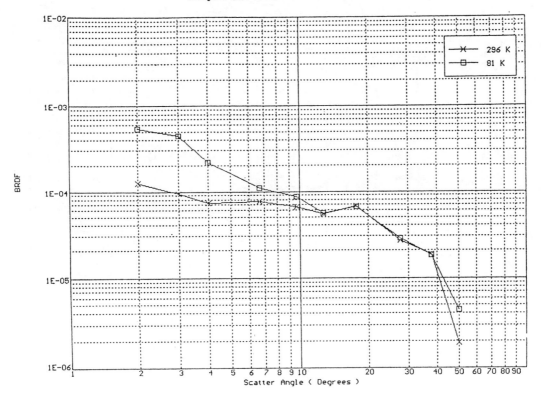

Figure 7
Cryo-Scatter of O-50 Beryllium
Sample #B4 Measured in VICS-2

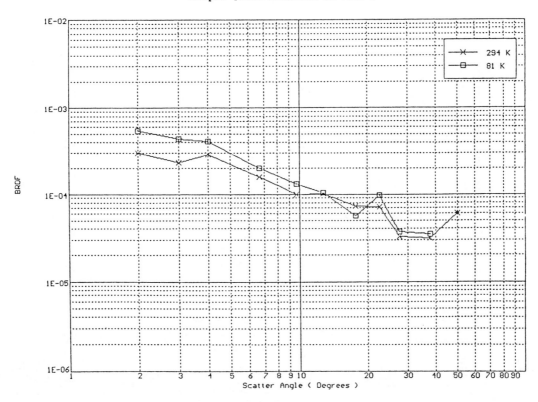

Figure 8
Cryo-Scatter of Spherical Powder Beryllium
Sample #S4C1 Measured in VICS-2

Figure 9
Cryo-Scatter of I-70 Beryllium
Sample #733 Measured in VICS-1

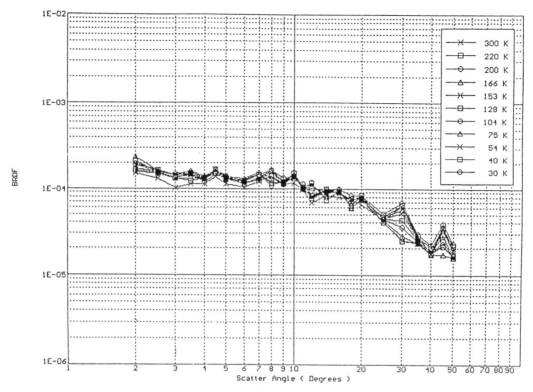

Figure 10
Cryo-Scatter of Be on Be
Sample #U-46B Measured in VICS-2

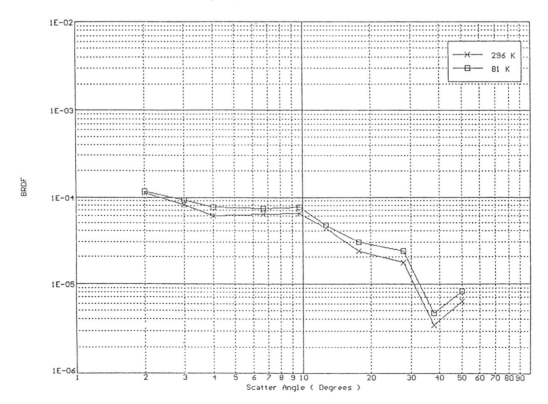

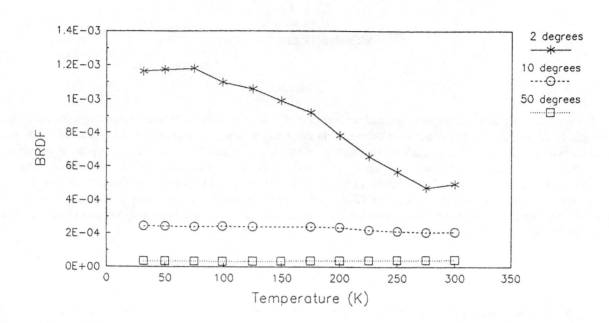

Figure 11
Typical Cryo-Scatter of O-50 Beryllium
As a Function of Temperature and Scatter Angle

Material Characterization of Beryllium Mirrors
Exhibiting Anomalous Scatter

C. M. Egert

Optics MODIL

Oak Ridge National Laboratory

Abstract

To understand the origin of anomalous scatter it is necessary to consider materials-related features which might be responsible for this anomalous behavior. In this study a variety of material characteristics of a sub-set of the beryllium mirrors used in an earlier investigation of anomalous scatter by Stover et. al.[1] is presented. Material characteristics investigated include near-surface chemical composition, grain size, surface particulate density, and ultraviolet-visible reflectance. These material characteristics are compared with the anomalous scatter level reported by Stover for the same mirrors. One mirror (B-85), which exhibits a high level of anomalous scatter, was found to have larger grain size (~28 µm) and a high density of localized surface porosity compared to other mirror samples with lower levels of anomalous scatter.

1. Introduction

A significant lack of wavelength scaling, sometimes referred to as anomalous scatter, has been reported for low scatter beryllium mirrors.[1] This non-topographic contribution to the total scatter in the IR region of the spectrum presents an important producibility issue in the manufacture of high quality beryllium optics. The apparent non-topographic nature of this phenomena suggests an underlying material-related cause for anomalous scatter. This report presents results of material characterizations of beryllium mirrors which have been previously included in investigations of anomalous scatter. The purpose here is to identify those material features which are characteristic of beryllium mirrors exhibiting high levels of anomalous scatter.

Stover et. al.[1,2] has reported anomalous scatter measurements for a variety of mirror samples including a molybdenum mirror considered free of significant anomalous behavior. Scatter measurements were taken at four wavelengths from 0.633 µm to 10.6 µm, the power spectral density (PSD) was calculated and integrated to obtain an estimate of the roughness for each wavelength. Anomalous scatter can be parameterized by calculating the standard deviation divided by the average of the four roughness values:

$$S = \Delta\sigma/\sigma_{avg}$$

In this form, higher values of S represent higher levels of anomalous scatter, while smaller values of S indicate low anomalous scatter. The anomalous scatter parameters for the beryllium mirrors calculated from the data presented in Ref. 1 and included in this study are presented in Table 2.

Not all beryllium mirrors exhibit unambiguous evidence of high anomalous scatter. For example, Vernold and co-workers have presented evidence that a thin, polished beryllium coating on a beryllium substrate does not exhibit anomalous scatter.[3] Also Stover found that some beryllium mirrors have lower levels of anomalous scatter than a molybdenum mirror which is often used as an example of a pure topographic scatterer. There are several possible explanations for this observation: first, random experimental noise in the scatter measurements themselves might lead to differences in PSD which would translate into non-zero S values. A second factor might be variation in location of scatter measurements on samples with varying surface quality. This would naturally lead to variations in roughness calculated from scatter measurements made at different locations on the same sample. Finally, it may be that some beryllium mirrors simply do not exhibit significant levels of anomalous scatter, or perhaps all mirrors

0-8194-0658-9/91/$4.00

exhibit some small level of anomalous behavior. A great deal more scatter experimentation is required to resolve this issue and determine which of these factors dominate. However, due to this uncertainty in the value of S we concentrate here on one beryllium mirror sample (B-85) which does exhibit a significantly high level of anomalous scatter, and compare its material properties with other beryllium mirrors which do not exhibit significant levels of anomalous scatter compared to the molybdenum mirror.

The material characterizations performed here indicate that sample B-85 does indeed exhibit some unusual material characteristics compared to the other beryllium mirrors included in this study. This sample has the largest grain size of the samples measured and a significant amount of open porosity was also observed. Recent theoretical advances suggest that both of these material characteristics may contribute to anomalous scatter.

2. Experimental Approach.

A set of beryllium mirrors listed in Table 1 were chosen from a larger set of mirrors included in Stover's investigation of anomalous scatter.[1] Both coated and uncoated hot isostatically pressed beryllium mirrors were included in this study. These mirrors were prepared by a variety of techniques including single point turning, conventional polishing, electrochemically assisted polishing, and non-contact polishing. Both material origin and fabrication methods are listed in Table 1. A variety of non-destructive material evaluation techniques were used to characterize these mirrors. Characterizations included grain size measurement, number density and composition of particulate or porosity sites at the surface, and UV reflectance (proportional to oxide thickness).

Table 1. Beryllium Mirrors Included In This Materials Characterization Study

Mirror Number	Material Type	Fabrication Notes
B-76	HIP Be	super-polished (POCT-33-9)
F-41	HIP Be	super-polished (POCT-33-3)
B-77	HIP Be	super-polished (POCT-18)
U-41	evaporated Be on Be	super-polished - low scatter
U-47	evaporated Be on Be	electrochemically assisted polish
B-85	HIP Be	CBN machined, non-contact polished

Grain size was determined by applying the line intercept method to polarized light metallographs of the beryllium surfaces. In this technique a series of randomly-oriented lines are drawn on the micrograph and the number of intersections of each line with a grain boundary is recorded. The total length of the line divided by the number of intercepts gives an estimate of the grain size. Line intercept measurements were performed at least four times on each of four metallographs taken at 400X magnification. The presence of considerable amounts of worked metal on the surfaces of some samples sometimes made accurate determination of grain size difficult.

While making grain size measurements a significant number of localized defects, either porosity or particulate, were observed on the surfaces of some mirrors. An attempt was made to quantify the areal density of these localized features by assembling a collage of six bright field metallographs at 100X magnification representing a region of the surface 2-mm by 2-mm. The number of particulate within this region were then counted and divided by the total area of the metallographs resulting in an estimate of the number density of particulate or pits on each surface. The relatively low magnification used to produce these collages restricted the minimum size particulate included in this estimate of number density to those with dimensions greater than ~5 μm. Smaller particulate could not be clearly resolved and were not included in this determination of particulate density. A few selected mirror samples were further analyzed using Scanning Auger microscopy to determine the composition of particulate at the surface.

Specular reflectance measurements from the ultra-violet through visible were also made to determine the existence of any variation in oxide thickness or optical properties which might be correlated to anomalous scatter behavior. Reflectance measurements were made using a Cary 2300 Spectrophotometer over a range of wavelengths from 0.2 to 2.0 µm. Three measurements were made on each sample in order to assess the magnitude of experimental noise or fixturing errors on the reflectance measurements. These repeated reflectance measurements were in good agreement with each other, with a total variation of less than 2% for each sample. For some samples IR reflectance was also measured using a Perkin-Elmer 938G Spectrophotometer over the wavelength range from 2.0 to 50 µm.

3. Material Characterization Results

Grain Size. A summary of the grain size measurements for the beryllium mirrors is given in Table 2 which also summarizes the other material characterization results. Polarized light micrographs at a magnification of 400X are shown in Fig. 1 and 2. These micrographs illustrate the considerable variation in microstructure observed for the different mirror samples. Some samples exhibited a clean grain structure, while in other micrographs the beryllium grain structure is much less clear due to the presence of worked or smeared metal apparently produced during the finishing operation. Sample B-85 also showed evidence (Fig. 1a and 2) of a worked surface with twinning of grains at the surface. However, several other mirror samples (Fig. 1c and 1d) also show worked surface structures with twinning, but not high anomalous scatter levels.

Most of the samples had an average grain size near 10 µm as expected for hot isostatically pressed beryllium material. For evaporated beryllium coatings (samples (U-41 and U-47) the near surface microstructure is columnar with column size increasing with increasing thickness. A 10 µm column size is also typical of thick (~250 µm) beryllium coatings.[4] One sample, U-41 has a smaller average grain size indicating a much thinner coating remained on the surface after polishing. Sample B-85 has a larger grain size of 28 µm compared to the other samples. Interestingly, sample B-85 exhibits a high level of anomalous scatter and sample U-41 a low level of anomalous scatter suggesting a possible connection between large grain size and high anomalous scatter.

Table 2. Summary of Material Characterization Results for Beryllium Mirrors.

Mirror Number	Anom Scatter ($\Delta\sigma / \sigma$)	Grain Size (µm)	Particulate Density (number/ µm^2)
B-76	0.33	9.0	0.023
F-41	0.40	10	0.030
B-77	0.43	11	0.085
U-41	0.52	5.0	0.117
U-47	0.53	10.5	0.048
B-85	1.03	28	0.228

Specular Reflectance. Specular reflectance was found to vary greatly in the UV region of the spectrum for the different samples in this study. Fig. 3 shows reflectance as a function of wavelength for two beryllium mirror samples which exhibit the greatest variation of those tested compared to calculated reflectance values obtained from the literature.[5] The greatest variation in specular reflectance was observed in the UV region of the spectrum. Reflectance measurements from all samples were in agreement with each other and with the reflectance obtained from the literature for wavelengths beyond 1 µm.

Reflectance of a clean mirror in the UV region is strongly dependent on oxide thickness and optical properties, and so this characteristic was used to assess oxide thickness. The characterization results summary in Table 2 shows UV reflectance values at 0.2 µm wavelength, a region of the spectrum in which the beryllium mirrors exhibited the greatest variation in reflectance. Lower UV reflectance is an indication of a thicker surface oxide layer and hence the beryllium mirrors examined exhibit some variation in oxide thickness. Despite this interesting variation, there is no correlation between the UV specular reflectance and anomalous scatter. In fact, beryllium mirror samples U-41 with a low anomalous scatter level, and B-85 with the highest anomalous scatter, both exhibit the same high UV reflectance.

Particulate Density. Results of the particulate density measurements on the beryllium mirror samples are also summarized in Table 2. Fig. 4 show bright field metallographs of the beryllium mirror surfaces for mirrors B-76 and B-85. An order of magnitude variation in the number density of particulate was found for the mirrors tested as Fig. 4 illustrate. Values of particulate density ranged from 0.023 per square μm for sample B-76 to 0.228 per square μm for sample B-85. Sample B-85 also exhibits the highest level of anomalous scatter. Mirror samples with lower levels of anomalous scatter also exhibit correspondingly lower particulate density values. For example, B-76 with the lowest measured particulate density also exhibits the lowest anomalous scatter level of the mirrors tested.

Besides counting particulate on the surface an attempt was also made to characterize the composition of particulate found on the surface of the mirror samples using Auger electron spectroscopy. Most particulate found on the mirror samples consisted of beryllium oxide or silicon based particulate (i.e. oxides or carbides). However, similar analysis of sample B-85 indicated that most of the "particulate" observed by light microscopy were actually localized regions of porosity. An example of this porosity is shown in Fig. 5 which is a scanning electron micrograph of two larger porosity sites on B-85. These porosity sites are not "clean" pits or voids in the surface but instead appear highly textured compared to the smooth surrounding region. Auger analysis of the beryllium within the site showed it to be essentially similar to the composition of the surrounding area. Significant porosity was not observed on any other mirror tested in this study.

Automated scanning electron microscopy was used to count these porosity sites on sample B-85. An area 1.15 cm^2 was scanned automatically and the mean projection (diameter) of the porosity sites was determined. Only porosity larger than 4 μm in mean diameter was included in this analysis. A frequency distribution as a function of porosity size is presented in Fig. 8. A total of 3,730 sites were counted in this analysis giving a ratio of porosity area to total area covered (the fractional area) of $A_f = 0.002$. The porosity size distribution for B-85 appears to peak at 4 or 5 μm with a tail extending out beyond 12 μm size. This distribution shows a significant amount of porosity in the size range equivalent to the wavelengths used in the anomalous scatter measurements. Higher magnification scanning electron microscopy suggests that this porosity does not extend much below 2 μm size, although small (< 1 μm diameter) silicon containing particulate similar to those found in other samples tested were also found on this sample.

4. Discussion.

Of the beryllium mirrors tested, sample B-85 with the highest level of anomalous scatter, also exhibited several interesting material characteristics which may contribute to its anomalous behavior. B-85 had the largest average grain size of the samples tested at 28 μm, roughly three times larger than most other samples tested. Theoretical models suggest that anomalous scatter may be related to the anisotropic optical properties of the crystalline grains at he beryllium surface.[6] Thus, it is not unreasonable to expect that grain size may play an important role in determining the level of anomalous scatter from a beryllium surface. This work suggests that a smaller grain size material might exhibit a lower level of anomalous scatter. This hypothesis is supported other recently reported results in which very fine grain beryllium coatings did not exhibit significant levels of anomalous scatter.[3]

Perhaps the most striking characteristic of sample B-85 when examined microscopically is the number and size of porosity sites at the surface. No other sample in this study exhibited such extensive porosity, and it is tempting to attribute the high level of anomalous scatter observed for this sample to this porosity. Size of the porosity sites on B-85 varied over a range which is expected to have the greatest effect on IR scatter. Theoretical analysis has suggested that a synergistic interaction between surface roughness (porosity) and grain orientation might explain anomalous scatter behavior.[7] Thus, the combination of porosity and large grain size might explain the high anomalous scatter level observed for some beryllium mirrors.

The material characterization results presented here have not definitively identified the material characteristics most responsible for anomalous scatter. Both porosity and large grain size were identified as unusual characteristics of a beryllium mirror exhibiting a high level of anomalous scatter.

Acknowledgements

A great many people assisted in the various characterizations presented in this paper. Special acknowledgement must go to Valerie Newman for metallography, Bill Bollinger for scanning electron microscopy, and Hugh Richards for scanning Auger microscopy. This work was sponsored by the Strategic Defense Initiative Organization through the Optics Manufacturing Development and Integration Laboratory (Optics MODIL) under U. S. Department of Energy Interagency Agreement No. 1855-1562-A1.

References

1. J. C. Stover, Marvin L. Brendt, Douglas E. McGary, Jeff Rifkin, *An Investigation of Anomalous Scatter from Beryllium Mirrors.*, Proc. SPIE 1165, 100 (1989)

2. J. C. Stover, *Multiple Wavelength Predictions of BRDF from Beryllium Mirrors*, Proc. SPIE 1530, (1991)

3. C. L. Vernold and J. E. Harvey, *Scattering Characteristics of Beryllium*, Proc SPIE 1530, (1991)

4. C. M. Egert and J. B. Arnold, *Sputtered Beryllium Coatings for the Precision Fabrication of Optical Components*, Proc. ASPE Spring Topical Meeting, (1991)

5. E. T. Arakawa, T. A. Calcott, and Y. Chang, *Optical Properties of Beryllium*, in Handbook if Optical Constants of Solids II; E. D. Palik, ed., p.421, Academic Press, New York, 1991.

6. E. L. Church, *Scattering by Anisotropic Grains in Beryllium Mirrors*, Proc. SPIE 1331, 12 (1990).

7. J. M. Elson, *Anomalous Scattering from Surfaces With Roughness and Optical Constant Perturbations*, Proc. SPIE 1530, (1991)

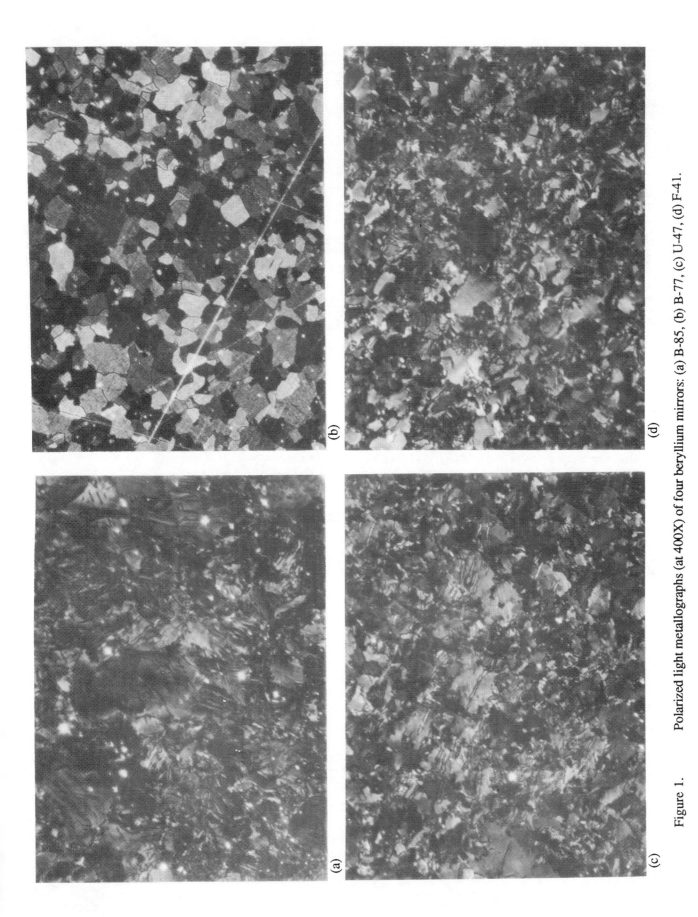

Figure 1.   Polarized light metallographs (at 400X) of four beryllium mirrors: (a) B-85, (b) B-77, (c) U-47, (d) F-41.

Figure 2.   Polarized light micrograph of grain structure of two mirrors showing the largest (B-85) and smallest (U-41) grain size of the samples studied.

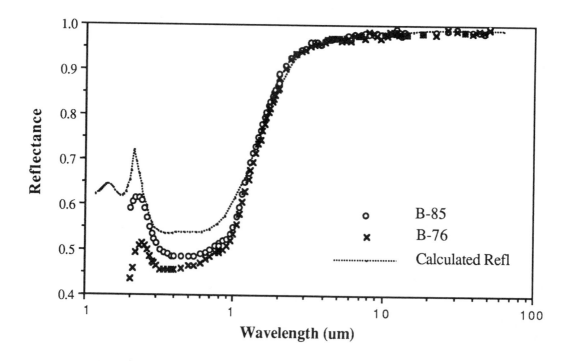

Figure 3.   Reflectance curves for beryllium mirrors B-85 and B-76 compared to calculated reflectance from Arakawa et. al.[5]

200 μm

Figure 4.   Bright field metallographs (at 100X) of mirrors B-85 (top) and B-76 (bottom) showing surface particulates and porosity.

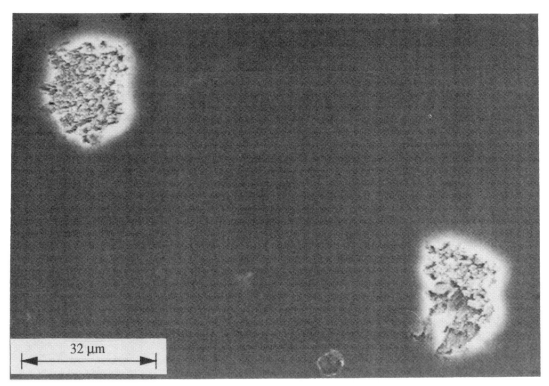

Figure 5.   Scanning electron micrograph of sample B-85 at 1130X magnification illustrating localized porosity on the surface.

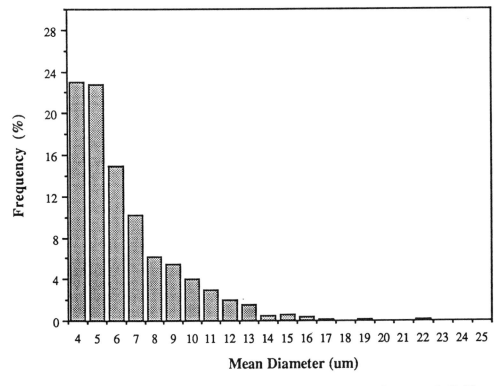

Figure 6.   Frequency distribution as a function of size for porosity sites on mirror sample B-85.

Scattering from slightly rough crystal surfaces

E. L. Church, USA ARDEC, Picatinny NJ 07806-5000

ABSTRACT

This paper discusses the reflection and scattering from slightly-rough crystal surfaces using first-order vector Fresnel-Kirchhoff diffraction theory. Results are given in the conventional, Stokes- and Mueller-matrix representations for cubic and uniaxial crystals in particular orientations.

1. INTRODUCTION

Beryllium and silicon carbide mirrors show "anomalous" scattering; that is, their measured BRDF's fall off more slowly with increasing radiation wavelength than the $\lambda^{(-4)}$ behavior expected for smooth topographic scattering [1].

Surface topography, however, is not the only source of scattering. Significant contributions can also come from fluctuations in the optical properties of the mirror surfaces, and since that scattering falls as $\lambda^{(-2)}$, it is a natural candidate for explaining the observed anomalies [2].

In the case of optically isotropic materials such scatter comes from spatial variations of their optical constants. Expressions for the BRDF of this type of scattering are well known [2-5].

In the case of optically anisotropic materials such as Be and SiC, scattering can also come from fluctuations in the orientations of the optic axis of the surface material [6]. This mechanism would appear to be particularly important in HIP'd beryllium mirrors since they show distinct anisotropic grain structure under polarized light. The theory of this type of scattering does not appear to have been discussed previously.

In an earlier paper we examined this mechanism using simple scalar scattering theory, and showed that the relative importance of anisotropic to topographic scattering is measured by the parameter

$$\alpha = \frac{TIS\,(anisotropic)}{TIS\,(topographic)} = \frac{1}{2}\left|\frac{n_o - n_e}{n_o + n_e}\right|^2 \Big/ \left[4\pi\frac{\sigma}{\lambda}\right]^2 \tag{1}$$

where TIS stands for total integrated scatter, r_o and r_e are the ordinary and extraordinary amplitude reflection coefficients of the material at normal incidence, and σ is the root-mean-square (rms) topographic roughness [6].

Figure 1 is a plot of the numerator on the right of Eq 1 for a beryllium crystal derived from the measurements of Weaver et al. [6,7]. Unfortunately, values are available only for relatively short wavelengths, and we cannot predict the value of Eq 1 at CO2 wavelengths, where the largest scattering anomalies are observed. However, if we take the numerator to have as low a value as 10(-5), and evaluate the denominator for a surface with an rms roughness of 10 Å, this ratio is 7.1. This relatively large number suggests that anisotropic scattering can contribute significantly to the observed anomalies, and justifies more detailed analysis. Hence this paper.

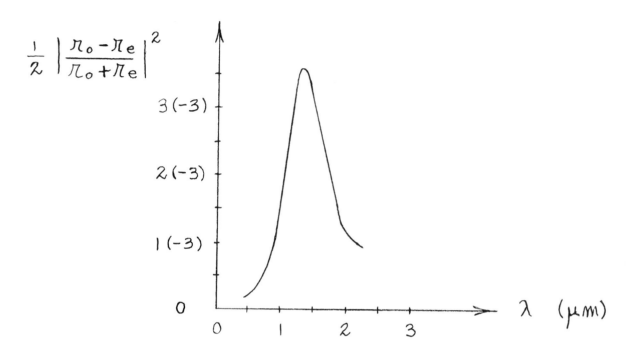

Figure 1. Experimental values of the numerator in Eq 1 for beryllium as a function of the radiation wavelength [6,7].

2. THE ROLE OF OPTICAL ANISOTROPIES IN SCATTERING

Scattering arises from any spatial variations in the optical properties of the surface which upset the delicate interference effects required for specular reflection. If these variations are small the BRDF has the form

$$BRDF = \frac{A}{\lambda^m} S_2(\vec{f})$$

(2)

where A is a reflectivity-obliquity-polarization factor, n is a number, $S_2(\vec{f})$ is the two-dimensional spectral density of the fluctuating quantity, and $\vec{f}$ is the spatial frequency, which is related to the scattering geometry. In the plane of incidence, for example, it has the familiar form

$$\xi_X = \frac{1}{\lambda} \left[\text{SIN}\, \theta_s - \text{SIN}\, \theta_i \right] \tag{3}$$

where θ_i and θ_s are the angles of incidence and scattering. Equation 2 provides a convenient framework for comparing different scattering mechanisms:

In the case of topographic scattering, n = 4 and S is the spectrum of the surface-height fluctuations. In the case of fluctuations in materials properties, n = 2 or 3 and S is the spectrum of their spatial distribution [2].

In our first beryllium paper we modeled the surface as a collection of grains with varying reflectivity [6]. In that case n = 2 and S was the spectrum of the distribution of grain boundaries. The role of the optical anisotropies is to give the grains different reflectivities depending on their orientation, and the degree of anisotropy then appears in the coefficient A in Eq 2. In particular, A is proportional to the numerator on the right-hand side of Eq 1, which vanishes for an isotropic material.

In the present paper we consider a different but related problem -- the effect of anisotropies on the A factor in the expression for topographic scattering from a a slightly rough single crystal. In this case n = 4 and S is the spectrum of the height fluctuations, as expected for ordinary topographic scatter, but if the crystal is not cubic the polarization properties of the scattering may be modified. In particular, the factor A may contain cross-polarization effects which are absent for an isotropic material.

We explore these polarization effects below using first-order Fresnel-Kirchhoff vector scattering theory, and present results in both the conventional and Stokes-matrix representations. The expressions obtained have immediate use in the analysis of reflectivity and ellipsometric measurements of beryllium crystals designed to extend the Weaver data to longer wavelengths. They are also a necessary first step in the vector calculation of the scattering from a HIP'd beryllium mirror, modeled as an ensemble of randomly oriented crystallites.

3. COORDINATE SYSTEM AND NOTATION

Figure 2 sketches the coordinate system and shows some of the notation used.

The planes of incidence and reflection coincide with the X-Z plane, and the crystal surface lies in the X-Y plane. Incident, reflected and scattered quantities are denoted by the subscripts "i", "r" and "s". States of polarization are indicated by the subscripts "s" or "p" ("h" or "TE" and "v" or "TM" in radar notation) when the electric vector is perpendicular to or lying in the plane of interest. Matrix elements follow the radar (and quantum-mechanical) convention of reading from right to left. That is, "ba" denotes the transition from state a to state b.

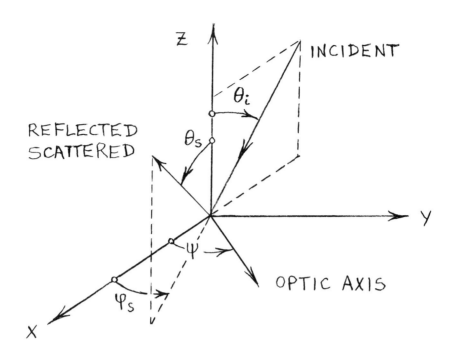

Figure 2. Sketch of the geometry and notation used in this paper.

We consider three types of crystals: 1) an isotropic (cubic) crystal, 2) a uniaxial crystal with its optic axis perpendicular to the surface plane, and 3) a uniaxial crystal with its optic axis lying in the plane of the surface, as shown in the Figure. Case 3 is the one of greatest interest since it displays the anisotropic effects to their fullest.

4. JONES REFLECTION MATRIX

The reflection of polarized light from an arbitrary surface is described by the 2 X 2 matrix equation

$$
\begin{pmatrix} E_p \\ E_s \end{pmatrix}_r = \begin{pmatrix} r_{pp} & r_{ps} \\ r_{sp} & r_{ss} \end{pmatrix} \begin{pmatrix} E_p \\ E_s \end{pmatrix}_i .
\tag{4}
$$

where the matrix elements are functions of the reflection geometry, the crystal type and orientation, and its indices of refraction: $N = n + i k$. (Sign conventions are discussed in Appendix 1.)

4.1 Isotropic crystal

In this case the off-diagonal terms in Eq 4 vanish and the diagonal terms are are the textbook Fresnel amplitude reflection coefficients [8]:

$$\mathcal{R}_{pp} = \frac{N_1^2 \cos\theta_i - N_0(N_1^2 - N_0^2 \sin^2\theta_i)^{1/2}}{N_1^2 \cos\theta_i + N_0(N_1^2 - N_0^2 \sin^2\theta_i)^{1/2}} \qquad (5)$$

and

$$\mathcal{R}_{ss} = \frac{N_0 \cos\theta_i - (N_1^2 - N_0^2 \sin^2\theta_i)^{1/2}}{N_0 \cos\theta_i + (N_1^2 - N_0^2 \sin^2\theta_i)^{1/2}} \qquad (6)$$

where N_0 is the index of the ambient (air, $N_0 = 1$) and N_1 is the index of the surface. At normal incidence these reduce to

$$\mathcal{R}_{pp} = -\mathcal{R}_{ss} = \frac{N_1 - N_0}{N_1 + N_0} \qquad (7)$$

4.2 Uniaxial crystal with its axis perpendicular to the surface

In this case the off-diagonal terms in Eq 4 are again zero but the diagonal terms are functions of both the ordinary, N_{1o}, and extraordinary, N_{1e}, indices of the surface [8]:

$$\mathcal{R}_{pp} = \frac{N_{1o} N_{1e} \cos\theta_i - N_0(N_{1e}^2 - N_0^2 \sin^2\theta_i)^{1/2}}{N_{1o} N_{1e} \cos\theta_i + N_0(N_{1e}^2 - N_0^2 \sin^2\theta_i)^{1/2}} \qquad (8)$$

and

$$\mathcal{R}_{ss} = \frac{N_0 \cos\theta_i - (N_{1o}^2 - N_0^2 \sin^2\theta_i)^{1/2}}{N_0 \cos\theta_i + (N_{1o}^2 - N_0^2 \sin^2\theta_i)^{1/2}} \qquad (9)$$

At normal incidence the dependence on the extraordinary index drops out and

$$\mathcal{R}_{pp} = -\mathcal{R}_{ss} = \frac{N_{1o} - N_0}{N_{1o} + N_0} \qquad (10)$$

4.3 Uniaxial crystal with its axis in the surface plane

In this case the off-diagonal terms in Eq 4 do not vanish. They and the diagonal terms are rather complicated functions of the geometry and crystal indices in general, and are given in Appendix 2. At normal incidence, however, they reduce to the simple expressions:

$$r_{PP} = -r_{SS} = r_0 \cos^2 \psi + r_e \sin^2 \psi \qquad (11)$$

and

$$r_{PS} = r_{SP} = (r_0 - r_e) \sin \psi \cos \psi \qquad (12)$$

where ψ is the orientation of the crystal axis (c.f. Fig 1), and r_0 and r_e are the ordinary and extraordinary Fresnel amplitude reflection coefficients of the surface at normal incidence:

$$r_0 = \frac{N_{1o} - N_0}{N_{1o} + N_0} \qquad\qquad r_e = \frac{N_{1e} - N_0}{N_{1e} + N_0} \qquad (13)$$

These quantities correspond to the conventional definitions of the ordinary and extraordinary reflectivity: r_0 is the normal-incidence reflectivity when the electric vector is perpendicular to the optic axis ($\psi = 0$) and r_e is its value when the vector is parallel to the optic axis ($\psi = pi/2$).

The 2 X 2 reflectivity matrix, Eq 4, is a building block for the 4 X 4 Stokes and Mueller matrices discussed below. First, however, we extend the above considerations to scattering.

5. JONES SCATTERING MATRIX

The scattering matrix plays a role similar to that of the reflection matrix discussed above [3]:

$$\begin{pmatrix} E_P \\ E_S \end{pmatrix}_s = \frac{ik}{4\pi} \cdot \frac{e^{ikr}}{r} \cdot \begin{pmatrix} F_{PP} & F_{PS} \\ F_{SS} & F_{SP} \end{pmatrix} \begin{pmatrix} E_P \\ E_S \end{pmatrix}_i \qquad (14)$$

The first-order Fresnel-Kirchhoff expressions for the matrix elements, $F_{\beta\alpha}$ are (Eq 15):

$F_{PP} = [-(1 + r_{PP}) \cos \theta_s + (1 - r_{PP}) \cos \theta_i] \cos \varphi_s - r_{PS} (1 + \cos \theta_s \cos \theta_i) \sin \varphi_s$,

$F_{PS} = [+(1 + r_{PP}) - (1 - r_{PP}) \cos \theta_s \cos \theta_i] \sin \varphi_s - r_{PS} (\cos \theta_s + \cos \theta_i) \cos \varphi_s$

and

$$F_{SP} = [-(1 + r_{SS}) + (1 - r_{SS}) \cos\theta_s \cos\theta_i] \sin\varphi_s - r_{SP}(\cos\theta_s + \cos\theta_i) \cos\varphi_s$$

$$F_{SS} = [-(1 + r_{SS}) \cos\theta_s + (1 - r_{SS}) \cos\theta_i] \cos\varphi_s + r_{SP}(1 + \cos\theta_s \cos\theta_i) \sin\varphi_s$$

Each of these elements is the sum of two terms: the first involves only the diagonal elements of the reflection matrix, Eq 4, and the second is proportional to its off-diagonal terms. When the off-diagonal terms vanish, Eq 15 reduces to textbook results [3].

5. PLANE-WAVE BRDFs

The conventional plane-wave BRDFs can be expressed directly in terms of the scattering matrix elements:

$$BRDF_{ba} = \frac{16\pi^2}{\lambda^4} \cdot \cos\theta_s \cos\theta_i \cdot Q_{ba} \cdot S_2(\vec{f}) \tag{16}$$

where

$$Q_{ba} = \left(\frac{\cos\theta_s + \cos\theta_i}{4 \cos\theta_s \cos\theta_i} \right)^2 |F_{ba}|^2 \tag{17}$$

and $S_2(\vec{f})$ is the power spectral density of the surface height fluctuations, Z:

$$S_2(\vec{f}) = \lim_{A \to \infty} \left\langle \frac{1}{A} \left| \int_A d\vec{x}\, e^{i2\pi \vec{f} \cdot \vec{x}} Z(\vec{x}) \right|^2 \right\rangle \tag{18}$$

Here $\vec{f}$ is the surface spatial frequency

$$\begin{pmatrix} f_x \\ f_y \end{pmatrix} = \frac{1}{\lambda} \begin{pmatrix} \sin\theta_s \cos\varphi_s - \sin\theta_i \\ \sin\theta_s \sin\varphi_s \end{pmatrix} \tag{19}$$

In the case of isotropic materials Eq 16 is essentially identical with the corresponding first-order vector-perturbation-theory or Rayleigh-Rice (RR) result [9,10]. There are two differences: The first is minor -- the Q values have different dependencies on the angles and surface indices -- but each reduces to the appropriate Fresnel intensity reflection coefficient in the specular direction:

$$Q_{ba}(\text{Specular}) = |\mathcal{R}_{ba}|^2 \tag{20}$$

The second difference is more interesting -- the first-order RR results in the literature only apply to isotropic materials, and predict zero cross-polarization effects in the plane of incidence, while the first-order FK results reported here apply to both isotropic and anisotropic materials, and do predict non-vanishing cross-polarization effects in the second case.

The present FK calculation predicts the ratio of the cross- to co-polarization intensities to be:

$$\beta = \frac{BRDF(cross-pol)}{BRDF(co-pol)} = \left| \frac{F_{ba}}{F_{aa}} \right|^2 \approx \left| \frac{\pi_{ba}}{\pi_{aa}} \right|^2 \tag{21}$$

where the approximation applies to near-specular scatter in the plane of incidence. If we evaluate this using the normal-incidence expressions for the reflection elements in Eqs 11,12, we find

$$\beta \approx \left| \frac{(\pi_0 - \pi_e) \sin\psi \cos\psi}{\pi_0 \cos^2\psi + \pi_e \sin^2\psi} \right|^2 \tag{22}$$

This vanishes for $\psi = 0$ or pi/2 but has the value of

$$\beta_{MAX} \approx \left| \frac{\pi_0 - \pi_e}{\pi_0 + \pi_e} \right|^2 \tag{23}$$

when ψ = pi/4. This quantity is twice the numerator in Eq 1 since the quantity appearing there involves an averaging over ψ. According to Fig 1, Eq 23 is ~ 0.7 % for Be at $\lambda \sim 1.2$ μm.

6. STOKES MATRICES

The full polarization properties of reflection and scattering are described by:

$$\begin{pmatrix} I \\ Q \\ U \\ V \end{pmatrix}_{\pi,s} = \begin{pmatrix} S_{11} & S_{12} & S_{13} & S_{14} \\ S_{21} & S_{22} & S_{23} & S_{24} \\ S_{31} & S_{32} & S_{33} & S_{34} \\ S_{41} & S_{42} & S_{43} & S_{44} \end{pmatrix} \begin{pmatrix} I \\ Q \\ U \\ V \end{pmatrix}_i \tag{24}$$

where the elements of the Stokes vector are:

$$I = I_p + I_s = |E_p|^2 + |E_s|^2$$
$$Q = I_p - I_s = |E_p|^2 - |E_s|^2 \tag{25}$$

$$U = 2 \ \text{Re} \ [E_P E_S^*]$$

$$V = 2 \ \text{Im} \ [E_P E_S^*] \qquad (25)$$

In the case of reflection, the 16 elements of the Stokes matrix are

$$S_{11} = 0.5 \ [|r_{PP}|^2 + |r_{SS}|^2 + |r_{PS}|^2 + |r_{SP}|^2]$$

$$S_{12} = 0.5 \ [|r_{PP}|^2 - |r_{SS}|^2 - |r_{PS}|^2 + |r_{SP}|^2]$$

$$S_{13} = + \ \text{Re} \ [\ r_{PP} r_{PS}^* + r_{SS} r_{SP}^* \ ]$$

$$S_{14} = - \ \text{Im} \ [\ r_{PP} r_{PS}^* - r_{SS} r_{SP}^* \ ]$$

$$S_{21} = 0.5 \ [|r_{PP}|^2 - |r_{SS}|^2 + |r_{PS}|^2 - |r_{SP}|^2]$$

$$S_{22} = 0.5 \ [|r_{PP}|^2 + |r_{SS}|^2 - |r_{PS}|^2 - |r_{SP}|^2]$$

$$S_{23} = + \ \text{Re} \ [\ r_{PP} r_{PS}^* - r_{SS} r_{SP}^* \ ]$$

$$S_{24} = - \ \text{Im} \ [\ r_{PP} r_{PS}^* + r_{SS} r_{SP}^* \ ]$$

$$S_{31} = + \ \text{Re} \ [\ r_{PP} r_{SP}^* + r_{SS} r_{PS}^* \ ]$$

$$S_{32} = + \ \text{Re} \ [\ r_{PP} r_{SP}^* - r_{SS} r_{PS}^* \ ] \qquad (26)$$

$$S_{33} = + \ \text{Re} \ [\ r_{PP} r_{SS}^* + r_{PS} r_{SP}^* \ ]$$

$$S_{34} = - \ \text{Im} \ [\ r_{PP} r_{SS}^* - r_{PS} r_{SP}^* \ ]$$

$$S_{41} = + \ \text{Im} \ [\ r_{PP} r_{SP}^* - r_{SS} r_{PS}^* \ ]$$

$$S_{42} = + \ \text{Im} \ [\ r_{PP} r_{SP}^* + r_{SS} r_{PS}^* \ ]$$

$$S_{43} = + \ \text{Im} \ [\ r_{PP} r_{SS}^* + r_{PS} r_{SP}^* \ ]$$

$$S_{44} = + \ \text{Re} \ [\ r_{PP} r_{SS}^* - r_{PS} r_{SP}^* \ ]$$

Appendix 1 discusses the sign convention regarding the Im terms.

When these elements are substituted into Eq 24 they give the Stokes vector of the reflected intensities. The corresponding Stokes vector of the quantity Q which appears in the expression for the BRDF, Eq 16 , can be obtained in the same way by making the replacement

$$\pi_{ba} \longrightarrow - \frac{\cos\theta_s + \cos\theta_i}{4\cos\theta_s \cos\theta_i} \cdot F_{ba} \qquad (27)$$

In the case of an optically isotropic surface, or a uniaxial crystal with its optic axis perpendicular to the surface plane, the off-diagonal matrix elements in the 2 X 2 reflection matrix vanish and the Stokes matrices for reflection and scattering-in-the-plane-of-incidence have the general form:

$$
\begin{pmatrix}
A & B & 0 & 0 \\
B & A & 0 & 0 \\
0 & 0 & C & D \\
0 & 0 & -D & C
\end{pmatrix}
\tag{28}
$$

where the four quantities A ... D are generally non-vanishing. In the case of a uniaxial crystal with its optic axis in the surface plane, the zero elements do not necessarily vanish, and provide a handle for detecting and measuring crystal anisotropies.

7. MUELLER MATRICES

An equivalent but simpler representation uses the Mueller matrix:

$$
\begin{pmatrix}
I_P \\
I_S \\
U \\
V
\end{pmatrix}_{n,s}
=
\begin{pmatrix}
M_{11} & M_{12} & M_{13} & M_{14} \\
M_{21} & M_{22} & M_{23} & M_{24} \\
M_{31} & M_{32} & M_{33} & M_{34} \\
M_{41} & M_{42} & M_{43} & M_{44}
\end{pmatrix}
\begin{pmatrix}
I_P \\
I_S \\
U \\
V
\end{pmatrix}_{i}
\tag{29}
$$

In this case the elements of the reflection matrix are:

$$
\begin{aligned}
M_{11} &= |r_{pp}|^2 \\
M_{12} &= |r_{ps}|^2 \\
M_{13} &= + \mathrm{Re}\,[\, r_{pp} r_{ps}^{*} \,] \\
M_{14} &= - \mathrm{Im}\,[\, r_{pp} r_{ps}^{*} \,] \\
M_{21} &= |r_{sp}|^2 \\
M_{22} &= |r_{ss}|^2
\end{aligned}
\tag{30}
$$

$$M_{23} = + \text{Re} [r_{ss} r_{sp}^*]$$

$$M_{24} = + \text{Im} [r_{ss} r_{sp}^*]$$

$$M_{31} = + 2 \text{Re} [r_{pp} r_{sp}^*]$$

$$M_{32} = + 2 \text{Re} [r_{ss} r_{ps}^*]$$

$$M_{33} = + \text{Re} [r_{pp} r_{ss}^* + r_{ps} r_{sp}^*]$$

$$M_{34} = - \text{Im} [r_{pp} r_{ss}^* - r_{ps} r_{sp}^*]$$

$$M_{41} = + 2 \text{Im} [r_{pp} r_{sp}^*]$$

$$M_{42} = - 2 \text{Im} [r_{ss} r_{ps}^*]$$

$$M_{43} = + \text{Im} [r_{pp} r_{ss}^* + r_{ps} r_{sp}^*]$$

$$M_{44} = + \text{Re} [r_{pp} r_{ss}^* - r_{ps} r_{sp}^*]$$

(30)

Again, the Stokes vector for the Q's and the BRDF can be obtained using Eq 27.

When the off-diagonal elements in the reflection matrix, Eq 4, vanish, the Mueller matrices for reflection and scattering-in-the-plane-of-incidence have the general form:

$$\begin{pmatrix} A+B & 0 & 0 & 0 \\ 0 & A-B & 0 & 0 \\ 0 & 0 & C & D \\ 0 & 0 & -D & C \end{pmatrix}$$

(31)

where the quantities A...D are the same as in Eq 28. If the off-diagonal matrix elements do not vanish, as in the case of interest here, the zero elements in Eq 31 need not vanish, and again provide a means for detecting and measuring crystal anisotropies.

8. FOLLOW ON

The next step is to evaluate the reflection and scattering matrix elements as functions of the angles of incidence and scattering and the radiation wavelength, using available data on the crystal indices [7]. These results can then be used to design efficient reflection/ellipsometry experiments for determining the indices at the longer wavelengths of interest.

The next step in the theory is to calculate the scattering from a surface modelled as a collection of randomly-oriented crystallites. That calculation will give the BRDF as the sum of three terms: 1) the topographic term -- the only one appearing in the present single-crystal calculation -- which varies as $\lambda\wedge(-4)$; 2) the grain-boundary term -- already considered approximately in [6] -- which varies as $\lambda\wedge(-2)$; and 3) an interference term between the two scattering mechanisms which varies as $\lambda\wedge(-3)$, as discussed in [2]. Elson has recently predicted significant interference effects for the analogous case of roughness plus inhomogenity scattering for beryllium under certain conditions [11].

It would be interesting to compare the cross-polarization terms, calculated here as a first-order effect for an anisotropic crystal, with the corresponding cross-polarization terms calculated as a second-order effect for an isotropic material. Such "second-order" calculations appear in the literature in both the Fresnel-Kirchhoff and the Rayleigh-Rice approximations, but their complexity hinders immediate interpretation [3,12,13].

These analytic results, combined with the yet-to-be-measured long-wavelength optical constants, will help identify the precise mechanisms responsible for the scattering from these mirrors and indicate manufacturing improvements required to improve their performance.

9. APPENDIX 1. SIGN CONVENTIONS

There is a sign convention in the literature having to do with whether time moves from right to left or from left to right. The practical difference is that imaginary quantities change sign in going from one system to the other. The possibilities are indicated in the following table:

Time factor	Index	Sign of Im terms in Eqs 25, 26, 30	Representative Authors
exp(-iwt)	n + i k	unchanged	Born and Wolf, Jackson, Stratton
exp(+iwt)	n - i k	opposite	Azzam and Bashara van de Hulst, Solimeno et al.

We use the first version. To convert our results to the second, the sign of the imaginary part of the index of refraction must be changed from positive to negative, as indicated, and the signs of the Im terms in Eqs 25, 26 and 30 must be reversed. Practical predictions are independent of the convention used.

10. APPENDIX 2. REFLECTION MATRIX ELEMENTS FOR OPTIC AXIS IN THE SURFACE PLANE

The following results are taken from Azzam and Bashara, quoting Sosnowski, who derived them from an obscure 1905 textbook by Curry:

$$r_{PP} = (F_1 G_4 + F_2 G_3)/(F_1 + F_2)$$

$$r_{PS} = F_1 F_2 (G_2 - G_1)/(F_1 + F_2)$$

$$r_{SP} = (G_4 - G_3)/(F_1 + F_2) \tag{32}$$

$$r_{SS} = (F_1 G_1 + F_2 G_2)/(F_1 + F_2)$$

where

$$F_1 = J/((N_0 \sin^2 \theta_i + J \cos \theta_i) \tan \psi)$$

$$F_2 = N_{10} \tan \psi (I + N_0 N_{10} \cos \theta_i)/(I N_{10} \cos \theta_i + N_0 J^2)$$

$$G_1 = (N_0 \cos \theta_i - J)/(N_0 \cos \theta_i + J)$$

$$G_2 = (N_0 N_{10} \cos \theta_i - I)/(N_0 N_{10} \cos \theta_i + I)$$

$$G_3 = (N_{10}^2 \cos \theta_i - N_0 J)/(N_{10}^2 \cos \theta_i + N_0 J) \tag{33}$$

$$G_4 = (I N_{10} \cos \theta_i - N_0 J^2)/(I N_{10} \cos \theta_i + N_0 J^2)$$

$$I^2 = N_{10}^2 N_{1e}^2 - (N_0^2 \sin^2 \theta_i)(N_{10}^2 \sin^2 \psi + N_{1e}^2 \cos^2 \psi)$$

$$J^2 = N_{10}^2 - N_0^2 \sin^2 \theta_i$$

Here ψ is the angle of the optic axis shown in Fig 2 and θ_i is the angle of incidence. At normal incidence, $\theta_i = 0$, these expressions reduce to Eqs 11,12 in the text.

11. ACKNOWLEDGEMENTS

This work was supported in part by the US Air Force, Rome Air Development Center, under contract F30602-85-C-0294. We also thank Dr L. Tsang and Dr. P. Z. Takacs for helpful discussions and encouragement.

12. REFERENCES

1. J. C. Stover, M. L. Bernt, D. E. McGary and J. Rifkin, "Investigation of anomalous scatter from beryllium mirrors", Proc SPIE 1165 100-106 (1989), and private communications.

2. E. L. Church and P. Z. Takacs, "Subsurface and volume scattering from smooth inhomogeneous materials", Proc SPIE 1165 31-41 (1989).

3. L. Tsang, J. A. Kong and R. T. Shin, "Theory of Microwave Remote Sensing", John Wiley and Sons, New York, 1985.

4. F. T. Ulaby, R. K. Moore and A. K. Fung, "Microwave Remote Sensing: Active and Passive": Vols 1 and 2, Addison-Wesley, Reading MA, 1981; Vol 3, Artech House, Inc, Dedham MA, 1986.

5. J. M. Elson, "Theory of light scattering from a rough surface with an inhomogeneous dielectric permittivity", Phys. Rev. B30 5460-5480 (1984).

6.   E. L. Church, P. Z. Takacs and J. C. Stover, "Scattering by anisotropic grains in beryllium mirrors", Proc SPIE 1331 12-17 (1990).

7.   J. H. Weaver, D. W. Lynch and R. Rosei, "Optical properties of Be from 0.12 to 4.5 eV", Phys. Rev. B7 3537-3541 (1973).

8.   R. M. A. Azzam and N. M. Bashara, "Ellipsometry and Polarized Light", North Holland, New York, 1977.

9.   E. L. Church, H. A. Jenkinson and J. M. Zavada, "Relationship between surface scattering and microtopographic features", Opt. Eng. 18 125-136 (1979).

10. D. E. Barrick, Chapter 9 in "Radar Cross Section Handbook", McGraw-Hill, New York, 1970.

11. J. M. Elson, "Anomalous scattering from optical surfaces with roughness and permittivity perturbations", Paper 1530-20 presented at the Annual Meeting of the SPIE held in San Diego, July 1991.

12. R. Schiffer, "Reflectivity of a slightly rough surface", Appl. Opt. 26 704-712 (1987).

13. J. C. Leader, "Analysis and prediction of laser scattering from rough surface materials", J. Opt. Soc. Am. 69 610-628 (1979).

A study of anomalous scatter characteristics

John C. Stover
Marvin L. Bernt
Tim D. Henning

TMA Technologies Inc.
PO Box 3118
Bozeman MT 59715

ABSTRACT

Anomalous, or non-topographic, scatter from beryllium optics at IR wavelengths has been recognized as a source of poor performance for space based IR imaging systems. This paper reports the results of experiments designed to discover clues as to the source of anomalous scatter so that it can be eliminated with new manufacturing techniques. Polarization, wavelength dependence and surface pitting were examined. None of these can be confirmed as the primary source of anomalous scatter based on the work completed to date.

2. Introduction

This document reports a continuation of the anomalous scatter investigation that TMA has been making with a team formed by the Oak Ridge Optics MODIL. Background material, such as definitions of BRDF, PSD, etc., and the work which established the existance of anomalous scatter are found in earlier reports of our progress [1-3].

Non-topographic scatter so completely dominates the mid–IR BRDF of many beryllium optics that making surface profile measurements is of little value for determining mirror quality. The effect makes it difficult to meet scatter specifications for beryllium IR imaging systems. The purpose of this continuing investigation is to learn more about the causes of non-topographic scatter so that production techniques can be devised to reduce it, or even eliminate it.

The relationship between surface roughness and scatter from smooth, clean, front surface reflectors is well understood. "Smooth" implies that surface height fluctuations are much less than a wavelength. This condition is met if you can see your face in the mirror and holds for this sample set. "Clean" implies that the BRDF is not dominated by scatter from surface contaminants. The samples were cleaned prior to each measurement. The "front surface" descriptor implies that scatter is due to just surface topography, and not other mechanisms, such as bulk scatter or position dependent changes in the optical constants. This condition is the one that is in some manner violated at IR wavelengths by a majority of beryllium mirrors. The measured BRDF from "topographic" reflectors can be used to find the surface power spectral density function (PSD) and the root mean square roughness (rms).

Data was taken on ten samples at wavelengths ranging from 0.325 to 10.6 micrometers. Nine of the samples were beryllium. The tenth, a molybdenum mirror with little non-topographic scatter, was used as a standard. All but one of the Be mirrors had been used in the previous study and are starting to show the effects of continued examination. They have been measured with a variety of instruments (including stylus profilometers), at several different laboratories. They did not come to the original study in new condition, and their manufacturing history is not well documented.

The scatter measurement work reported here concentrated in three areas. These are polarization effects, wavelength dependence effects, and the role played by small point defects that are present on many beryllium mirrors. In each case there was reason to believe that these issues should be examined for possible clues as to the cause of non-topographic scatter. Our approach, data and results are covered in the next

$$F_{total} = F_t + F_a = \frac{(const.)\, S_t\, COS\, \theta_s}{\lambda^4} + \frac{S_a}{\lambda^2} \qquad (2)$$

$$F_{total} = 1/\lambda^4 \; + \; 0.1/\lambda^2 \qquad (3)$$

The table below evaluates the two terms and their ratio as a function of wavelength. The topographic term dominates in the visible and the anomalous term dominates in the mid-IR.

λ	F_t	F_a	$F_a \setminus F$
0.1	10^4	10	10^{-3}
1.0	1	0.1	0.1
10	10^{-4}	10^{-3}	10

The BRDF data was analyzed for the presence of such a term in the following manner. First, the surface PSD was found using the 0.325 BRDF data. Examination of the PSD calculations (Figures 3) shows essentially the same result is obtained for both the 0.325 and 0.633 scans, so it is reasonable to assume that this calculation is correct. Then the PSD was used to calculate topographic scatter at 10.6 micrometers. This was subtracted from the measured 10.6 scatter and the remainder assumed to be anomalous scatter caused by the hypothetical inverse squared term. The constant value in this term was then found. The resulting two term equation was then used to calculate the BRDF at 3.39 and 0.633 micrometers.

In order to ease the calculation load, each measured BRDF was fit with a straight line in log/log space. Plots are produced after entering two points from any two BRDF data sets. The results for two samples are shown in Figures 4 and 5. The measured scatter is above and the calculated scatter below. In the complete data set a reasonable correspondence is sometimes found; however, in general the predictions are poor. The conclusion is reached that non-topographic scatter does not follow a simple inverse square relationship.

There is still some hope for this approach. The existence of an inverse square term implies that a cross term (inverse wavelength cubed) exists. The assumption here was that this third term is small. It may not be. Thus the door is still open to produce a power series fit that will provide clues to the general nature of the source of the measured anomalous scatter.

5. Point Defect Contributions

Another possible source of anomalous scatter that has received considerable attention within the beryllium optics community is the presence of small (about 10 micrometer) defects found in many mirror surfaces. They generally take the form of small surface pits and are violations of the smooth surface condition required of Equation (1). BRDF maps were made of each sample, using a 1.06 micormeter laser source, to quantify the presence of these defects. The data was obtained by imaging the sample with a CCD camera, calculating BRDF and color plotting the results. Figure 6 is a grey scale version of one map. We have looked at the pits in two ways.

First, as suggested in a MODIL Scatter Workshop, we tried treating the pits as sources of Lambertian scatter. We converted the map to a histogram of BRDF level by number of pixels. We picked a threshold of three standard deviations above the mean as a defect threshold and found the resulting fractional defect area, A_d. Values ranged from almost nothing to 1.56% with 0.5% typical. Under the Lambertian assumption the BRDF was calculated as

$$F_l = A_d \, R/\pi \qquad (4)$$

where R, the measured specular reflectance of the mirror, varies from about 0.4 to 0.96 over the wavelengths in question. Typical values at 10.6 micrometers are about 10^{-3} sr^{-1}. Comparison of these values to the measured high angle scatter data reveals that the computed Lambertian BRDF is generally at least an order of magnitude than the measured BRDF. The conclusion is that the pits on these samples do not contribute as Lambertian sources of scatter.

Pit contribution was also examined by looking for a correlation between anomalous scatter and the fractional pit area, A_d, as shown in Figure 8. The anomalous scatter parameter was generated from the calculated PSD's by taking the quadrature difference of the associated rms roughness values ($[\sigma^2_{10.6} - \sigma^2_{.633}]^{1/2}$) for the 10.6 and 0.633 micrometer data. There is no apparent correlation. It should be remembered that the samples used in this study come from varied manufacturing processes. They are not a controlled suite of samples. The apparent lack of correlation could be caused by other sample differences. The A and B data sets are an exception to this as they are taken from two locations on the same sample. "A" was a very low pit density location and "B" a much higher density location. These points are marked in Figure 8. Although the pit densities differ, the anomalous scatter levels do not. We are forced to conclude, from this limited data, that there is little effect from the pits.

6. Conclusions

This research has resulted in reducing the list of possible causes for anomalous scatter rather than identifying one or more prime contenders. Each possible source (polarization mechanisms, inverse wavelength squared terms and pits) has produced little or no correlation to anomalous scatter. There are a couple of qualifiers to these results. The inclusion of an inverse wavelength cubed term might change that result, and a better controlled sample set will reduce the number of unknown variables.

A number of experiments remain for the coming year. If a pitted sample is measured after a thin layer of aluminum is applied (so that the pits are preserved) and the anomalous scatter disappears, then this would probably end any speculation about pits as a major source of anomalous. We plan to measure several samples on the full polarization control instrument that we will have available in the Fall of 1991. We may yet see some polarization effects by watching for variation in all sixteen Mueller elements. We also hope to measure some single crystal samples in the coming year and check for wavelength depence on a more controlled set of samples.

7. Acknowledgements

The authors wish to recognize the support of the Optics MODIL at Oak Ridge National Laboratory for funding and technical collaboration. Dr. Eugene Church, of Picatinny Arsenal, and Dr. Peter Takacs of Brookhaven National Laboratory performed a parallel study on different aspects of the problem, and we thank them for many stimulating discussions.

8. References

1.   Stover, J.C., M.L. Bernt, D.E. McGary, J. Rifkin; "An Investigation of Anomalous Scatter from Beryllium Mirrors", ORNL/Sub/89-VK810/1, Oak Ridge National Laboratory, (1989).

2.   Stover, J.C., M.L. Bernt, D.E. McGary, J. Rifkin. "Investigation of Anomalous Scatter from Beryllium Mirrors." Proc. SPIE 1165-43 (1989b).

3.   Stover, J.C., Optical Scattering: Measurement and Analysis, page 150, McGraw-Hill New York, NY 1990.

4.   Church, E.L., P.Z. Takacs. "Subsurface and Volume Scattering From Smooth Surfaces." Proc. SPIE, 1165-04, (1989).

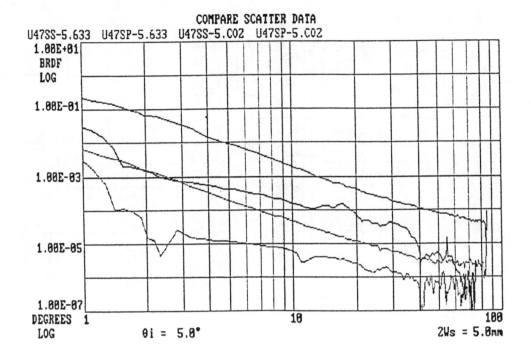

Figure (1)

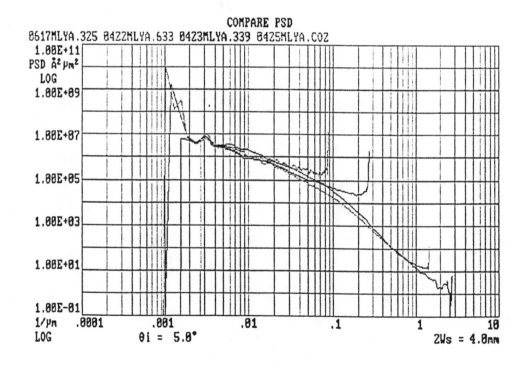

Figure (2)

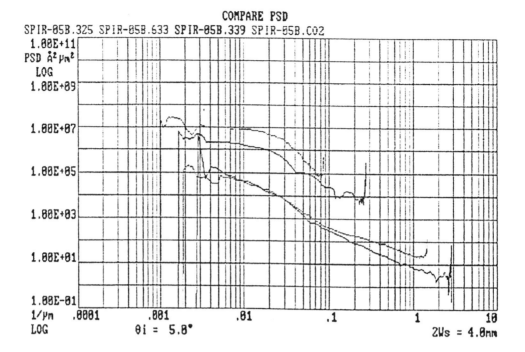

Figure (3)

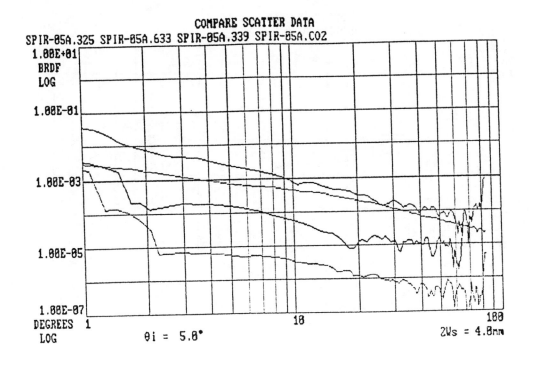

COMPARE SCATTER DATA

SPIR-85A.325 SPIR-85A.633 SPIR-85A.339 SPIR-85A.CO2

$\theta i = 5.8°$

$2Ws = 4.0mm$

Calculated BRDF

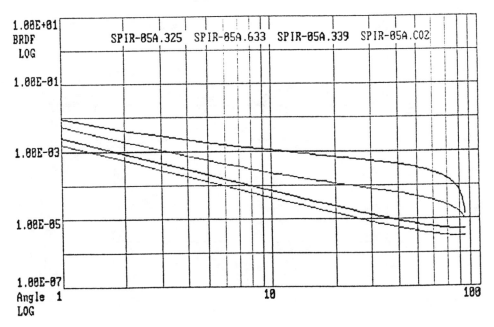

SPIR-85A.325   SPIR-85A.633   SPIR-85A.339   SPIR-85A.CO2

Figure (4)

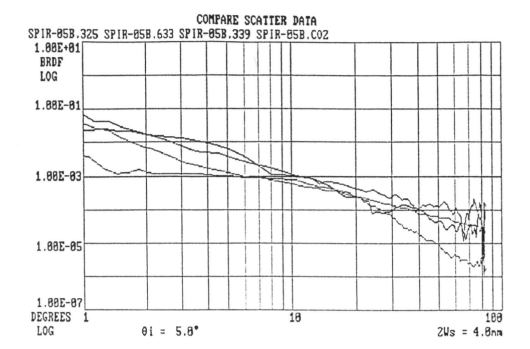

COMPARE SCATTER DATA
SPIR-85B.325 SPIR-85B.633 SPIR-85B.339 SPIR-85B.CO2

1.00E+01
BRDF
LOG

1.00E-01

1.00E-03

1.00E-05

1.00E-07
DEGREES   1   10   100
LOG   θi = 5.8°   2Ws = 4.0nm

Calculated BRDF

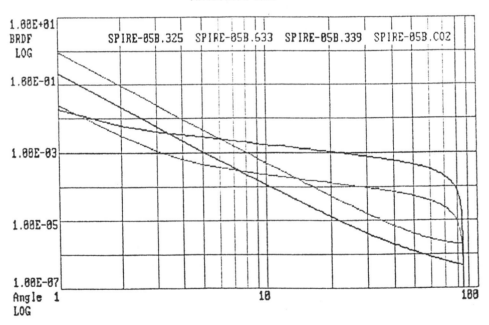

1.00E+01
BRDF
LOG

1.00E-01   SPIRE-85B.325   SPIRE-85B.633   SPIRE-85B.339   SPIRE-85B.CO2

1.00E-03

1.00E-05

1.00E-07
Angle 1   10   100
LOG

Figure (5)

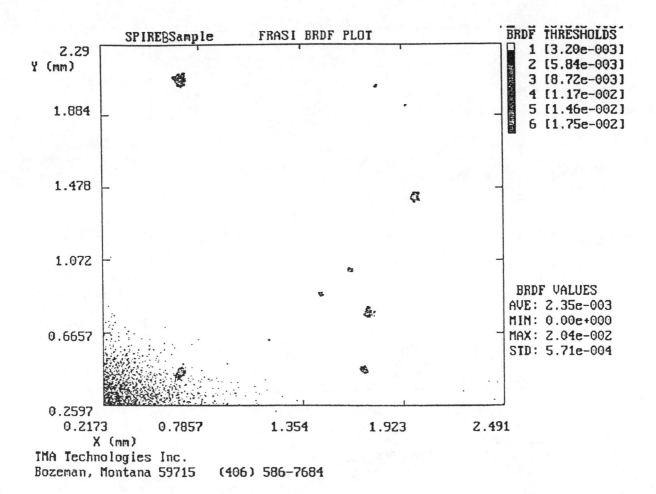

Figure (6)

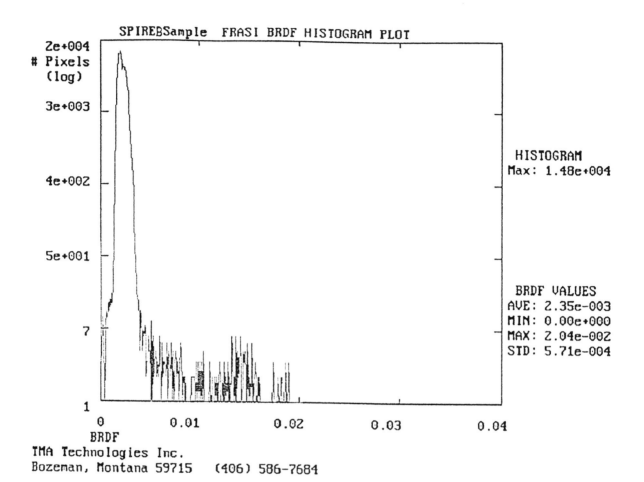

Figure (7)

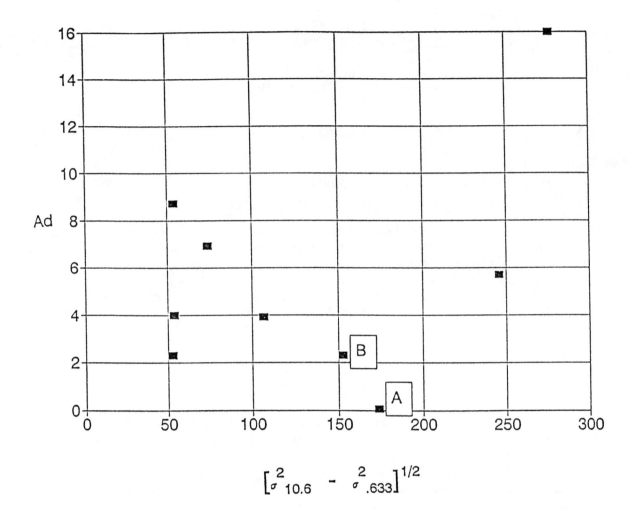

$$\left[\sigma^2_{10.6} - \sigma^2_{.633}\right]^{1/2}$$

Figure (8)

Anomalous scattering from optical surfaces with roughness and permittivity perturbations

J. M. Elson

Research Department, Physics Division
Naval Weapons Center, China Lake, CA 93555

ABSTRACT

This paper discusses the wavelength dependence of angle-resolved scattering (ARS) from optical surfaces that have surface roughness and dielectric permittivity fluctuations. Such surfaces are assumed to be of "optical quality" in that the rms roughness is much less than the incident wavelength. Of particular interest is scattering from beryllium optical components. Depending upon fabrication methods, it has been found that ARS measurements from beryllium at visible through infrared wavelengths do not scale as wavelength to the inverse fourth power. This is contrary to first-order roughness scattering theory that predicts a Rayleigh-like wavelength dependence. This indicates that other nontrivial sources of scattering are also in effect. The results of this work show that a surface that scatters due to surface roughness and dielectric permittivity perturbations along with a statistical correlation between these sources can yield a non-Rayleigh wavelength dependence. This can explain the anomalies seen in beryllium scattering and other materials.

1. INTRODUCTION

The basis of the analysis presented here was originally applied to anomalies in angle-resolved scattering (ARS) as measured from silver surfaces.[1] The anomalies in question refer to differences between scattering theory and measurements for transverse-electric (TE) and transverse-magnetic (TM) polarization of the incident beam. Wavelength dependence was not considered. In this earlier work,[1] it was shown that (1) differences between roughness scattering theory and ARS measurements could be accounted for when perturbations about the nominal optical constant values were included and (2) there was a statistical correlation between the roughness and optical constant perturbations. In fact, this cross-correlation or interference between scattering sources was the major contributor to the anomalies. When evaluating mirrors designed for infrared (IR) applications, it is convenient to measure ARS at visible wavelengths and then, using scaling laws, infer scattering levels at IR or other wavelengths. This is a viable idea assuming that scattering scales with the inverse fourth power of the wavelength or some other known wavelength dependence. Also, a common parameter for determining the quality of optical finish is the rms roughness. This is a valid concept only if roughness is the sole source of scattering.

Stover *et al.*[2] and Egert[3] have recently reported order of magnitude discrepancies in consistency of inferring the power spectral density of beryllium surfaces from measurements of scattered light at 0.6328-, 1.06-, 3.38-, and 10.6-μm wavelengths. Church *et al.*[4,5] have provided an analysis of possible causes of such anomalies that includes effects due to material birefringence. In the work presented here, birefringence is not considered; rather, other possible explanations for these reported anomalies are given.

0-8194-0658-9/91/$4.00

2. BACKGROUND AND THEORY

Most details of the first-order theory used here are found elsewhere,[1] where many questions may be answered. Sources of scattering include (1) surface roughness, which perturbs the phase of the scattered field, (2) subsurface dielectric perturbations, which perturb the nominal reflectance of the scattered field, and (3) supersurface particulates, which are independent scatterers on the surface. Scattering from sources as in (3) are not considered here. Figure 1 shows some notation and indicates the scattering sources as roughness $\Delta z(x,y)$ and dielectric $\Delta\varepsilon(x,y;\lambda)$ perturbations, where $\varepsilon_1(\lambda)$ is the nominal dielectric constant of the substrate. Unlike Ref. 1, any z-dependence of $\Delta\varepsilon(x,y;\lambda)$ is neglected here.

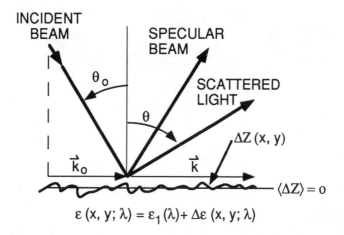

$$\varepsilon(x,y;\lambda) = \varepsilon_1(\lambda) + \Delta\varepsilon(x,y;\lambda)$$

Fig. 1. Schematic of the model for a scattering surface showing surface roughness $\Delta z(x,y)$ and dielectric perturbations $\Delta\varepsilon(x,y;\lambda)$. The numerical results of this work are given in the plane of incidence, where the angles of incidence and scattering are θ_0 and θ, respectively.

2.1. Phase perturbation scattering: wavelength dependence

This section considers only scattering from surface roughness. The differential scattered power per unit solid angle normalized to the incident power (BRDF $\cos\theta$, where θ is the polar scattering angle) is

$$\frac{1}{P_o} \left< \frac{dP_r}{d\Omega} \right> = F_r \, g_r \, (\vec{k}_o - \vec{k}) \quad , \tag{1}$$

where $k = (2\pi/\lambda)\sin\theta$, $k_o = (2\pi/\lambda)\sin\theta_o$, and θ_o is the polar angle of incidence. F_r is an algebraic factor dependent on angle, wavelength, etc., as discussed in Ref. 1. Also,

$$g_r(\vec{k} - \vec{k}_o) = <|\Delta z (\vec{k}_o - \vec{k})|^2>/L^2 \tag{2}$$

is the power spectral density of the surface roughness where $<...>$ denotes ensemble average and

$$\Delta z(\vec{K}) = \int d^2\rho \, \Delta z(\vec{\rho}) \, e^{i\vec{K} \cdot \vec{\rho}} \tag{3}$$

By letting $|\epsilon_1(\lambda)| >> 1$, the wavelength scaling of Eq. (1) can easily be evaluated for two particular cases of wavelength dependence on $\epsilon_1(\lambda)$. These are (1) $\epsilon_1(\lambda)$ and $\Delta\epsilon(x,y;\lambda)$ independent of wavelength and (2) $|\epsilon_1(\lambda)| \sim \lambda^2$ and $|\Delta\epsilon(x,y;\lambda| \sim \lambda^2$ (metallic-like). These two cases both yield

$$\frac{1}{P_o} <\frac{dP_r}{d\Omega}> \sim \lambda^{-4} \tag{4}$$

While there may be some cases where these limiting results may be valid, neither of these examples are of general use since $\epsilon_1(\lambda)$ is obviously material dependent. Nevertheless, in a crude way this shows that roughness scattering does not seem to suggest any surprise behavior in wavelength dependence.

2.2. Reflectance perturbation scattering: wavelength dependence

This section considers only scattering resulting from perturbations on the nominal Fresnel reflectance that are due to fluctuations in the subsurface dielectric permittivity. Again, the differential scattered power per unit solid angle normalized to the incident power is (BRDF $\cos\theta$, θ is polar scattering angle)

$$\frac{1}{P_o} <\frac{dP_d}{d\Omega}> = \frac{F_d g_d(\vec{k}_o - \vec{k})}{|q + q_1|^2} , \tag{5}$$

where $q = (2\pi/\lambda) [\epsilon_1(\lambda) - \sin^2\theta_o]^{1/2}$ and $q_1 = (2\pi/\lambda) [\epsilon_1(\lambda) - \sin^2\theta]^{1/2}$. The power spectral density of dielectric perturbations is

$$g_d(\vec{k}_o - \vec{k}) = <|\Delta\epsilon(\vec{k}_o - \vec{k})|^2>/L^2 , \tag{6}$$

where $\Delta\epsilon(K)$ is analogous to Eq. (3) with Δz replaced by $\Delta\epsilon$. The factor F_d is similar to F_r and is also given in Ref. 1. For the same limiting cases as discussed for Eq. (1), different results are seen:

(1) $$\frac{1}{P_o} <\frac{dP_d}{d\Omega}> \sim \lambda^{-2} \tag{7a}$$

(2) $$\frac{1}{P_o} <\frac{dP_d}{d\Omega}> \sim \lambda^{-4} \tag{7b}$$

For case (1), where $\varepsilon_1(\lambda)$ and $\Delta\varepsilon(x,y;\lambda)$ do not vary with wavelength, the ARS is seen to vary as the inverse square of the wavelength; whereas, for the metallic-like dependence, case (2), the inverse fourth power prevails. This result suggests that ARS from wavelength-independent dielectric perturbations can behave differently on wavelength than ARS predicted from roughness sources. However, beryllium certainly does not have a wavelength-independent dielectric function for the visible through IR. Hence, these limiting cases do not provide an explanation for a material like beryllium. Close inspection of Eq. (5) reveals the physical basis of how Eqs. (7) are obtained. The $|q + q_1|^{-2}$ term of Eq. (5) accounts for the depth of the scattering volume. For a $\varepsilon_1(\lambda)$ independent of wavelength, the $|q + q_1|^{-2}$ varies as wavelength squared; whereas, for $|\varepsilon_1(\lambda)| \sim \lambda^2$, this depth term is roughly constant with wavelength. In other words, for a dielectric permittivity that has little dependence on wavelength, the scattering volume depth effectively increases as the square of the wavelength, which results in the ARS intensity falling off only as inverse wavelength squared. For a metallic-like material, the scattering volume remains approximately constant with wavelength and thus the wavelength scaling remains as inverse fourth power. That Eq. (5) results from scattering due to fluctuations in the nominal reflectance is rather easy to see. It is a simple exercise to let the dielectric permittivity $\varepsilon(x,y;\lambda) = \varepsilon_1(\lambda) + \Delta\varepsilon(x,y;\lambda)$ and expand the Fresnel reflectance about $\Delta\varepsilon(x,y;\lambda)$, yielding

$$ R(x,y;\lambda) = R_o(\lambda) + \frac{\partial R}{\partial \varepsilon} \Big|_o \Delta\varepsilon(x,y;\lambda) \quad , \tag{8} $$

where the subscript 0 refers to $\Delta\varepsilon(x,y;\lambda) = 0$. Comparing this result for the appropriate polarizations with Eq. (5) shows very close functional dependences when $\vec{k} \approx \vec{k}_o$, i.e.,

$$ \frac{1}{P_o} \left< \frac{dP_d}{d\Omega} \right> \rightarrow \frac{\cos\theta_o}{\lambda^2} \left| \frac{\partial R}{\partial \varepsilon} \right|_o^2 \frac{<|\Delta\varepsilon|^2>}{L^2} \quad . \tag{9} $$

2.3. Phase plus reflectance perturbation scattering: wavelength dependence

When roughness and dielectric perturbations are considered together, both Eqs. (1) and (5) contribute to the overall scattering intensity but there is a potentially influential additional term as well. Equations (1) and (5) are based on independent scattering from roughness and dielectric perturbations, respectively. The ARS is dependent upon the statistical nature of these perturbations. However, the statistical correlation between the roughness and dielectric perturbations also must be taken into account. The general expression for ARS with both sources in effect may be written as

$$ \frac{1}{P_o} \left< \frac{dP_d}{d\Omega} \right> = \frac{1}{P_o} \left[\left< \frac{dP_r}{d\Omega} \right> + \left< \frac{dP_d}{d\Omega} \right> + \left< \frac{dP_{rd}}{d\Omega} \right> \right] \quad , \tag{10} $$

where the "new" term represents the interference term generated by the nature of the cross-correlation between the roughness and dielectric perturbations. In analogy with Eqs. (1) and (5), the last term of Eq. (10) is written as

$$\frac{1}{P_o} < \frac{dP_{rd}}{d\Omega} > = \text{Re}\left[F_{rd}\, g_{rd}\, (\vec{k}_o - \vec{k}) \right] \quad , \tag{11}$$

where the cross-power spectral density is written as

$$g_{rd}(\vec{k} - \vec{k}_o) = \frac{<\Delta(\vec{k} - \vec{k}_o)\, \Delta\varepsilon^*(\vec{k}_o - \vec{k})>}{L^2} \quad . \tag{12}$$

Again, the F_{rd} term is discussed in Ref. 1. If the perturbation variables are statistically independent, then Eq. (12) and this interference term would vanish. As in Eqs. (4) and (7), the wavelength dependence of the interference term for cases 1 and 2 is as follows:

$$(1) \qquad \frac{1}{P_o} < \frac{dP_{rd}}{d\Omega} > \sim \lambda^{-3} \tag{13a}$$

$$(2) \qquad \frac{1}{P_o} < \frac{dP_{rd}}{d\Omega} > \sim \lambda^{-4} \quad . \tag{13b}$$

2.4. Auto- and cross-power spectral density assumptions

In order to look for anomalous effects in ARS from the optical surface being modeled here, some assumptions regarding the statistical nature of the roughness and dielectric perturbation random variables are needed. The power spectral density expressions as shown in Eqs. (1), (5), and (12) are assumed here to be a Fourier transform of their respective correlation function. This is written as

$$g_x(\vec{K}) = \int d^2\tau\, G_x(\vec{\tau})\, e^{i\vec{K}\cdot\vec{\tau}} \quad , \tag{14}$$

where X is r(roughness), d(dielectric) or rd(roughness-dielectric). For X = r or d, the autocorrelation function is written as the sum of an exponential and a Gaussian

$$\begin{aligned} G_x(\tau) &= \delta_{lx}^2 \exp(-|\vec{\tau}|/\sigma_{lx}) + \delta_{sx}^2 \exp(-\tau^2/\sigma_{sx}^2) \\ &= G_{lx}(\tau) + G_{sx}(\tau) \end{aligned} \quad , \tag{15}$$

where the δ and σ denote rms and correlation length values, and the l and s subscripts refer to long and short range, respectively. For the cross-power spectral density, Eqs. (14) and (15) are used along with the assumption that the cross-correlation function is given by

$$G_{rd}(\tau) = e^{i\psi(\lambda)}\left\{ \left[G_{lr}(\tau)\, G_{ld}(\tau) \right]^{1/2} + \left[G_{sr}(\tau)\, G_{sd}(\tau) \right]^{1/2} \right\} \quad . \tag{16}$$

This expression is clearly an attempt to combine the statistical properties of both the roughness and dielectric perturbations; also, it inherently assumes that the roughness and dielectric perturbations are indeed correlated. The phase term $\exp(i\psi(\lambda))$ is included because the correlation of roughness and dielectric variables can be complex, where

$$\psi(\lambda) = \tan^{-1}\left[\text{Im } \varepsilon_1(\lambda)/\text{Re } \varepsilon_1(\lambda) \right].$$

(17)

As it turns out, the cross-correlation assumptions above are crucial in the results to follow in that certain materials can exhibit anomalous wavelength dependence of ARS.

3. NUMERICAL RESULTS

Numerical results are given for ARS from optical surfaces with beryllium and silver optical constants. It will be seen that beryllium can exhibit anomalous wavelength scattering whereas, silver does not. The random variable statistical parameter assumptions for the surface roughness include

Long-range correlation length (σ_{lr}) = 6.0 µm
Short-range correlation length (σ_{sr}) = 0.3 µm
Long-range rms (δ_{lr}) = 20 Å
Short-range rms (δ_{sr}) = 30 Å.

These rms values combine to yield a total rms surface roughness of about 36 Å. These variables are used in Eqs. (15) and (16). The random variable statistical parameter assumptions for the dielectric permittivity fluctuations include

Long-range correlation length (σ_{ld}) = 6.0 µm
Short-range correlation length (σ_{sd}) = 0.3 µm
Long-range rms (δ_{ld}) = $0.2\,|1 - \varepsilon_1(\lambda)|$
Short-range rms (δ_{sd}) = $0.2\,|1 - \varepsilon_1(\lambda)|$.

Note that the corresponding correlation lengths for the roughness and dielectric variables are taken to be the same. Further, the long- and short-range dielectric rms values are equal and are taken to be essentially proportional to the magnitude of the nominal dielectric constant. In other words, the rms variation of the dielectric perturbations is assumed to be on the order of 20% of the nominal dielectric magnitude as well as depending on wavelength. Compared to comparable ratios of roughness to wavelength percentages, this 20% value seems quite a large variation for a perturbation parameter. However, it will be assumed that the important consideration is the change in nominal reflectance. To illustrate, Fig. 2 shows the beryllium reflectance versus wavelength for the zero- and first-order terms of Eq. (7). It is seen that the perturbation on the nominal reflectance is very small, and thus the magnitude assumption of δ_{ld} and δ_{sd} is justified. Figures 3 through 6 show predicted beryllium ARS curves for wavelengths of 0.6328, 1.06, 3.39, and 10.6 µm, respectively. The optical constants of beryllium were obtained from Ref. 6. Both TE and TM polarizations are given. The angle of incidence is 5°, and the scattering is shown in the plane of incidence. Each plot has three curves corresponding to different sources of scattering: solid (roughness only), dot (dielectric only), and dash (roughness plus dielectric). Looking at Figs. 3 through 6, the relative position of the solid curve (roughness) to the dash curve (roughness plus dielectric) can be seen to change by about an order of magnitude! From Fig. 3 at 0.6328 microns, the solid curve indicates more ARS intensity than the dashed curve. From Fig. 6 at 10.6 µm, the solid curve is far below the dashed curve. This indicates that far more ARS occurs at IR wavelengths than predicted by roughness scattering theory, and this is consistent with ARS measurements from some beryllium components.[2-3] However, not all beryllium seems to exhibit such strong anomalies, and this fact supports the idea that surface preparation could be a major factor.

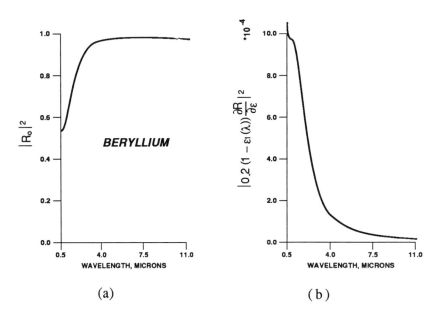

(a)   (b)

Fig. 2. (a) Zero-order reflectance and (b) first-order correction to the p-polarized Fresnel reflectance versus wavelength using beryllium optical constants. For a dielectric perturbation on the order of $0.2 \, |1 - \varepsilon_1(\lambda)|$, it is seen that the first-order correction to the intensity reflectance is much smaller than the nominal unperturbed reflectance.

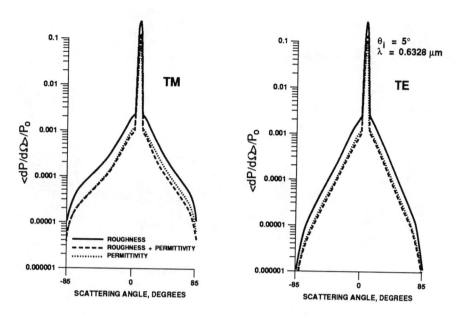

Fig. 3. TM and TE ARS curves versus polar scattering angle for λ = 0.6328 μm and an angle of incidence of 5° for beryllium. The three curves represent scattering due to roughness-only (solid), dielectric-only (dot), and roughness-plus-dielectric (dash) perturbations.

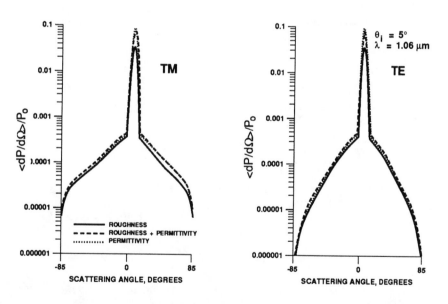

Fig. 4. TM and TE ARS curves versus polar scattering angle for λ = 1.06 μm and an angle of incidence of 5Å for beryllium. The three curves represent scattering due to roughness-only (solid), dielectric-only (dot), and roughness-plus-dielectric (dash) perturbations.

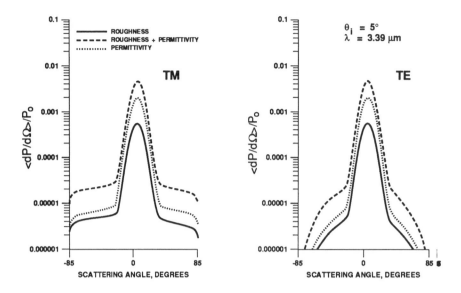

Fig. 5. TM and TE ARS curves versus polar scattering angle for λ = 3.39 μm and an angle of incidence of 5° for beryllium. The three curves represent scattering due to roughness-only (solid), dielectric-only (dot), and roughness-plus-dielectric (dash) perturbations.

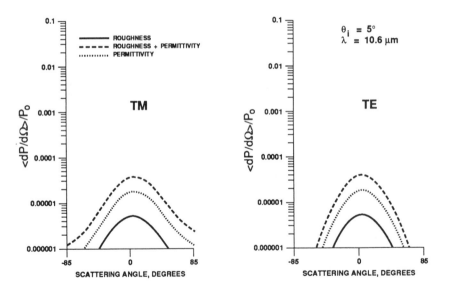

Fig. 6. TM and TE ARS curves versus polar scattering angle for λ = 10.6 μm and an angle of incidence of 5° for beryllium. The three curves represent scattering due to roughness-only (solid), dielectric-only (dot), and roughness-plus-dielectric (dash) perturbations.

For the sake of comparison, similar curves for TM polarization are shown in Figs. 7 and 8 for silver surfaces. The only difference between these two figures and Figs. 3 through 6 is that silver optical constants are substituted for beryllium. In Figs. 7 and 8 it is seen that the relative difference between the solid and dash curves does not change greatly for the four wavelengths shown. This indicates that silver does not have strong ARS wavelength anomalies.

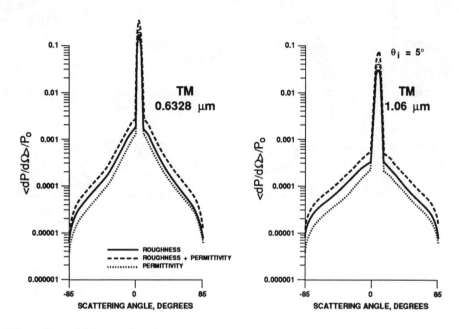

Fig. 7. TM ARS curves versus polar scattering angle for $\lambda =$ 0.6328 and 1.06 μm and an angle of incidence of 5° for silver. The three curves represent scattering due to roughness-only (solid), dielectric-only (dot), and roughness-plus-dielectric (dash) perturbations.

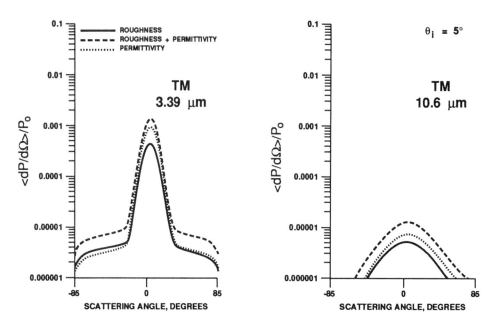

Fig. 8. TM ARS curves versus polar scattering angle for $\lambda = 3.39$ and 10.6 μm and an angle of incidence of 5° for silver. The three curves represent scattering due to roughness-only (solid), dielectric-only (dot), and roughness-plus-dielectric (dash) perturbations.

4. CONCLUSIONS

Based on the modeling of the scattering surface as given in this work, the numerical results given above indicate that wavelength scaling of angle-resolved scattering (ARS) from beryllium can deviate by an order of magnitude from that predicted by roughness-only scattering theory. On the other hand, all things being equal except the scattering medium, some materials do not have such severe wavelength anomalies, e.g., silver. This seems to indicate that dielectric scattering is more important in some materials than others or is fabrication dependent.

Looking in detail at the numerical results given here, it turns out that the biggest factor in the beryllium anomalous behavior is the phase factor $\exp(i\psi(\lambda))$ in Eq. (16). For beryllium from 0.6328 to 10.6 μm, the real part of $\varepsilon_1(\lambda)$ changes from positive to negative, which causes the interference term in Eq. (10) to change sign. This changes the interference term contribution to ARS from destructive in the visible to constructive in the infrared. In silver, there is no similar contribution from $\exp(i\psi(\lambda))$.

The major conclusion is that dielectric scattering is always present, but certainly to different degrees depending upon the material and fabrication history. Also, based on the ARS formulas, there are a number of different characteristics between roughness-only scattering and dielectric-only scattering. These characteristics could perhaps be used to identify one type of scattering versus another.

5. ACKNOWLEDGEMENTS

This work was performed with Independent Research funding. The author wishes to thank Gene Church, Peter Takacs, John Stover, and Charles Egert for helpful discussions and reference material.

6. REFERENCES

1. J. M. Elson, "Theory of Light Scattering From a Rough Surface With an Inhomogeneous Dielectric Permittivity," *Phys. Rev.* B30, pp. 5460-5480, November 1984.

2. C. M. Egert, "Anomalous Scatter Workshop Review," Proceedings of the Eighth Industrial Briefing, ORNL/OMIS-80/18, Oak Ridge, TN, pp. 123-134, November 1990.

3. J. C. Stover, M. L. Bernt, D. E. McGary and J. Rifkin, "An Investigation of Anomalous Scatter From Beryllium Mirrors," *Proc. Opt. Soc. Am.* 1165, pp. 100-106, August 1989.

4. E. L. Church and P. Z. Takacs, "Subsurface and Volume Scattering From Smooth Surfaces," *Proc. Opt. Soc. Am.* 1165, pp. 31-38, August 1989.

5. E. L. Church, P. Z. Takacs and J. C. Stover, "Scattering by Anisotropic Grains in Beryllium Mirrors," *Proc. Opt. Soc. Am.* 1333, pp. 205-220, July 1980.

6. E. T. Arakawa, "Optical Properties of Beryllium," Proceedings of the Fifth Industrial Briefing, ORNL/OMIS-88/6, Oak Ridge, TN, pp. 237-250, November 1989.

Characterization of hot isostaticly pressed (HIP'ed) optical quality beryllium

J. L. Behlau
M. Baumler

Eastman Kodak Company, Federal Systems Division
121 Lincoln Avenue, Rochester, New York 14653-8101

ABSTRACT

Producibility of low scatter, HIP'ed beryllium optics requires the manufacturer to know more than just the surface roughness or BRDF of his parts in work. The limitations of his test apparatus (spatial frequency range, height resolution, steepness of slopes, polarization sensitivity, available wavelengths) require that "overlapping" data be taken. This doesn't just mean that the "same" data needs to be taken on similar instruments. Instead, a collection of both quantitative and qualitative data from different types of analysis equipment must be combined to form a more complete picture of the interactions between the material and the wavelengths of interest. This paper discusses the results from several different tests which (when combined) give the manufacturer enough information to determine whether or not there is any more that can be done in his shop to improve the scatter function. We demonstrate that a variety of objective and subjective testing is necessary to determine the "true" characteristics of uncoated HIP'ed beryllium mirrors. We show results of testing, give a discussion on the interpretation of the data and demonstrate how it was used to optimize production results.

1. INTRODUCTION

In the course of coming on line, the Kodak beryllium facility was tasked with producing high quality beryllium optics which would meet the surface roughness and scatter specifications required by our potential customers. Our goal then was to develop our polishing techniques to provide very low surface roughness, thereby reducing the level of scatter. We soon discovered that for rms surface roughness values less than 30Å, BRDF remained the same for individual parts (figure 1) and was inordinately high overall.

In an effort to discover why surface roughness told us one story, but scatter told us quite another, we began a variety of experiments aimed at finding two things, a way to accurately quantify the surface roughness of beryllium with a microsurface profiler and, a way of quantifying some surface feature(s), the volume of which tended to correspond to the level of scatter.

2. ANALYSIS DETAILS

2.1 MISLEADING DATA

Early in the development of our processes, we relied heavily on the topographic data from the (phase sensitive) Wyko Topo-2D. However, when this data couldn't be supported by the scatter measurements, its usefulness on bare beryllium surfaces became questionable. In an effort to verify the Wyko data (or find an instrument that supported our BRDF results) we tried two other devices, the Chapman Instruments MP2000, and a Taylor-Hobson form Talysurf. Both the stylus instrument and the Chapman device gave very high numbers for rms surface roughness. In fact, if we had used these numbers the BRDF would not have seemed out of line. However, once you take into account the spatial frequency range of the Talysurf and the fact that the data was taken from an acetate replica, you can see how the stylus numbers may have been based on high frequency data that was collected from the replica. Unfortunately some questions regarding these measurements still remain unanswered. The results from the MP2000 were intriguing, and believable (if you only wanted correlation with BRDF data). At this point we started to rely more upon the MP2000 than the Topo. It would take us several months and much data before we would correct (refine) our thinking.

0-8194-0658-9/91/$4.00

BRDF COMPARISONS

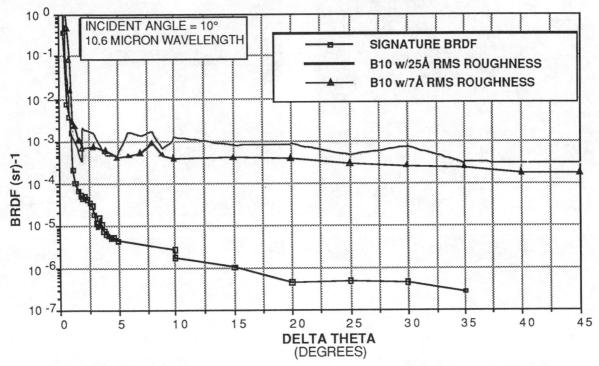

Figure 1. Comparison of the forward scatter from HIP'ed beryllium part B10 (polished to two different rms roughness results) demonstrates that the surface roughness as measured by the Topo-2D profilometer is not the key contributor to these levels of scatter.

2.2 QUALITATIVE ANALYSIS: NOMARSKI / POLARIZATION MICROSCOPY

We have analyzed customer furnished samples which had been polished by other companies and found that some of the labs probably did not have a polarizing microscope available to them. If they had used one they would have realized that, even though the BRDFs of their samples were low, the samples were worthless as stable cryogenic optics because the plastically deformed layer had not been removed. As a standard practice, we record (photographically) the images from our samples with a Nomarski style microscope. The Nomarski gives the manufacturer his first "view" of surface roughness. The instrument in the beryllium processing area offers 400X, 200X and 50X magnifications with a Nomarski prism attachment, and a 400X magnification without a prism so that with crossed polarizers we get a good view of the optically active beryllium grains. Using the microscope to analyze surfaces, the polisher looks for removal of the plastically deformed layer (an unstable layer resulting from machining damage), scratches, differential grain heights (some grains polish faster than others due to their orientation), gouges, twinning (machining damage to individual crystal structures), and residual etching damage (See figure 2). With the microscope as the sole analysis tool, a variety of beryllium polishing techniques were analyzed and honed. Though this data is qualitative, it has played a large part in the development of our processing and understanding of beryllium.

2.3 MP2000 VS. TOPO-2D

When the Wyko Topo-2D is used for surface roughness measurements, visual examination of the *fringe* (phase) data allows the polisher to easily see when measurements are being made across differential grains, or through pits. The video system of the MP2000 gives the same data, and there is some evidence that one can see how an area is being interpretted by the MP2000 visually, since contiguous grains in the plain reflect light differently. Using the Nomarski style microscope, we attempted to predict surface roughness and BRDF. The Topo system appeared to be in agreement with the qualitative evaluations from the Nomarski photos, with rms following differential grain volume, and peak-to-valley dependent upon whether or not the scans passed through pits. The MP2000 data didn't appear to fall in line with any of our predictions.

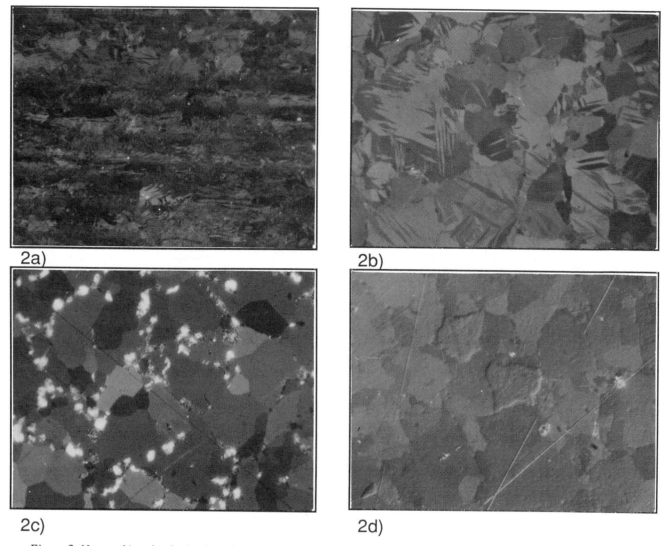

Figure 2. Nomarski and polarization photomicrographs (400X) of HIP'ed beryllium displaying a) a plastically deformed layer, b) twinning, c) etching damage d) scratches and differential grain heights.

After a period of time, two things forced us to re-evaluate the use of the MP2000 microsurface profiler on bare HIP'ed beryllium. The first was agreement between the Wyko and Nomarski microscope, and the second, was learning that from grain to grain, beryllium is an anisotropic, *optically active*[1,2] material. At .6328 μm and normal incidence, beryllium grains alter the polarization of the incident beam (figure 3). Once you understand that the grains alter polarization (depending on grain orientation) another interesting observation can be made; viewed with polarization microscopy, differentially polished grains tend to be the same shade of gray, supporting the idea that specific crystal *orientation* allows the differential polishing to occur.

Beryllium grain to grain anisotropy explains why the (polarization sensitive) MP2000 rms numbers for microsurface roughness were as much as an order of magnitude higher than what the (phase sensitive) Wyko had reported for the same samples. The difference was greater for the 20X Nomarski II option than for the 10X Nomarski I. This may have been due to the difference in maximum slope sensitivity (Max Slope: Nom. I w/10X = 2.78°, Nom. II w/20X = 20.1°). On an optically active material such as beryllium, where the microstructure is anisotropic from grain to grain, the MP2000 *currently* runs into problems. Problems with using the MP2000 to measure the surface roughness of HIP'ed beryllium arise when the test beam and reference beam move from a grain in one crystal lattice orientation onto a grain of another. The difference in orientation of the grains is important because different orientations cause the reflected beams to have a change in polarization (from grain to grain). These changes in polarization are interpreted as slope differences by the MP2000, when in

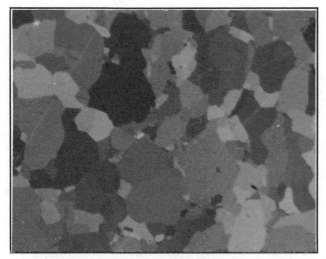

Figure 3. 400X Polarization photomicrographs of bare beryllium with the analyzer rotated to two different positions, demonstrating the "optically active" nature of beryllium.

actuality, the grains have the same slope and height.

To verify our reasoning, we coated one half of a HIP'ed beryllium sample with a thin layer of aluminum. As seen in figure 4, Nomarski images show both halves have the same topography. Both halves of the sample were tested with three instruments, the MP2000 (with both magnification and prism options), the Topo-2D (with a 20X objective) and a stylus instrument. The Topo and the MP2000 (20X Nomarski II) evaluations were performed at Kodak. The 10X Nomarski I option was used to make the MP2000 measurements at Chapman Instruments. The stylus work was performed by Jean Bennett of the Naval Weapons Center in China Lake. Table 1 below, lists the numerical results of the evaluations.

Chapman Instruments has been very responsive to our needs, and is currently developing software which will enable the MP2000 to measure accurately across materials with different optical constants (complex indecies of refraction). This will be valuable to anyone who needs to measure the height differences across two different materials (i.e. emulsion thickness, coating thickness, semi-conductors etc.).

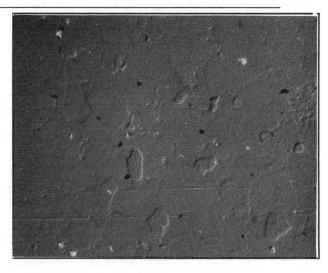

4a)   4b)

Figure 4. 200X Nomarski photomicrographs of a) the uncoated half of a HIP'ed beryllium puck and b) the coated half, demonstrating that the thin layer of aluminum should not have an affect on the measureable rms surface roughness.

		Coated	Uncoated	Scan Length
Wyko Topo-2d		25.3Å	27.5Å	600 µm
Chapman	10x, Nomarski I	22.2Å	45.3Å	1 mm
MP2000	20x, Nomarski II	29.3Å	276.9Å	1 mm
Stylus		26.56Å	25.48Å	1 mm

Table 1) A HIP'ed beryllium puck was coated on one half of its polished surface. Both halves were tested for RMS microsurface roughness. The results above compare both of the surfaces as well as the types of measurement apparatus. The Topo data was an average of 12 measurements at different locations. The stylus data is an average of 3 readings.

It is clear that one must know the attributes of the material to be tested and test set limitations when selecting instrumentation. The Chapman device has trouble measuring roughness on uncoated beryllium. However, it has been utilized to measure grain diameters. This has been done by observing spikes in the power spectral density, by analyzing the autocovariance function (figures 5a) and observing the plateaus in the plotted output of slopes (figures 5b) all of which reveal the typical grain diameters if care has been taken to set the scan length to pass through the centers of several grains.

2.4 KODAK BRDF MEASUREMENTS

Scatter measurements are performed on the Kodak BRDF device, which has compared well with the TMA CASI™ system (figure 6) and other systems around the country.[3,4] The Kodak BRDF device is capable of full angular coverage for plane of incidence (PLIN) BRDF measurements at 10.6 µm, and full spherical coordinate BRDF measurements at .6328 µm (with step sizes of .002° available). The system has a standard 7mm diameter beam at the sample surface, adjustable focus for the testing of spherical optics, and can handle optics 18" in diameter (no weight limit).

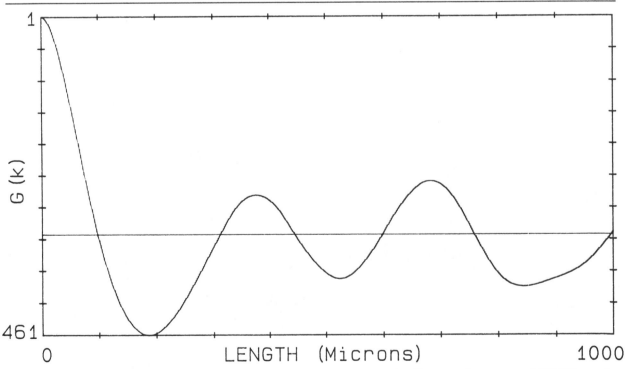

Figure 5a. The Autocovariance function of properly filtered data from the Chapman Instruments MP2000 can be used to determine typical beryllium grain sizes.

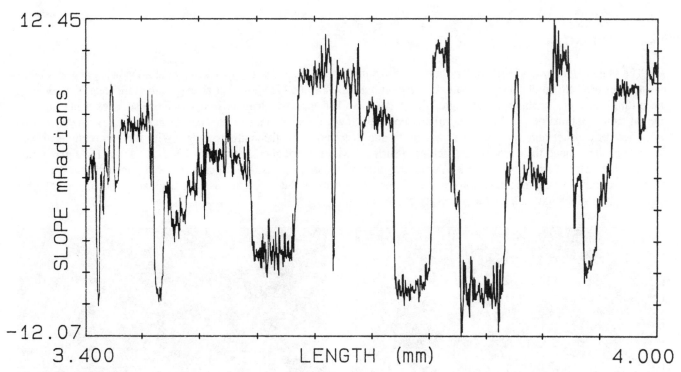

Figure 5b. A plot of bare beryllium slope data, collected from the polarization sensitive Chapman Instruments MP2000 can also be used to determine beryllium grain sizes. The apparent change of slope is due to differences in reflected polarization from contiguous grains with different crystal lattice orientations.

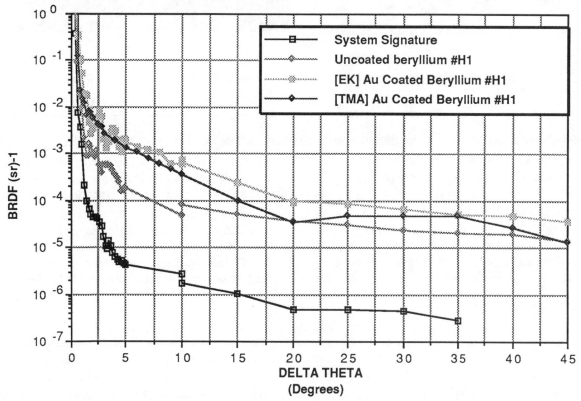

Figure 6. BRDF of a gold coated beryllium sample as measured by TMA Technologies and Eastman Kodak Company. Notice that BRDF drops to the same level for this component, with or without the gold coating. Testing was done at 10.6μm, and an incident angle of 10°. Delta Theta = ABS(specular angle - detector angle).

2.4.1 PITTING

The scatter from typical HIP'ed beryllium (at .6328 μm) is easy to see with the naked eye at all viewing angles. When the samples are held so that an illumination source with a dark surround is viewed at near specular reflection, a haze (from the pitting) becomes visible. Up through July of 1989, we had visually characterized beryllium surfaces using the high magnification options on our Nomarski microscope. By revising our Nomarski analysis to include a low magnification view, we were better able to determine the density and distribution of differential grains. With this information in hand Kodak polishers tuned their process and achieved surfaces with rms roughness less than 10Å. However, for a given part, varying the surface roughness below 30Å rms had little or no affect on BRDF. Furthermore, by including a dark field photomicrograph at 50X magnification (figure 7), the area density of the pitting is easier to evaluate. The BRDF is higher on those parts which have a higher area density of the pitting.

7a)   7b)

Figure 7. 50X Nomarski photomicrographs showing a) differential grans and scratches, b) darkfield illumination showing pit density.

When we first realized the severity of the pitting in the beryllium surfaces we proposed that the pits be treated as local Lambertian scattering sites. A Lambertian scatterer would have a BRDF of .31 sr^{-1} which would be constant for all angles. If a significant proportion of the surface were acting as a Lambertian scatterer, the BRDF curve would bottom-out at a constant level. By considering the measured area density of the pits to be 1% of the surface, we expect to see a leveling off of the BRDF data to a constant of 0.0031 sr^{-1}. This quick calculation corresponded to the BRDF value measured from a beryllium sample which displayed a pit area density of 1% (measured at 10.6 μm, 10° angle of incidence, 98% reflectivity). As for why there is more scatter at longer wavelengths,[5] there are simply fewer scattering sites with the correct diameters for the shorter wavelengths (in our data, there are more pits in the 4 μm to 12 μm range, with an average of approximately 8 μm). For wavelengths smaller than these pit diameters, there is structure in the pits which may still need to be considered.

2.5 SEM AND AUGER ANALYSIS

Once we determined that the pitting was the likely source of *most* of the anomalous scatter, we performed further analysis of these areas using SEM and Auger chemical analysis on typical beryllium pucks polished with alumina abrasives. The high magnification revealed most of the pits to have diameters of approximately 10 μm (note that for a wavelength of 10 μm this would be a perfect scatterer). The Auger system has a tunable (diameter) ablation beam, capable of analyzing materials in an area of .1 μm^2. Practical application of the laser to the pit area allowed us to use a larger beam. The pit material proved to be beryllium-oxide (BeO$_2$). There was no evidence of any other material within the pits. We then tested a diamond polished beryllium part. Many of the pits in the surface of the diamond polished part were found to have inclusions. Chemical examination of the pits using Auger analysis revealed the primary pit material to be high in oxygen content (beryllium-oxide). In figure 8 you can see scratches emanating from the pit, showing how the shattered Be_2 was dragged from the site. The inclusion in the pits turned out to be carbon. We don't know what form this carbon is in.

Why did carbon show up when we polished with a diamond abrasive? Could the carbon we saw be from our diamond polishing process? No, the abrasive used is roughly .1 μm in diameter versus the 2 μm to 3 μm diameters of the particles measured in the SEM data. Also, closeup views of the carbon particle show oxide material ramped up on the sides of the particle. Though this is not conclusive evidence, we don't believe the carbon is a result of the polishing process. It is likely that the formation of the oxide and carbon was due to a hydrocarbon inclusion in the base powder of the beryllium prior to HIP'ing. Whether the formation of the beryllium oxide takes place during HIP'ing or annealing hasn't yet been determined.

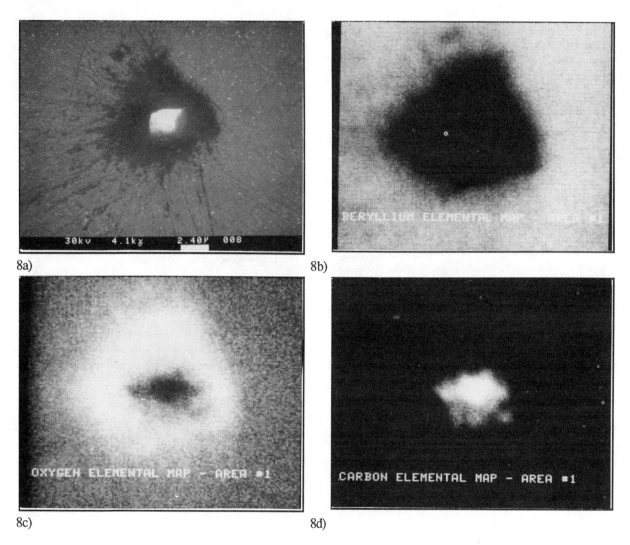

8a) 8b)
8c) 8d)

Figure 8. SEM and Auger photographs displaying a) a carbon particle (bright region) in a beryllium-oxide 'pit' b) area of high beryllium content c) areas of high oxide content and d) areas of high carbon content. Auger analysis took place after 7 minutes of sputtering. The glow of the carbon particle in the SEM photo (a) is a common response for organic materials with electron stimulus.

Why hadn't the carbon shown up in the alumina polished parts? Earlier scanning electron microscopy on *alumina* polished surfaces showed only deep pits with no inclusions, this is due to the tumbling of the abrasive between the pitch lap and the beryllium surface. Any inclusions were ripped out by the loose .3 μm diameter abrasive. The specially prepared surface had been diamond polished. Diamond embeds itself in the pitch material, or "charges" the lap, instead of tumbling loosely between surfaces. This allowed a more gradual erosion of surface material, so that the BeO_2 areas were not as deep, there were not as many scratches emanating from the pits, roughly half of the pits contained a carbon particle embedded in the center of the BeO_2, and these particles appear to have been polished to the level of the surrounding beryllium plain.

Note: A chemically enhanced polishing method has been successfully used to polish the oxide material to a smoother finish. This has done nothing to reduce the scatter because the complex index of refraction is different in the high oxide areas than in the pure beryllium areas.

3. SPUTTERED BERYLLIUM

In order to separate surface roughness from a material property, we had a 2" diameter beryllium part sputter overcoated with a 125 μm layer of beryllium (which remained significantly thicker than a "skin depth" after polishing). The sample was polished to a rms surface roughness of 3Å. Its BRDF was measured to be the same as that of our aluminum coated BK7 reference mirror (at 10.6 μm and an incident angle of 10°). The psd plot of the sputtered beryllium tested at .6328 and 10.6 μm shows that the material is significantly closer to wavelength scaling than bare HIP'ed beryllium (Figure 9). The flaw in this experiment (remember we are looking for scatter properties of beryllium without pits) is the potential isotropism of the microsurface due to the coating layer. We no longer have the affect of large anisotropic grains, which according to Church[6], will scale as λ^{-2} in our scattering function, and will probably be the dominant scattering mechanism for bare beryllium once the pitting is eliminated. The lack of total wavelength scaling may be due to a very fine anisotropic grain structure or contamination, or simply a different surface roughness due to being tested in a different location (for the sputtered beryllium sample, the two wavelengths were tested roughly a year apart).

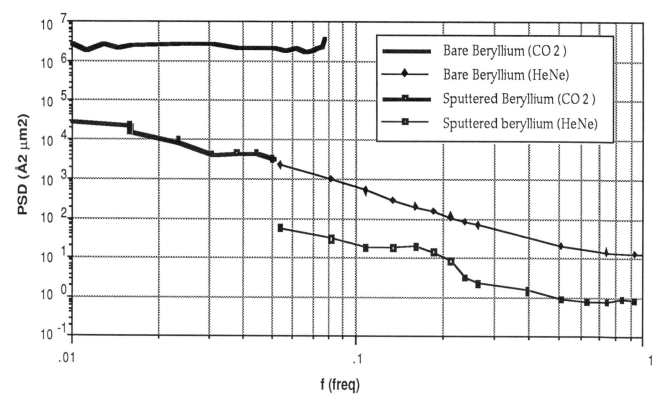

Figure 9. The PSD plots of sputtered beryllium at .6328 μm and 10.6 μm versus bare HIP'ed beryllium shows that the beryllium overcoated beryllium demonstrates an improvement in wavelength scaleing. The remaining scatter may be due to fine grain structure of the coating or contamination.

4. FOLLOW-ON

It is our belief that the beryllium polishing facilities have reached the limit of their abilities to provide a HIP'ed beryllium surface which is cryogeniclly stable, has low rms roughness, and the best achievable scatter function. We are now at the point where scatter is material dependent and preparation dependent. The first order of business then, is the elimination or prevention of the agglomerated oxide material which forms the pitting in the surface of beryllium optics (not to be confused with typical 100Å thick surface oxidation). The possibilities for its elimination at this time depend on when the beryllium oxide is formed. Is there a means of altering the HIP'ing process to block the formation of the oxide, without disturbing the other properties? Since carbon was present in the oxide, was it somehow acting as a *seed* in the formation of the oxide? If so, how can the carbon (or hydrocarbon) be removed from the bulk material? If the oxide forms during annealing (post machining) can the annealing process be altered to block the formation of the oxide? Can parts be manufactured without annealing? What role do the anisotropic grains play (contribution to scatter)? Much of our current effort has been aimed at resolving these questions through empirical means.

Working with Rome Laboratories on the Beryllium Scatter Analysis (BSA) program, we have laid out an aggresive materials, manufacturing and test plan aimed at answering these questions. The ultimate goals of the BSA program are; the creation of a model (so that scatter predictions may be made accurately), the creation of material specifications and processing and testing criteria for the manufacture of high quality large (>1 m) beryllium optics. That effort is reported on at this conference in a paper titled "The Beryllium Scatter Analysis Program."

4. ACKNOWLEDGEMENTS

The measurements and insights provided by Jean Bennett', Chapman Instruments, and TMA Technologies have aided us in the development of our understanding of beryllium. We would also like to thank Tom Leonard for including us in the BRDF round-robins, and for being an apt teacher.

5. REFERENCES

1. Private correspondence with James Marder of Brush Wellman in verifying the techniques of color metallography and the conclusions drawn from observations, especially in regards to materials in which there are different optical constants for different grain orientations (tin, uranium, *beryllium*...etc.).

2. Color Metallography, *Vol. 9, 9th ed.*, pgs.135-138. 389. ASM (1985).

3. Thomas A. Leonard, M. Pantoliano, "BRDF Round Robin," *Proc. SPIE* 967, 226 -235 (August 1988).

4. Thomas A. Leonard, M. Pantoliano, J. Reilly, "Results of a CO_2 BRDF Round Robin," *Proc. SPIE* 1165, 444- 449 (August 1989).

5. John. C. Stover, M. Bernt, D. McGary, J. Rifken, "An Investigation of Anomalous Scatter from Beryllium Mirrors," *Proc. SPIE* 1165, 136-150 (1989).

6. E. L. Church, "Scattering by Anisotropic Grains in Beryllium Mirrors," *Proc. SPIE* 1331, 12-16, (1990).

7. Jean. M. Bennett, L. Mattsson, Introduction to Surface Roughness and Scattering, OSA, Washington (1989).

The Beryllium Scatter Analysis Program

J. L. Behlau, E. M. Granger,
J. J. Hannon, M. Baumler
Eastman Kodak Company, Federal Systems Division
121 Lincoln Avenue, Rochester, New York 14653-8101

J. F. Reilly
Rome Laboratory, Griffiss AFB, Rome, New York 13441-5700

ABSTRACT

Many groups today are researching the characteristics of beryllium, in an attempt to find ways of producing high quality (low scatter) stable beryllium optics. This paper discusses a two part study in which **1)** an attempt is being made to determine the best, raw beryllium mixture and preparation, machining and polishing processes, test and analysis methods, and **2)** a proposed model for the prediction of scatter from beryllium surfaces (based on a knowledge of surface and subsurface interactions with incident wavelengths) will be refined against empirical data. We discuss design of the experiment, the model, and some of the early results.

1. HISTORY

Kodak Beryllium Research

In December of 1989, Kodak collected research data on a beryllium mirror which could explain the formation of beryllium-oxide (BeO) inclusions in beryllium optics[1]. SEM photographs and Auger analysis revealed carbon inclusions approximately 3 μm in diameter in the center of the BeO. Kodak proposed that the carbon was in the original powdered beryllium prior to hot isostatic pressing (HIP'ing), and that it acted as a seed in the formation of the oxide. In another development, Kodak polished and tested a beryllium puck which had been coated with sputtered beryllium. The coated part had a surface roughness (rms) of 3 Å and a scatter function as low as that of an aluminum coated BK7 reference mirror (two orders of magnitude below the typical bidirectional reflectance distribution function (BRDF) of uncoated beryllium optics [Figure 1]).

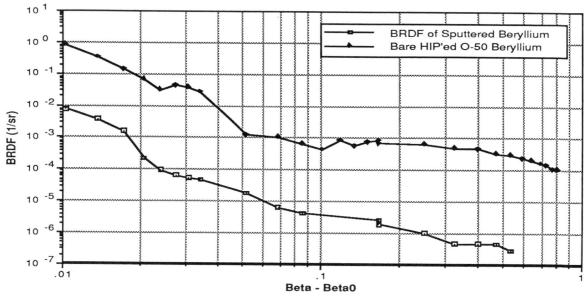

Figure 1. The BRDF function of a beryllium substrate, sputter coated with beryllium, and polished to a 3Å rms surface roughness (Wyko Topo-2D measurements) vs. a typical bare beryllium scatter function (18Å rms roughness). The coating was 125 μm thick prior to polishing.

0-8194-0658-9/91/$4.00

2. THE BERYLLIUM SCATTER ANALYSIS (BSA) PROGRAM

2.1 The Materials and Processing Matrix

The "Materials and Processing Matrix" (Figure 2) is a layout out the number of samples of each material required, and which processing steps each sample must undergo.

2.1.1 Quantities

The number of samples required was driven by three needs; keeping the program costs reasonable, keeping the one year contract life achievable, and having at least two samples per parameter change so that statistics would be more meaningful. In researching which process parameters have the greatest effect on scatter, process changes would not be required on all materials. That would have required a great many more than the 36 samples the program currently employs. Instead, Kodak chose a "standard" material, and varied all the processes only on *this* material. The material chosen for the standard was O-50.

2.1.2 Materials

The materials in the matrix are from Brush Wellman and represent both off-the-shelf beryllium materials and special blends specifically produced for this program. The off-the-shelf materials are O-50, I-70, and Spherical Powder. The blends are of O-50 with high carbon content, and O-50 with high BeO content. The O-50 material, recently developed by Brush Wellman, has approximately the same chemistry as the spherical powder but is manufactured in the same manner as I-70 (impact ground). Kodak data also shows the O-50 beryllium to be more sensitive to subtle process changes, which makes it more desirable as a process monitor.

The O-50 material was viewed by Brush Wellman to be the best candidate for high quality beryllium optics, surpassing I-70 and Spherical Powder because of its larger grain size[3] and lower BeO content. The I-70 material has been used extensively throughout the industry as an optical quality material and must be included for several reasons (historical cross referencing, grain size, chemistry, etc.). The spherical powder beryllium was developed by arrangement between the Air Force and Brush Wellman. It was anticipated that its spherical shape would help reduce the problems which are associated with grain orientation. Table 1 below gives some chemistry results from the raw powder and the HIP'ed billets used in the BSA program. Taking into consideration the possibility that contaminants in the beryllium may migrate towards the evacuation stem of the HIP can during processing, the decision was made to polish those surfaces furthest from the evacuation stem.

The special blends of O-50 material were added to the matrix because of the role that carbon appears to play in the creation of BeO deposits, which result in pits on the sample surface. Carbon is most likely being introduced to the base material as a hydrocarbon, and being reduced during a later process. There are many mechanisms by which a hydrocarbon can be mixed into the beryllium powder (vessels where the powder is stored, rubber hoses used for high pressure air, etc.). As a datum to indicate whether or not the volume of BeO inclusions followed the volume of carbon, consideration was given to lowering the carbon content of the beryllium powder. However, removing or reducing the level of carbon inclusions is costly, and beyond the scope of this program. Instead, carbon has been added in the form of a graphite flake to achieve a mixture that has a 300% increase in carbon content over off-the-shelf O-50, at nearly the limit allowable for the material.

The O-50 with high BeO content was selected to determine whether or not adding oxide would cause more pitting in the polished samples. Reducing the BeO content of the material was ruled out because a low level of oxide stabilizes the material, making it safer to handle. Brush Wellman increased the oxide content by placing the powdered form in a controlled environment of higher temperature and humidity.

2.1.3 Machining, Annealing, and Etching

Because there is a possibility that BeO sites are formed during annealing, the parameter option of annealing versus not annealing samples, is included within the matrix. This created a potential problem since annealing repairs some deep surface damage induced by machining. Therefore, a method of machining the samples without inducing deep damage had to be utilized for this program. The damage that did occur needed to be etchable to depths easily removed by polishing. All parts within the matrix were machined to a diameter of 5 cm and a thickness of 1 cm. Each part was beveled, and three 8-32 x 1/4" holes were drilled and tapped on the back surface. The tapped holes are for use in bolting the parts to a "cold finger" during sputter coating. No one expects to coat all the samples, however if in the future an experiment requires a coating, drilling and tapping the holes (post polish) would have been an expensive operation.

Part #	Spherical Powder	I-70	STD O-50	Δ Carbon	Δ BeO	w/o Anneal	With Anneal	.3μm pre polish	2μm coating	125μm coating	.3μm polish	1μm polish	Diamond Polish	Replicate
1			•			•		•	•					
2			•			•		•	•					•
* 3			•			•		•	•					
4			•				•	•	•					
5			•				•	•	•					
* 6			•				•	•	•					
7			•				•			•	•	•		
8			•				•	•		•	•	•		
* 9			•				•	•		•				
10			•				•	•		•	•			•
11			•				•	•		•	•			
* 12			•				•	•		•				
13			•				•	•			•		•	
14			•				•	•			•		•	
* 15			•				•	•		•				
16			•	•							•			
17			•	•							•			
18			•				•				•			•
19			•				•				•			
20				•		•					•			
21				•		•					•			
22				•			•				•			
23				•			•				•			
24					•	•					•			
25					•	•					•			
26					•		•				•			
27					•		•				•			
28			•				•				•	•		
29			•				•				•	•		
30			•				•				•		•	
31			•				•				•		•	
32		•					•				•			
33		•					•				•			
34	•						•				•			
35	•						•				•			
36			•								•			

Figure 2. The BSA programs Materials and Processing Matrix. Those sample numbers with asterisks are coating witness samples.

Beryllium Material	Grain Size (µm)	PRE-HIP BeO (%)	PRE-HIP Carbon (%)	POST-HIP TOP BeO (%)	POST-HIP TOP Carbon (%)	POST-HIP BOTTOM BeO (%)	POST-HIP BOTTOM Carbon (%)
O-50	15.9	.31	.024	.26	.013	.25	.012
O-50 + Carbon	13.7	.28	.075			.24	.068
O-50 + BeO	14.5	.46	.010			.42	.012
I-70	8.8	.64	.030	.66	.026	.64	.028
Spherical Powder	12.2	.42	.092			.43	.089

Table 1. Chemical analysis provided by Brush Wellman for pre and post HIP. The analysis also includes samples from both the top (towards HIP can evacuation stem) and bottom of the HIP cans for I-70 and O-50.

2.1.4 Coating

Because Kodak had achieved a high quality optical surface with a 125 µm thick sputtered beryllium coating, a thick, very fine grain beryllium coating was included in the matrix to demonstrate a possible means of getting a high quality optic utilizing todays materials and technology. By utilizing the thick coating (polished), all surface features of the original polished substrate are eliminated, and Kodak is able to achieve a surface rms roughness of 2 to 3 Å. Thin (2 µm) coatings were added to mask the effects of large anisotropic grains while still retaining the reflectivity of beryllium and the approximate surface roughness of the uncoated substrate. These coatings will aid us in determining the level of contribution to scatter by subsurface effects, and pitting.

2.1.5 Polishing

Several polishing techniques were required to accomplish a variety of surface finishes. Kodak has available conventional dual oscillating polishing (DOP) machines and continuous polish (CP) machines. The DOP machines are used primarily for curved optics and diamond polishing. Diamond polishing will be used for achieving both the 40Å rms surface roughness and the 3Å rms surface roughness requirement. Diamond polishing will also be used to demonstrate pit chemistry, since the diamond charged lap will polish the pit inclusions instead of ripping them out. The CP machines, with a .3 µm alumina grit, will be used for producing optics with an rms surface roughness of approximately 25Å (as measured on the Wyko Topo-2D).

2.1.6 Replication

If the anomalous scatter[4] is due to grain boundaries, non-cubic lattice structure and other intrinsic properties, then by changing the material (to aluminum) but retaining the surface topography, we can verify these effects. Several experiments using aluminum overcoatings on beryllium substrates have been attempted, but the results may be contaminated with artifacts caused by; "show through", the surface roughness of the coatings, and pinholes in the coating. In order to totally isolate the surface topography from any potential subsurface effects, Kodak selected a replication method used by manufacturers of diffraction gratings.

Hyperfine, Inc., manufacturer of precision-ruled and replicated gratings and echelles of very high efficiency, will replicate three beryllium samples using this unique method of replication. The replica material will be aluminum bonded to glass by a space qualified epoxy. Three completely characterized samples will be replicated. The replication process will yield three first generation replicas (think of these as negatives) and six second generation replicas (positive copies of the beryllium originals).

2.1.7 Testing

Brush Wellman will provide; chemical analysis of the beryllium materials (pre- and post HIP), density measurements, thermally induced porosity (TIP) response, grain sizes and metallurgic samples. The TIP testing is accomplished by measuring the density of a sample before and after a thermal soaking for three hours at 788°C (1450°F).

Kodak will provide: Nomarski photomicrographs of each sample, surface roughness measurement and analysis (Wyko Topo-2D and Taylor-Hobson form Talysurf), BRDF, SEM photos, ESCA and Auger analysis.

Chapman Instruments will provide non-contact surface micro-roughness profiling and analysis of coated optics and replicas, utilizing the MP2000.

TMA Technologies will provide measurements BRDF at two angles and three wavelengths, isotropic analysis of each material, low resolution 10.6 μm Raster scans and high resolution 1.06 μm Fraster scans.

Battelle Pacific Northwest will provide coating tests on the witness samples. These tests include a Sebastian pull test, coating thickness analysis, and thermal shock tests.

2.1.8 The Scatter Model

Kodak will develop a model of surface scatter that deals with surface topography, subsurface nontopographic components of scatter and Lambertian scattering produced by surface pitting.

Measurements on beryllium mirrors by C. L. Vernold[5] and J. C. Stover[4] have shown that the BRDF varies as a function of incident angle and wavelength. These studies have shown that the amount of scattering produced by 10.6 μm light is greater than would be predicted from BRDF measurements made at .6328 μm. As indicated in the literature, this lack of wavelength shift-invariance indicates that the scatter is not just a function of surface topography.

Many phenomena could contribute to this effect. First, shift-invariance requires that the surface be a clean front surface reflector, and that the rms roughness be much less than the wavelength of interest. Therefore, the simple model will not predict scatter if the surface has large pits, inclusions, or subsurface damage caused by mirror fabrication techniques. Second, there are many subsurface features such as grain size, oxide content, grain orientation and dielectric anisotropy that may produce scattering.

Measurements made at Kodak, on an uncoated 17-inch beryllium mirror indicated that this lack of shift invariance could be attributed to pits that were found to cover about one percent of the mirror surface. The pits were 6 to 10 μm in diameter at the surface and were 6 to 10 μm deep. The size of these defects is well outside the range of topographic errors allowed by the wavelength independent BRDF model. Both the .6328 and 10.6 μm BRDF measurements resulted in BRDF values of approximately .003 sr[-1] for large scattering angles. This result is significantly higher than should be expected given the measured surface roughness of 26Å.

An example of surface pits is shown in Figure 3. Kodak's hypothesis is that these pits are acting as scattering sites, adding a Lambertian scattering component to the overall scattering characteristic of the surface. A 100% reflective Lambertian scatterer would have a BRDF of 0.31 sr[-1], which would be constant for all angles. If a significant proportion of the mirror surface were acting as a Lambertian scatterer, the BRDF curve should asymptotically approach a constant level. A cursory analysis of the surface indicates that approximately 1% of that surface is accounted for by the pits. If this represents the whole surface and the pits are acting as Lambertian sources, then the BRDF should level off at 0.0031 sr[-1], which is consistent with the value found in Kodak's BRDF measurements.

The physical mechanism, which is the basis of this model, is that each pit acts as a small light trap at short wavelengths. Therefore, we would expect Lambertian scatter from the pit to increase with wavelength. This scattering hypothesis has gained additional support from a test in which a beryllium mirror was overcoated by sputtering beryllium on the surface, then polished with standard methods. This process resulted in a surface that was nearly shift invariant. In addition, this surface oxidized in normal fashion, indicating that, after the pits are eliminated, there is little effect from the beryllium oxidation layer.

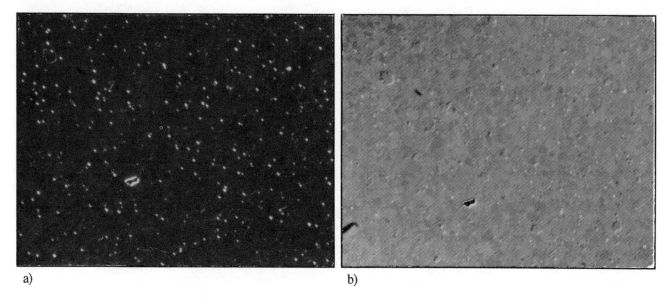

a)   b)

Figure 3. a) A 200X darkfield Nomarski photomicrograph showing clearly the pit density of the sample surface, b) the same area in brightfield view.

There is yet another source of scatter to be considered: that resulting from deep subsurface effects. Church[6] has shown that for subsurface effects such as grain size, oxide content, grain orientation, grain boundaries, pits, inclusions, dielectric anisotropies on the grain structure, and such can be modeled by modifying the shift invariant model using a λ^2 factor to weight the BRDF measurement in order to produce an LSI scatter power spectral distribution for subsurface effects.

The most common scatter theory, Rayleigh vector perturbation,[7] takes into account the full nature of the electromagnetic field and includes the scattering process as a function of polarization. For smooth surfaces, the scalar diffraction theory is sufficient to describe scattering. The scalar theory implies that the scattering (BRDF) can be determined from the Fourier transform of the surface profile. This simplification implies that the scattering behavior is linear shift-invariant (LSI) relative to the direction cosines of the incident radiation. In this form, the intensity of the scattered light is given by the power spectral density function (PSD), which is the Fourier transform of the autocorrelation of the surface profile.

To develop the PSD, the Fourier transform of the surface microroughness must be expanded in term of the spatial frequency,[4]

$$f = \frac{[sin(\Theta_S) - sin(\Theta_I)]}{\lambda}$$

(1)

where λ is the wavelength of the incident light. The directions of the incident and the scattered radiation are denoted by θ_I, ϕ_I and θ_S, ϕ_S respectively, as shown in Figure 4 ($\phi_I = 0°$ in this example).

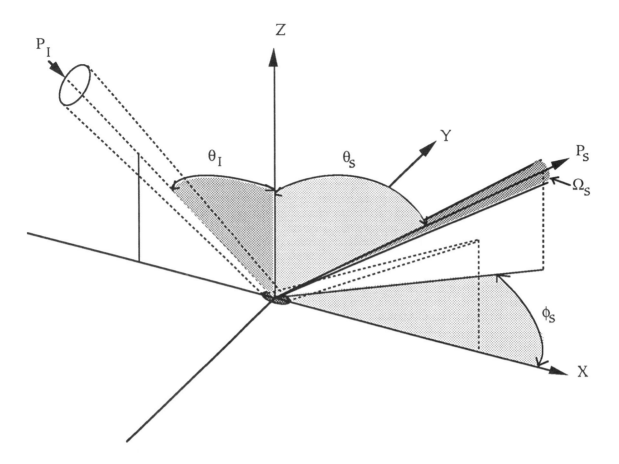

Figure 4. Directions of incident and scattered radiation.

The BRDF is defined as

$$P_I \ / \ P_S \ \ [\Omega \ cos \ (\Theta_s)] \ \ sr^{-1}$$

(2)

where P_I and P_S represents the incident and scattered power and Ω is the solid angle of the collection system. Under the condition that the surface is clean and has an rms surface roughness much less than the wavelength of the incident radiation, the BRDF is related to the two dimensional PSD[8] of the scattering surface by

$$PSD \ (f_x, f_y) = \frac{BRDF \ (\ \lambda^4 \ )}{16 \pi^2 \ cos \ (\Theta_I \ ) \ cos(\Theta_S \ )Q} \qquad \mu m^4$$

(3)

where f_x and f_y are the spatial frequency component in the x and y direction and Q is the material polarization factor and is approximately equal to the surface reflectivity.

Once the PSD is measured, the scattering function for any wavelength or incidence angle can easily be calculated if the surface conforms to the shift-invariant scattering model. As noted, however, beryllium does not conform to the shift-invariant model. The following section develops a model that describes the scattering radiation as a function of wavelength and angle that includes contributions from topographic and non-topographic features of the beryllium surface.

The model for the PSD of light scattered from beryllium can be evolved from the linear shift-invariant model. The formulation and test of the model depends on the LSI properties of metals such as aluminum. Figure 5a shows the LSI property holds for aluminum for wavelengths of .325 and 10.6 μm. In contrast, Figure 5b shows that beryllium apparently is not an LSI material.

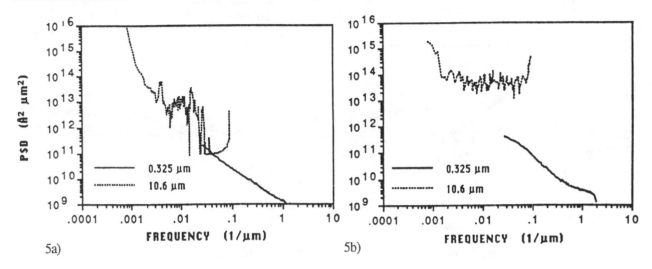

5a)   5b)

Figure 5. LSI properties for a) aluminum coated fused silica and b) beryllium, using wavelengths of .325 μm and 10.6 μm (from C. L. Vernold, "Application and Verification of Wavelength Scaling for Near Specular Scatter Predictions," SPIE, 1165, 29 [1989]).

Kodak's hypothesis for developing the model for uncoated beryllium is that approximately 99% of the surface obeys the LSI property and the remaining one percent of the surface has the properties of deep (many wavelengths) non-topographic defects. Figure 6 shows how the scatter from the deep pit type of defect changes as a function of wavelength. At short wavelengths, the incident light tends to be absorbed in the surface void. At long wavelengths, where the wavelength is nearly equal to the diameter of the pit, the light is scattered over a wide angle. Using the simple assumption that the long wavelength scatter is Lambertian, the difference between the classic LSI result and the measured scatter can be approximately explained by the Lambertian scatter from the deep pit defects.

The model describing the total PSD from the three primary scattering contributors can be written as a function of wavelength.

$$PSD_T = \alpha(\lambda) PSD_S + B(\lambda) PSD_N + \gamma(\lambda) PSD_L \tag{4}$$

Where PSD_S is the shift-invariant contributor, PSD_N is the deep-defect contributor, and PSD_L is the Lambertian scatter portion produced by pits and large-particle surface contaminants. Each contributes to the total scatter at the measurement wavelength in the proportion $\alpha(\lambda)$, $B(\lambda)$, and $\gamma(\lambda)$ for the shift-invariant, deep-defect, and Lambertian sources. Since the shift-invariant is proportional to λ^4, the deep defect to λ^2, and the pitting (Lambertian scatterer) is constant, we can expand equation 3 to produce a model to describe the total PSD:

$$PSD(f_x, f_y) = \frac{BRDF[\alpha(\lambda) \lambda^4 + B(\lambda) \lambda^2 + \gamma(\lambda)]}{16\pi^2 \cos(\Theta_I) \cos(\Theta_S) Q} \tag{5}$$

In this form, each primary contributor is shift-invariant with respect to scattering angle. The wavelength squared term produces an angular (frequency) invariance for the nontopographic deep defects in beryllium. The Lambertian contribution is assumed to contribute a constant amount of power at all spatial frequencies.

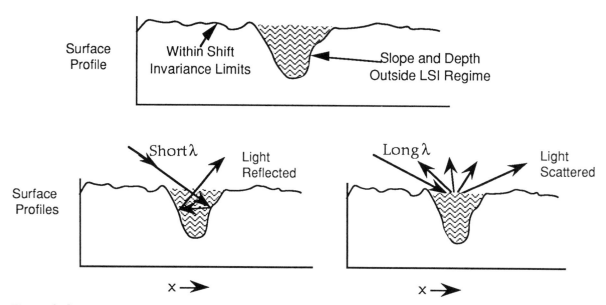

Figure 6. Scatter from deep-pit type defect changes as a function of wavelength.

Figure 7 shows the application of the model. The scatter from the surface pits or defects is not an LSI process; therefore, the PSD_L must be determined for each wavelength and as a function of scattering angle.

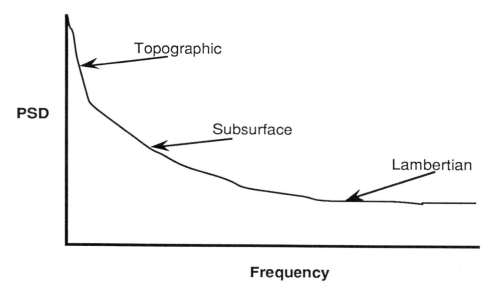

Frequency

Figure 7. Application of the BSA model.

Figure 8 is an illustration of how the model can be applied to determine the percent scatter that can be attributed to the Lambertian and λ^4 scaling of the PSD. Two wavelength scans yield the PSD_s of these three mechanisms through a regression technique to find the values for α, B, and γ that are required to map the wavelength and angular dependence of the measured scattered light. The results can be checked with a third wavelength scan.

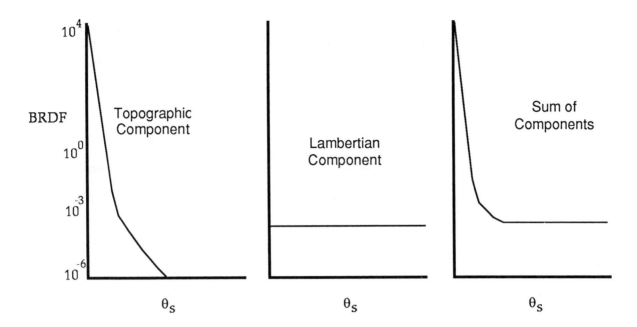

Figure 8. The model can be applied to determine percent scatter.

2.1.9 Testing The Model

Both machined and coated aluminum surfaces exhibit the LSI scattering property. This property is useful in testing the scatter due to voids or large defects. Using methods that are similar to those used to replicate diffraction gratings, we will replicate the surface of the beryllium optic (the material of the replica will be aluminum). The replica will be measured for BRDF. If the defect model is correct, the aluminum replica surface will demonstrate the same PSD properties as beryllium. Since aluminum normally displays LSI properties, the non-LSI nature of the aluminum replica would have to be a function of the area of the large defects. We hypothesize that the aluminum replica will behave in this way. If we do not observe this response, then the deviation from LSI will have to be attributed to non-topographic defects in beryllium. The model needs to be tested at several levels of surface roughness to clearly demonstrate that the PSD_T has one component that is related directly to the measured surface roughness (minus the large defects).

2.1.10 Model Summary

Beryllium mirror samples do not usually exhibit the linear shift-invariance observed in materials such as aluminum and copper. Beryllium overcoats on beryllium surfaces, after normal processing, have produced surfaces that very nearly obey the LSI rule. Since the only difference between the samples is surface pits and/or subsurface defects, Kodak proposes a new area-weighted model that handles not only the LSI part of the scatter but also the contributions from large surface defects and subsurface defects. The large angle constant scatter appears to be produced by a small area of deep surface pits or defects. Therefore, the scatter is not a pure LSI process and the non LSI scattering effects will have to be included in the model as a function of wavelength and scattering angle.

Kodak's hypothesis is that the aluminum replica will show the same lack of LSI as the original beryllium surface (as a result of pits and scratches). The model will be tested for several levels of surface roughness to ensure that the area-weighted scattering still does not reject the null hypothesis. If the hypothesis is rejected, then the source of the non LSI scattering is from some deep non-topographic features.

The aluminum replica test will help to separate the deep-defect wavelength-squared portion of the total scatter of beryllium. This is because the replica has only the topographic structure of the beryllium surface. Therefore, the replica scatter information is important in determining the percent of the scatter that is due to deep defects.

By using several beryllium mirrors manufactured with different pit densities, and surface roughness along with the aluminum replicas, we will be able to determine the percent of scatter attributable to surface topography, surface pitting, and deep surface defects. Using these weights, Kodak will be able to produce a model that will aid in the definition of the manufacture of new beryllium surfaces.

3.0 COMPARISON OF TEST RESULTS WITH THE MODEL

Kodak will use the measured data to evaluate constants in the model. The model has three major components:

o The traditional formulation in which the scattering is a function of the two-dimensional PSD of the surface (the topography) and the scattering angle

o The scattering caused by subsurface effects

o Scatter due to surface pits and voids.

Scatter sources could include contamination in the form of particulates or contaminant films. Both sources will be controlled during measurements and will only enter Kodak's analysis as sources of noise on the data. Other sources to be characterized by testing and, therefore, optimized by Kodak's analysis include
o Residual molecular films
o Trace elements in the beryllium
o Oxide layers on the surface
o Surface features with lateral dimensions as small as 0.2 μm
o Grain size and orientation
o Discontinuous grain boundaries
o Voids
o Inclusions
o Subsurface damage.

Residual molecular films, trace elements in the beryllium, and oxide layers on the surface are scatter sources that can be measured on the beryllium surface with Kodak's ESCA equipment. The depth of the oxide layer can be seen from the relative concentrations of beryllium and oxygen as a function of depth. Auger analysis also yields data on chemical constituents of the surface, however, the Auger analysis is best suited for high resolution analysis whereas the ESCA analysis covers relatively large areas (1 mm or larger).

Surface features with lateral dimensions as small as 0.2 μm can be characterized with Kodak's scanning electron microscope (SEM). An SEM photomicrograph of a beryllium pit is shown in Figure 9. Auger analysis of this pit indicated that the pit floor is composed of beryllium oxide, and that the particle at the center of the pit is carbon.

Grain size and orientation will be identified with a Nomarski microscope, utilizing the polarizer and analyzer of the system.

Grain boundary surface discontinuities will be found during microsurface roughness measurements on the WYKO Topo-2D optical profilometer. This examination requires a null fringe pattern in which a dark fringe is spread across the field of view. The presence of surface height discontinuities becomes apparent at the grain boundaries if it is present. The surface height difference can then be measured.

Voids and inclusions will be examined during WYKO Topo-2D testing as well as with Kodak's SEM when appropriate. With our model, Kodak will evaluate each of the three constants. There are only constant values for a given state of the surface. Each scatter source will affect the constants in the model. For example, Kodak expects that process parameters that yield surfaces with few voids will result in a Lambertian coefficient that is very small. Those producing little subsurface damage will result in a small λ^2 coefficient.

Our ability to predict scatter in the IR will be assessed with respect to each of these measurable scatter sources through a regression analysis of the data. Some of the parameters will prove to have no effect on the ultimate scattering of the surface. These will be eliminated from the model. Those that prove to be significant will be kept. The sample size in Kodak's study is small with respect to the number of parameters that we are examining. It may, therefore, be impossible to separate analytically the effects of all the sources. Some engineering judgment will have to be applied at times. Kodak expects that the results of the study will support their contention that the majority of anomalous scatter is arising from previously observed 6 to 10 µm diameter pits in the surface. Prior experience with a beryllium overcoat sputtered onto a beryllium substrate has indicated that beryllium surfaces that exhibit neither pits nor typical grains produce low BRDF values. The model and experimental results will allow us to identify the characteristic of the beryllium material that causes the unexpected scattering. Standard factor analysis will be used to determine the significant factors in the model. The model will be changed to best represent the different conditions that have been included in the test (such as beryllium oxide, carbon, pits, crystal structure, surface roughness and such). This model will then be used to optimize the beryllium processing.

4.0 EARLY RESULTS

1) One observation in the early analysis of data collected from the BSA samples is that in the comparison of annealed and unannealed O-50 samples there is strong evidence that fewer pits occur in the unannealed samples. IR scatter measurements representing both processes) are shown in figure 9 below.

2) Of the annealed samples, the spherical powder beryllium has achieved the best scatter function.

3) An analysis of the pit material of diamond polished bare beryllium yielded only 1 pit in 12 with a carbon inclusion, however, 8 of the 12 pits analyzed by X-ray stimulated secondary electron emissions contained grains of uranium and/or titanium. Other contaminants showing up consistently, in unagglomerated form were silicon and magnesium.

4) There is evidence that damage from annealing may not be deeper than 10 to 20 µm. Note that this is more material than is usually removed by polishing. If a manufacturer feels a need to remove that 20µm of material (i.e. to induce a sag) the part would usually be re-machined, re-etched and re-annealed, thereby re-inducing the annealing damage.

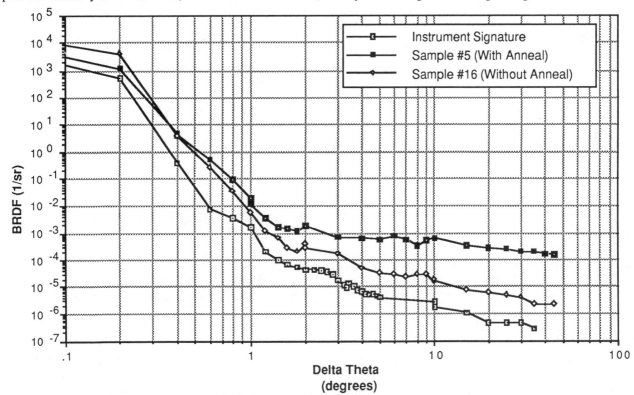

Figure 9. BRDF measurements comparing annealed and unannealed O-50 beryllium. Measurements were taken with 10.6 µm wavelength and a 10° angle of incidence.

5) Examination of the 10.6 µm BRDF data from the thin (2 µm) coated beryllium samples show an improvement in the BRDF function for annealed samples, and a degradation in scatter function for unannealed samples. Both now have approximately the same scatter function.

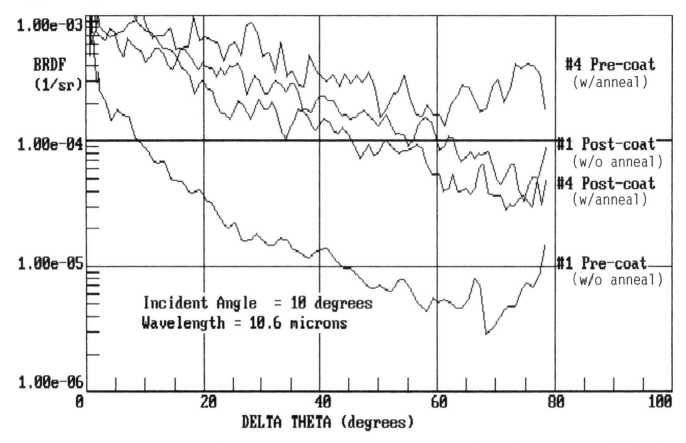

Figure 10. Pre- and post-coat (2 µm thickness) BRDF data comparing an annealed sample with an unannealed sample.

Note: data from the thick (125 µm) coatings, and replication data is not available at this time.

5.0 REFERENCES

1) J. L. Behlau, M. Baumler, "Characterization of Hot Isostaticly Pressed (HIP'ed) Optical Quality Beryllium," Proc. SPIE, *1530*, (1991).

2) RADC is now called Rome Laboratory (RL)

3) Note that all grain sizes are as measured and reported by Brush Wellman.

4) J. C. Stover, et al., "Comparison of Wavelength Scaling Data to Experiment, " Proc. SPIE *967*, 44 (1988)

5) C. L. Vernold, "Application and Verification of Wavelength Scaling for Near Specular Scatter Predictions," Proc. SPIE, *1165*, 18 (1989).

6) E. L. Church, H. A. Jenkinsons, J. M. Zavada, "Relationship Between Surface Scattering and Microtopographic Features," Opt. Eng. *18*, 125 (1976).

7) F. Toigo, et al., "Optical Properties of Rough Surfaces; General Theory and the Small Roughness Limit," Phys. Rev. B15, No. 12, 5618 (1977)

8) F. E. Nicodemus, et al.; Geometrical Considerations and Nomenclature for Reflectance, NBS Monograph 160, U.S. Dept. of Commerce, Oct. 1977.

OPTICAL SCATTER:
APPLICATIONS, MEASUREMENT, AND THEORY

Volume 1530

SESSION 4

Instruments and Techniques

Chair
Cynthia L. Vernold
Hughes Danbury Optical Systems, Inc.

Mapping of imbedded contaminants in transparent material by optical scatter

Donald A. Rudberg, John C. Stover and Douglas E. McGary

TMA Technologies
PO Box 3118
Bozeman, MT 59715

ABSTRACT

The paper reviews the design of an instrument designed to detect and map internal defects found in PMMA plastic. The material is used to manufacture intraocular lenses and must be free of defects larger than about 20 micrometers for a variety of safety, manufacturing and cosmetic reasons. The instrument detects scatter from a diode laser with a CCD array camera to map defect location. This information is used to avoid manufacturing lenses that will eventually be rejected.

2. INTRODUCTION AND PROBLEM STATEMENT

Quality control of raw material used in production of optical components is a matter of concern from the viewpoint of performance, cosmetics, and safety.

The raw optical material, called PMMA, used in the manufacture of hard intraocular lenses invariably contains small, bulk defects as a result of the material production process. These defects impair the optical quality of the finished lens, but it is of greater concern that they can pose a safety hazard. Defects range from opaque, carbonaceous inclusions, through moderately translucent inclusions, and to bubbly voids. Most of them have a generally spherical to oval character, although a very few can be found that are stringlike. Some bubble defects also have small carbon inclusions embedded in them.

In the finished product, inclusions are a potential source of device failure, whereas voids are more a cosmetic concern. Ophthalmic procedures may require that laser energy be passed through the optics, creating the possibility that inclusions, which absorb energy, may undergo rapid heating when struck by the beam, causing thermal cracking or destruction of the optic. The bottom line is that both inclusions and voids are to be excluded from the finished product. Lens manufacturers desire to identify all defects larger than a given size (which is not yet a well-determined value), so a threshold detection procedure is appropriate.

3. PURPOSE OF THE INSTRUMENT

The instrument discussed here acts as a primary quality filter whose purpose is to identify locations of defects in the sheets of raw optical material. The present inspection process involves individuals scanning the plates, using 100 power microscopes. Quality of inspection varies according to inspector experience, visual acuity, and mental readiness. Depth of field is small with a 100 power microscope, so it is very easy for a particle to go undetected. Operators also have varying degrees of alertness, both among individuals and from time to time during the day. It was the desire of the lens manufacturer to have a fast, uniform process that would inspect the critical central region of PMMA plates.

This instrument is the result. It is the first unit in an automated production train that converts incoming blank sheets into finished optics. Along with other necessary machines, the train includes a coring robot that cuts circular blanks from the PMMA plates. The specific function of the TMA instrument is to create a catalog of defect coordinates and sizes, and to pass that catalog to the coring robot. Software in the robot then determines suitable core locations and controls the actual cutting process. Statistics on yield are also kept by the robot, although determination of core locations and of statistics could just as easily be done by the TMA instrument.

A final quality control inspection of the finished lens takes place later, but it is desirable to avoid any processing of lens blanks with defects. Costs are reduced by avoiding the processing of blanks that will ultimately be rejected, and inventory control requires less effort as a result of reduced re-ordering and tracking of replacement lenses.

For this application, quality control of intermediate processes is the driving concept of the instrument, rather than direct quality control of the finished product. However, by modification of mechanical procedures so that finished lenses can be handled, and modification of the imaging system so that it can look into a finished lens, a similar instrument could carry out the final inspection as well. Indeed, variants of the instrument would be suited to a wide variety of material inspection processes.

4. MATERIAL AND INSPECTION SPECIFICATIONS

PMMA is a clear plastic that is fully tolerated within the eye. It has an appropriate index of refraction for ophthalmic use (about 1.49) and it is readily machined. The production process of the bulk plastic finishes with the pouring of large sheets having a three millimeter thickness. The large sheets are later cut into twelve by five inch plates for ease of handling. Intraocular lenses are manufactured from the central two millimeters of a sheet, so surface defects (scratches, abrasions, etc.) and dust must be ignored during inspection. Nor should defects in either the top or bottom half-millimeter of the sheet be detected since they will not become part of the finished lens.

The specification for intraocular lens quality is that there shall be no visible defects in the lens. This is not a very well-defined specification. It does not indicate a precise defect size that is unacceptable, although 10 to 20 microns is often used. A typical sheet of PMMA material can have a thousand defects in the range from 2 to 10 microns, a few hundred from 10 to 20 microns, and fewer than a hundred defects larger than 20 microns. Experienced lens inspectors will see 10 to 20 micron defects if the defects are favorably illuminated and focused, but they might miss them under less favorable conditions. Larger defects are more easily detected by operators. The TMA instrument introduces repeatability and flexibility, the two important factors of quality control in this environment. It allows setting of a variable threshold of light scatter intensity so that all defects of that scatter intensity or greater are detected.

5. DESIGN CONSTRAINTS AND DECISIONS ABOUT THE INSTRUMENT

It is possible to emulate the procedure involving human operators in forming an image of the material for each frame of an inspection instrument, i.e., to use direct reflective or transmissive illumination with machine vision to detect defects. But such a procedure would display unwanted defects on the surfaces and those within 0.5 mm of the surfaces. Human operators are good at distinguishing surface defects from bulk defects, but it is computationally intensive to use machine vision for the task. It was immediately clear that any but the simplest processing would defeat the time limit imposed by the instrument user. For efficient production, a twelve by five inch PMMA sheet must be inspected within 30 minutes. Thirty minutes per plate translates to less than one second of total time for sample movement, image capture, and processing for each of the over 2000 small frames that would be taken of the plate.

A related constraint was that allowable instrument cost dictated use of a PC and a commercially available array processor hosted by the PC, rather than a more powerful processing engine that would allow more sophisticated image processing. So the processing algorithm had to be simple to be fast. Consequently, the image had to be quite simple in structure. Also, as much responsibility as possible had to be transferred to the image illumination and capture system. Fortunately, the population of significant defects in any given frame is usually quite sparse, so a simple, sparse-defect algorithm could be used. It allowed all of the array processor tasks to be performed in 180 milliseconds on a 512 by 480 pixel array. It is limited to calculation of defect scatter magnitude and centroid coordinates.

In addition to the above issue of surface defect recognition, it is time-consuming for either human operators or machines to distinguish between defects in the central two millimeters of a plate and those in the outer half-millimeter of each side. With these constraints on processing time and instrument cost, a different approach to image formation would be the key to achieving desired instrument performance. So the decision was made to illuminate only the region of the plate to be inspected, i.e., the central two millimeters, and to capture images of light scattered normally to the illuminating beam by the defects. Light scattered by the defects is used as a metric, rather than a direct measure of size. The array processor algorithm used works equally well either in calculating cross-sectional area of defect images, or in calculating summed light scatter.

This is a remarkably effective way to inspect transparent material. It eliminates surface defects from the image, as well as defects in the outer one-half millimeters of material. Rather than looking for specific size of defects in a brightfield image, defects stand out as bright points in a darkfield image. A much improved depth of field is achieved by using a relatively low-magnification optic system to scan the necessary frame area, rather than by using a microscope. The desired material volume can be inspected without changing focus.

Thus the most important issue becomes quality of illumination. In effect, the problems of separating surface defects and outer region defects from desired defects, and of exposing desired defects for imaging have been cast into this single question: What is the quality of the beam in illuminating the desired region?

6. INSTRUMENT CONCEPT

Microscopic viewing of light scattered by defects in the material presents an image similar to the Milky Way in microcosm. Large numbers of very small defects form a dim, diffuse background for lesser numbers of larger, brighter particles. Although there has been no directed effort at analysis of particle size statistics, they appear to follow a Rosin-Rammler distribution, in which the number of defects increases rapidly as defect size decreases. The same sort of size distribution function applies to aerosols and ungraded particulates, such as pulverized coal or gravel deposits. Thus, much of the practical application revolves around determination by the instrument user of an acceptance size threshold that is small enough to insure a good product, yet large enough to give a reasonable yield.

The detection process is conceptually very simple: Introduce a collimated, directional, uniform light beam into the plane of the material and view images of light scattered by the defects. When done against a black background, the result is a darkfield image in which the defects stand out as scatter points. The sample scatters enough light so that a CID camera and an eight-bit frame grabber can record the image. Subsequent processing is done to identify scatter values from defects and coordinates of the scatter centroids, which are passed to other processing stations along the production train.

For spherical objects that are large compared to a wavelength, light scatter varies as the object area, all other things being equal. Embedding of the area relation in software allows an approximate sizing of defects, the approximate nature being due to variable scatter characteristics of actual defects, e.g., a dark particle will scatter less than a light particle, and defects are rarely spherical. Instrument threshold setting is done by capturing an image of a known dark particle at the threshold size, and setting the threshold scatter value at the particle scatter. A dark particle is chosen because they are of most concern. They also have by far the greatest population. They almost always have a rough surface so that they give good scatter from the PMMA/particle interface. Bubble inclusions of the same size as a dark particle appear brighter and translucent inclusions are dimmer.

Now that system requirements are generally defined, and error sources considered, it is time to consider a real source. Ideally, it is collimated, directional, and uniform. Laser illumination is collimated, its direction is controllable, and, when expanded, its central region can have acceptable uniformity. A laser diode was selected for its small size and easy mounting. With suitable expansion optics, it yields a beam having a fairly uniform profile in a central region of approximately four millimeters wide and a little over two millimeters high. The beam is then passed through the first aperture of two millimeters high and some six millimeters wide. This aperture defines beam height without regard to beam width. Slight beam spread in the horizontal plane is not a problem in this application. The second aperture is on the far side of the plate and is used as a target to aim the beam in both the horizontal and vertical directions.

7. INSTRUMENT DESIGN AND OPERATION

As shown in Figure 1, the instrument consists of a set of orthogonal stages driven by stepper motors, with a platform for mounting the PMMA plate. Illumination is a 670 nm laser diode beam that enters the side of the plate rather than the top or the bottom. The beam has been enlarged, collimated and directed through an aperture so that it impinges on the central two millimeters with a precise two millimeter height and an approximate four millimeter width. It then passes through the short dimension of the plate, and finally through a second aperture with a built-in beam dump. The first aperture is used to define the size of the incident beam, while alignment of the beam is done by using both of them as targets. The laser diode and its optics are mounted on the lower, or X, stage so that they move with the stage, keeping the beam under the center of the camera field of view. The X-stage has a motion of five inches parallel to the laser beam. The upper, or Y, stage moves transversely to the beam. It has a motion of twelve inches relative to the X-stage.

The choice of camera was influenced by the character of the scattered-light image. The scattered light is generally of small area, but it can be of high intensity, especially as defect size increases. CCD cameras produce a smoother image than do CID cameras, but they can exhibit blooming and streaking that are unsuitable for this environment. So the decision was made to use a CID camera and to tolerate its higher noise.

The camera is fixed in location above the stages, looking down on the plate. The plate is moved by computer-controlled stages in a pre-defined pattern beneath the camera. The camera operates in conjunction with a commercially available frame grabber for image capture. The frame grabber and array processor have their own independent communication bus so that data transfers can take place directly between them without using the PC bus. At the end of the measurement, the instrument automatically sends the defect locations to the automated coring robot or stores the defect locations in a data file. Each optical blank that is inspected has a barcode sticker that is read by the instrument and is used for identification throughout the process and in subsequent data files.

Instrument operators are required only to load plates, enter identification and press a key. The instrument automatically scans a plate, and it produces a defect map plot and file with the plate identification. At any time during measurement or plate loading, the robot controller can interrupt the machine for transfer of any existing defect file. Measurement resumes after transfer.

Technicians and engineers have access to expanded menus that initialize system parameters and that test instrument functions. File transfer requests by the coring robot are not honored during initialization or test. The instrument is configurable, so instrument initialization software can be run to set:

1. The hardcopy output mode: Plot defect map, print defect listing, both or none.
2. The RS-232 output mode, automatically transfer defect file or just store the defect file on the hard disk.
3. The size of the defects to locate.

Testing software allows the user to separately test different parts of the instrument. These parts are:

1. The CID camera and frame grabber
2. The array processor
3. The stage movement hardware.
4. The printer/plotter.
5. The laser diode driver system
6. The bar code reader.

8. ERROR SOURCES AND CONTROL

Errors in defect detection come from three sources: (1) non-ideal illumination, (2) defect variability, and (3) system noise. We will treat the sources in order, beginning with a discussion of the consequences of not achieving the collimated, directional and uniform illumination mentioned earlier. An index of refraction of 1.50 is used in this and all subsequent discussions.

Collimation is required to insure that only the central region of the material is illuminated. Beam spreading in air of one milliradian half-angle yields a beam width of 2.4 millimeters at the exit interface. Beam spread of more than 2.6 milliradians half-angle in air will allow the beam to reach the distal surfaces, causing surface defects (scratches, abrasions, dust, etc.) to be incorrectly detected as inclusions.
An associated issue is failure to illuminate the same defect set when the plate is turned such that the entrance and exit faces are interchanged. Non-duplicated illumination occurs when (1) the beam is not aligned with the plane of the plate, or (2) the entrance surface is not normal to the beam so that refraction occurs. Alignment of the beam with the plane of the plate is relatively easy to achieve, requiring that the beam be lined up with the target apertures on either side. Alignment to around 0.5 milliradian can done readily.

Entrance surface normality of the PMMA plate is more difficult to control. A smooth surface oriented normally to the illuminating beam is necessary, making preparation of the plate very important. The lens manufacturer has experimented with several edge preparation procedures. The most suitable one is use of diamond-surfaced tools to mill a smooth edge, which is then lightly polished to minimize scatter. A 0.5 degree machining error at the entrance face, which is not an outrageous value, leads to lack of duplicate coverage that is 17 percent of the material. Correspondingly, assuming uniform distribution of defects, no system noise, and all other conditions ideal, 17 percent of defects will not be duplicated upon interchange. The condition of greater concern is that part of the bulk to be inspected is not illuminated. In this case (again assuming no beam spreading), 9 percent of desired volume is not scanned. Fortunately, the real condition of beam spreading will tend to cover the non-illuminated volume just discussed, so reality is a benefit in this case. At the extreme, a machining error exceeding 0.7 degrees will refract the beam to strike a surface of the plate, causing surface conditions to be interpreted as bulk defects.

Uniformity of the beam is desired because the instrument works by measuring light scatter rather than by direct measurement of defect size. The reason for doing it this way is that scatter measurement allows scanning of a much larger region than does image metrology, resulting in faster plate inspection. However, using scatter implies that identical defects located in different regions of a non-uniform beam will yield non-identical scatter. Lack of beam uniformity in the horizontal direction can be viewed with the imaging system and its effects on scatter can be removed by software. That still leaves a variation of signal-to-noise ratio across the width about which nothing can be done. Non-uniformity in the vertical direction cannot be determined when the imaging system is normal to the specimen and is thus not removable by software. It is essential to begin with as uniform a beam profile in the vertical direction as possible. For the application, beam expansion and a spatial filter were used to achieve the maximum beam uniformity. The beam is then windowed to a near top-hat profile. No problem has been noted as a result of the diffraction caused by windowing. A considerably less vexing problem is that beam spreading discussed above reduces intensity slightly along the length of the beam, resulting in a near-inconsequential decrease of scatter values.

System noise sources are (1) the camera, (2) laser diode output power variations, and (3) the motor driver system. Camera noise is dependent on temperature and integration time. Temperature dependence of CID camera noise is such that an approximate doubling occurs with each 6 degree Celsius rise. Dark noise increases linearly with camera integration period, but since signal also increases linearly with integration, basic camera quality is the important factor involved. It is changeable only by selection of a another camera. Laser diode power output is checked for operation within limits after each linear scan in the y-direction. Motor driver noise is minimized by proper placement and careful cable selection. With all noise sources operating normally, a minimum signal-to-noise ratio of 3:1 is achieved for 10 micron dark defects.

9. INSTRUMENT PERFORMANCE AND CONCLUSIONS

Although final performance figures are not complete, the instrument appears to be very good compared to human operators. Two factors can be cited: (1) Defect detection is repeatable within the limits of error sources, and (2) the entire plate volume in the central two millimeters is scanned without need for refocusing. There are no data regarding the accuracy and repeatability of human operators so a direct comparison is not possible. However, the instrument finds more defects that are missed by operators than is the reverse case.

The instrument is designed so that errors are made on the side of over-detection. While it is important that the instrument allow good material to be processed into finished lenses, it is paramount that the instrument not allow defective material to enter the manufacturing chain. The design insures blockage of defective material with an accepted risk of blocking some good material. Ideally, every defect in the central two millimeters that exceeds a certain size is identified, and no defects in the outside half-millimeter are flagged, regardless of size. Repeated measurements of a given plate would yield exactly the same set of flagged defects every time, regardless of plate orientation. This ideal state is not reached because of system noise and geometric imperfections of beam spreading, edge refraction, vertical misalignment, and plate thickness variation.

Repeatability of measurements of a given plate in a fixed orientation is affected by system noise. Under fixed geometric conditions, 90 to 95 percent of detected defects remain from measurement to measurement, while the rest, which are near threshold, move in and out from one measurement to another. A change of threshold does not eliminate the behavior since it only moves the variability to a new level. Generally, the defect count changes very little from one measurement to another, and the location of acceptable core locations changes only slightly.

Geometric imperfections will yield variation of results when a given plate is loaded in a new orientation. The basis for variability in defect identification has been established above in the section on error sources and error control. Experimental results are that about 70 percent of defects are duplicated when a plate is rotated such that the entrance and exit surfaces are interchanged. Again, the defects that move in and out from measurement to measurement are usually near threshold.

Some geometric imperfections of the PMMA plate are outside of TMA control. Because the plates are poured, they vary in thickness by as much as 0.5 mm. Turning the plate over so that the top and bottom surfaces are interchanged illuminates a somewhat different region of the plate, so duplication declines to around 60 percent. Bowed plates are not a problem because a vacuum chuck is used.

The conclusion is that use of light scatter to inspect optical material for embedded defects is an effective, low-cost alternative to other methods.

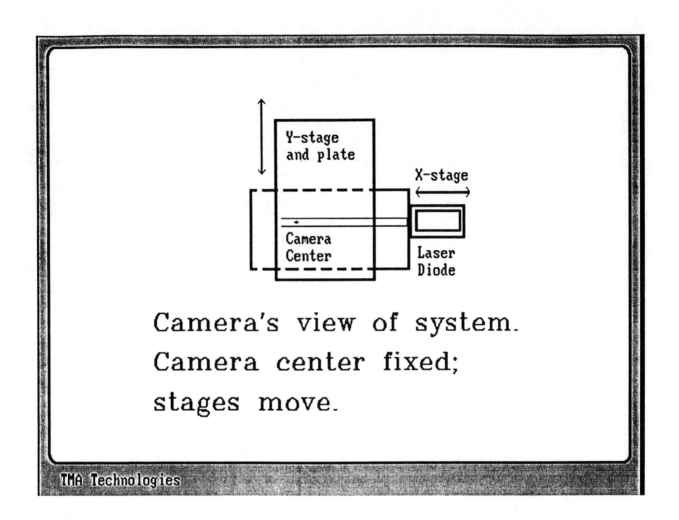

Figure 1

Experimental Study of the Laser Retroreflection of Various Surfaces

Liu Wen-Qing, Jiang Rong-Xi, Wang Ya-Ping and Xia Yu-Xing
Anhui Institute of Optics and Fine Mechanics, Academia Sinica
P. O. Box 1125, 230031-Hefei, Anhui, P.R. China

ABSTRACT

The development and use of an automatic system to measure the absolute retroreflectance of various surfaces without any reference standard is described in this paper. Preliminary results are obtained for several samples: defused SiC plates coated with gold, unworked aluminum sprayed with different thickness of paint and metal plate sprayed with different SiN powder and paint mixture. The results indicate that a correct choice of the parameters allows complementary information of the retroreflection. The accuracy of the system is confirmed to be from 1% to 9% of measured value (for CO_2 laser source) and higher accuracy may be obtained for other wavelength.

INTRODUCTION

The retroreflection properties of a surface play an important part in many fields[1] and have a broad range of applications[2,3]. The reflectance signatures, besides being the physical and chemical properties of the sample, is affected by irradiation, surface texture, surface flatness, angle of incidence and viewing[4]. To obtain reliable and reproducible measurement results, it is important that the measuring method be directional, simple and repeatable.

In this paper, the design of a simple double beam retroreflectometer enable the reflectance to be measured at different angles of incidence and viewing without using a reflectance standard. Several surface samples as mentioned above have been measured by this system. The results from this study can be used to define signature model of the surface, to analyze the surface roughness and to evaluate photometric properties of surfaces sprayed with various silicon carbide powder and silicon nitrogen powder.

APPARATUS AND METHODS

A schematic of the retroreflectance measurement system is shown in figure 1. The operational principle of the system is simple. B_1 and B_2 are two optically identical beam-spitters. In order to eliminate the measurement errors due to the fluctuations of the light source intensity and processing electronics with time, the double beam ratio recording system is used[5]. A linearly polarized beam from a CW 5W CO_2 laser of the sealed-off type is divided into a reference beam and a sample beam by beam-spitters B1. The reference beam is incident on detector D_2 through an integrating sphere, which is used as an

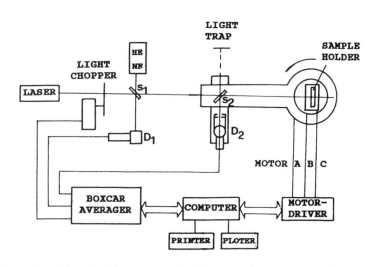

Fig.1. Block diagram of the system. Legend: B_1, B_2: Beam splitters, D_1, D_2: Detectors, Motor A, B, C for changing angle of incidence and viewing, and movement of traslation.

attenuator. The sample beam reflected by the sample optic returns to beam-splitter B_2 and has the same optical path lengths as the reference beam. The reflectance of the sample is the ratio of intensities of sample and the reference beams and is given by

$$R(\theta,\lambda) = K \frac{V_{D1} R_{B1}(\lambda)}{V_{D2}(1-r_{B1}(\lambda)(1-r_{B2})r_{B2}(\lambda)}$$

where V_{D1} and V_{D2} are the readout voltage of the detector D_1 and D_2, respectively. r_{B1} and r_{B2} are the reflectance of the beam-spitters B_1 and B_2, respectively. K is the system constant by taking into account the factor of the integrating sphere attenuator, the responsibility and the sensitivity of two detectors, obtained by experiment. A visible HeNe laser is utilized to alignment the system and to indicate CO2 laser beam position.

By considering the CW light beam and the pulsed light beam, the boxcar averager (EG&G model 162) is used[6]. The signals from two detectors (thermal electric type TGS) are amplified by two differential amplifiers (Tectronix AM502) and then their outputs are sent to two gated integrator (model 165) of the boxcar average. The ratio of the intensity of the sample beam to that of the reference beam is measured directly. In order to collect the ratio data, we developed an interface used to communicate between boxcar averager and a computer (IBM 286). The measured ratio is stored in the computer through this interface. Various angles of incidence and viewing are driven by stepping motors, which are controlled by the computer through a DI/DO interface, to move to the designated angle position within the accuracy of 0.01 degree. The linear movement of the sample is also performed by a stepping motor within the accuracy of 5 μm.

By proper selecting the AD and TS parameters of the boxcar averager, the signal-to-noise ratio can be improved. In order to further increase the SNR, the measurement is repeated ten times on one point in the sample at a repetition of 15 Hz (for CW CO2 laser source). The averaged value is assigned to that point in the software. Uncertainties of the retroreflectance measurements obtained with this system are evaluated by standard reference to be 0.2 - 9 % of the measured value.

EXPERIMENTAL RESULTS AND DISCUSSION

The reflecting distribution were first performed on the SiC plate made in our laboratory and SiC sandpaper with different grit size from 60 μm to 600 μm. Each sample has the same material weight and the sample surface was coated with 5 μm of gold by a vapor deposition technique. A strong dependence of the retroreflectance on the grit size was found at 10.6 μm wavelength, by comparison with other wavelength 0.6328 μm, 0.94 μm and 1.06 μm. For different incident angles, all samples demonstrate the specular peak with different peak intensity. The retroreflectance distribution was related to the surface structure, wavelength and the grit spacing[7]. A broad uniform diffuse reflectance and a small retroreflectance was obtained for SiC plate (not sandpaper) with 120 μm grit size as shown in figure 2. It seems that this sample has approximately Lambertian characteristics in IR range. All sandpaper

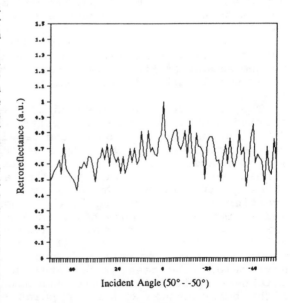

Fig. 2. Retroreflectance of SiC plate.

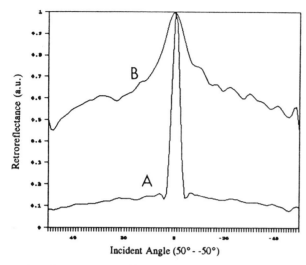

Fig.4. Retroreflectance of metal plate with SiN powder and white paint mixture. A, 1:45; B, 6:45, respectively.

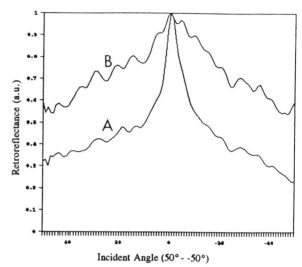

Fig. 3. Retroreflectance of metal plate with SiN poder and green paint mixture. A,1:45; B,6:45.

samples exhibit poor uniform distribution of the retroreflectance because of the loose grit spacing.

The measurement of the retroreflectance are performed on some metal plates sprayed with mixture of SiN supperfine powder and paint. The SiN powder are produced by the method of laser-induced vapor phase synthesis[8]. Figure 3 indicate the results for plates with the SiN powder and green paint mixture 1:45 and 6:45, respectively. The curves are normalized for easy comparison. Both shape and intensity of the retroreflectance exhibit difference. The intensity ratio of the retroreflectance between the 1:45-mixture and the 6:45 one was calculated to be 1.66. Since the mixture was sprayed on the same type of substrate, the difference can be attributed to the mixture properties.

Similarly, the measurement was also performed on other two samples with the same mixture ratio. Instead of the green paint, the white paint was used. As expected, besides the shape, the white paint mixture has higher retroreflectance intensity than the green one as shown in figure 4. The ratio between the white paint mixture and the green one, in this case, is 5.6 for 1:45-mixture and 1.98 for 6:45-mixture, respectively. The experimental results obtained seems to indicate that the distribution of the retroreflectance may be controlled not only by the amount of SiN powder, but by the paint's properties. By performed on the other wavelength, Similar results can be obtained with the different Mixture.

The experiment performed on one sample with different surface position was carried out to examine the paint thickness effects on the retroreflection distribution. Figure 5 shows the results foe a sample with 35 μm paint thickness on the base surface. The sample was placed with three position $\varphi=0°$, $\varphi=45°$ and $\varphi=90°$, respectively. φ, in this case, is defined as a angle between surface streak and incident beam plane. The retroreflection intensity is similar for all three positions. But the angle distribution strongly depend upon the sample position. The interesting is that when $\varphi=0°$, the distribution angle is 2 times and 4 times broader than those when $\varphi=45°$ and $\varphi=90°$, respectively, indicating a grating phenomena. The sample shows a sharp specular retroreflection angle for $\varphi=90°$. There is a limited paint thickness. In the case of the paint thickness under the limit, the retroreflectance of a sample is attributed both to the paint surface and the base (substrate) surface. Moreover, by increasing the paint thickness up to 100 μm, this phenomena almost disappeared. The influence of the light beam through the front paint surface and onto the base surface can be neglected, therefore the base retroreflection effect was not very significant.

In Conclusion, we have described a system for laser retroreflectance measurement at different angle of incidence and viewing. Of the samples evaluated, the unacceptance amount of retroreflectance must be considered when some diffused plate is used as a Lambertian surface. The characterization of the retroreflection distribution can be controlled as needed by selecting the sample surface material and/or its mixture. This properties may be used in analysis of target signatures[9].

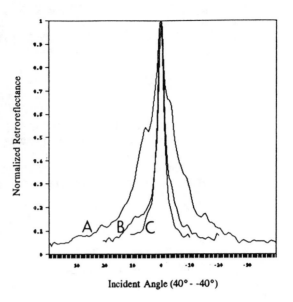

Fig. 5. Retroreflectance of plate with 35 μm thickness paint on the surface. A, φ=0°; B, φ=45°; C, φ=90°, (see text)

REFERENCE

1. Zu-Han Gu, Richard S. Dummer, Alexei A. Maradudin, and Arthur R. McGurn, "Experimental study of the opposition effect in the scattering of light from a randomly rough metal surface" Appl. Opt. vol.28(3), pp. 537-543, 1989.

2. W. Erb, "Computer-controled gonioreflectometer for the measurement of spectral reflection characteristics" Appl. Opt. vol.19(22), pp. 3789-3794,1980.

3. R.S. Ram, O. Prakash, J. Singh, S.P. Varma, "Absolute reflectance measurement at normal incidence" Optics & Laser Tech. vol.22(1), pp.51-55, 1990.

4. B. Hapke, "Bidirectional Reflectance spectroscopy" J. of Geophysical research, vol.86(B4), pp.3039-3060, 1981.

5. H. Takahashi, M. Kimura, R. Sano, "Automatic reflectivity map measurement of high power CO_2 laser optics" Optics & Laser Tech. vol.21(1), pp.37-40, 1989.

6. "Model 162 Boxcar averager operating and service manual" EG&G PARC, USA.

7. T.W. Stuhlinger, E.L. Dereniak, and F.O. Bartell, "Bidirectional reflectance distribution of gold-plated sandpaper" Appl. Opt. vol.20(5), 1981.

8. Li Daohuo, "Matching and optimizing of the experimental parameters for synthesis of Si_3N_4 by laser", Advances in laser science-III, pp. 613-615, American Institute of Physics, Confrence proceeding N. 172, New York 1988.

9. D.K. Killinger and A. Mooradian, (eds.), "Optical and laser remote sensing", Springer series in optical sciences 39, New York 1983.

Design considerations for multipurpose bidirectional reflectometers

John T. Neu and Martin Bressler

Surface Optics Corporation
P.O. Box 261602, San Diego, CA, 92196-1602

ABSTRACT

A bidirectional reflectometer designed for obtaining both signature data and optical component scattering information must satisfy a wide range of measurement requirements and sometimes conflicting requirements and criteria.

The basic definition of bidirectional measurement is reviewed; the design approaches, measurement geometries, measurement procedures and techniques, recommended for a multipurpose instrument are described and discussed.

1. BRDF DEFINITION AND USES

The bidirectional reflectance distribution function (BRDF) of a material, as illustrated in the flower-like graphic of Figure 1, is a very complete and useful description of the reflectance properties of a material. It describes how beams of a given spectral content, incident upon a plane sample at various angles, will be reflected into the hemisphere surrounding the sample.

Each measured reflectance value of the BRDF is a function of four plane angles - two plane angular quantities to describe the direction of the incident beam with respect to the sample center, and two to describe the direction of the measured or received beam reflected from the sample. The bidirectional reflectometer (BDR) instruments which are used to obtain BRDF data must therefore have at least four angular degrees of freedom.

BDRs have two major uses. They are used to obtain data for use in signature analysis (determining how aircraft, missiles, and other vehicles will appear to various observers and observing systems) and for measuring the scattering characteristics of the optical components of telescopes and other optical systems.

2. MEASUREMENT GEOMETRIES

Several different measurement geometries or instrument configurations are used in achieving the four angular degrees of freedom required for a BDR.

*   Goniometric — In the so called goniometric system, the sample remains fixed in a horizontal plane. The source and detector beams each have two degrees of freedom and may be pointed toward the sample from any position in the hemisphere surrounding the sample. The angular designations for the goniometric system are shown in Figure 2 and a conceptual schematic of a goniometric measuring device is presented in Figure 3. The source and receiver are mounted on separate arms, each arm has two degrees of freedom and both source and receiver can be moved to any point on the hemisphere surrounding the sample.

*   Modified Goniometric — A modification to the goniometric system is sometimes used, when large heavy sources are to be employed (e.g., high temperature blackbodies). Mounting such heavy sources on a moving arm would require unfeasibly large and massive arms and instruments. The source is therefore fixed and an arm mounted optical system allows one degree of freedom - the elevation angle at which the source radiation is directed towards the sample can be varied. The receiver arm remains unchanged providing degrees of freedom in both azimuth and elevation. The fourth degree of freedom is provided by rotation of the sample about a perpendicular axis through its center.

*   Fixed Source — In another common configuration, shown schematically in Figure 4, the source is fixed, and the source and receiver beams lie in a plane. The receiver arm rotates in this plane providing one degree of freedom. The other three degrees of freedom are provided by sample rotations - designated in Figure 4 as roll, pitch and yaw.

The distribution of the four degrees of freedom in these three BDR configurations are summarized in Table 1.

Table 1
Distribution of Degrees of Freedom in BDR Configurations

SYSTEM	DEGREES OF FREEDOM			
	SOURCE	SAMPLE	RECEIVER	TOTAL
Goniometric	2	0	2	4
Modified Goniometric	1	1	2	4
Fixed Source	0	3	1	4

The fixed source system provides mechanical simplicity and compactness. It is commonly used with laser sources in BRDF determination of scattering from telescope optical components. The goniometric and modified goniometric systems offer other advantages to general purpose instruments and these configurations are usually used for instruments used to provide signature data. The goniometric system advantages in obtaining signature data are:

a. The goniometric angular coordinate system is the same as that of the ERAS format which has become the standard for signature data and signature generation computer programs.

b. Although the fixed source configuration angles can be automatically transformed to goniometric system angles the transformation involves both magnifications and demagnifications of error. The angle setting accuracy of the fixed source system, must sometimes be less and sometimes much greater than the settings of the goniometric system in order to achieve the same accuracy in goniometric angle.

c. The angles are easily visualized in the goniometric system, and the instrument operator has a direct sense of what is being measured. Further only one angle needs to be changed for each successive measurement of a hemispheric mapping - in horizontal rings or vertical slices.

d. Polarized measurements are sometimes conducted for signature data, and require four complete BRDF measurements - one for each of the illuminator/receiver polarization combinations ($\parallel \parallel$, $\parallel \perp$, $\perp \perp$, $\perp \parallel$). In a goniometric system, only a simple 0 or 90° setting of the two polarizers is required for all the measurements made to map the hemisphere surrounding the sample and develop the BRDF. This simple adjustment can be done manually. The fixed source system, in contrast, would require both polarizers to be set to specific different angles for each separate measurement. Since hundreds of separate measurements are often required, polarized measurements with a fixed source system would have to be automated to be practical.

3. BEAM SAMPLE ARRANGEMENTS

BDR measurement requirements can vary widely depending on the purpose and use of the data to be obtained. For example, data to be used to develop vehicle signatures should include data from the fullest possible range of measurement positions over the hemisphere surrounding the sample. Measurements of the scatter from optical components, on the other hand, may not need the same range of measurement positions but must be made at very low background radiation levels. In addition, a general purpose BDR must be able to be used with different sizes of samples - small round samples as well as extended samples of indeterminate shape when it is impractical or unfeasible to obtain the former.

Several different modes or arrangements of illumination and detection beams and samples, used to meet these varying requirements, are illustrated in Figures 5, 6, and 7, and their relative advantages summarized in Table 2.

The stem mounted, round sample, shown in Figure 5, completely covered by both illuminating and detecting beams, is an example of an arrangement that has been commonly used for obtaining data for use in signature development because it permits a full hemispheric range of measurement positions. However, at certain retro angles, with a small angular spread between source and detector beams, there can be spurious signals developed by reflection and scattering from the sample edges, sample mounting plate, and stem. Such problems can arise only in a small portion of the total range of measurement positions and the spurious signals can be minimized, but not eliminated, through the use of absorbers, baffles and special fixtures.

Table 2

Comparison of Beam Sample Arrangements

I.	II.	III.		ADVANTAGES
				I. Source and detector beams slightly larger than sample diameter.
				II. Large source beam, small detector beam, extended sample.
				III. Source beam shaped with automatic aperture to project constant diameter illuminated circle on sample; detection beam diameter slightly larger than illuminated circle diameter.
X			1.	Allows full angular coverage - no measurement positions in which detector beam encompasses only a portion of the illuminated sample area.
X		X*	2.	The same illuminated sample area is measured at all measurement positions.
	X	X	3.	Eliminates possibility of spurious signals generated by reflection of radiation by sample edges and sample support structures.
	X	X	4.	Good arrangement for component scattering measurements.
X		X	5.	Good arrangement for signature data measurements.
	X	X	6.	Samples can be small fixed sizes or extended - need only be large enough to encompass source beam spot. Multiple measurements can be made on extended samples without need to cut out specimens, and destruction of parts for measurement purposes often avoided.

*   Usable at θ_i angles 0 to about 70°.

Scattering measurements can be made using the different beam sample arrangement shown in Figure 6. The use of an extended sample, eliminates the spurious signal problem of the previously discussed arrangement, but introduces a problem at certain measurement positions - at large elevation angles in particular where only a portion of the area intercepted by the detection beam is illuminated by the source.

The arrangement being used in current general purpose BDRs is shown in Figure 7. The use of automatically adjusting apertures which limit the illuminating area to within the sample boundaries, requires a well collimated beam and can be an effective tool up to illumination elevation angles of about 70°.

4. BEAM COLLIMATION CONSIDERATIONS

Bidirectional reflectance is defined in terms of perfectly parallel or collimated beams. Although parallel beams can be closely approximated they can never be perfectly parallel and any bidirectional reflectance measurement or BRDF should be qualified in terms of the parallelism or collimation of the incident source and measured reflection beams.

Both high beam power density and a high degree of collimation can be achieved with many laser sources. With incoherent blackbody sources, the degree of collimation that can be achieved varies inversely with the beam power density (see Figure 8). Since high beam power density relates directly to the signal strength and signal-to-noise ratio of a given measurement, and since low level reflectance values against blackbody sources are difficult to measure - particularly in the longer infrared wavelength where both source radiance and detector sensitivity are relatively low - good BDR designs allow for trading beam collimation for beam power density when necessary. (Such trades are easily achieved by means of a variable source aperture.)

5. FURTHER WAYS TO IMPROVE MEASUREMENTS

Several design and procedural methods have been developed for eliminating spurious signals and improving measurement accuracy.

- The alternate measurement of a known reflectance standard and the sample, under the same measurement geometries can provide checks and corrections for uncertain measurement arrangements. In addition, when the time span between standard and sample measurement is on the order of a few seconds or less, such alternate measurement can be used to automatically correct or eliminate any errors due to drift or change in the source output, detector temperature/responsivity, or electronics, which occur over time periods longer than the sample to standard interval. Rotating stages containing both sample and standard are included in systems currently being built to allow rapid alternate movement into and out of the measurement position.

- Some laser sources can exhibit intensity changes too rapid to be corrected by alternate sample and standard measurements, and a separate source monitor should be used to correct for such source variations.

- Automated data processing systems are being built in which the number of data samples or integration time is increased until a desirable or acceptable signal-to-noise ratio can be produced. This measurement time - sensitivity tradeoff feature can be particularly useful in long wavelength infrared measurements of low reflectance materials, and can establish an optimum use of measurement time over a wide range of material characteristics, measurement geometries and signal strengths.

6. SUMMARY

A number of design features and techniques have been recommended for multipurpose BDRs - measurement systems which will utilize blackbody and other continuous sources as well as laser sources and which will be used to obtain full hemispheric BRDF data for signature use as well as for measuring scatter of optical components. These are:

- the use of the modified goniometric mechanical arrangement and goniometric coordinate system.

- providing automatic control of illuminating beams to restrict illumination to within sample boundaries.

- using variable source or source image apertures to allow operational tradeoffs of beam collimation for sample irradiance and signal-to-noise ratio.

- rapid alternate measurement of samples and known standards to eliminate errors due to drift and change of system components.

- source monitors to correct for rapid source intensity changes.

- data acquisition designs which allow integration time or number of samples to increase until a given signal-to-noise ratio is achieved.

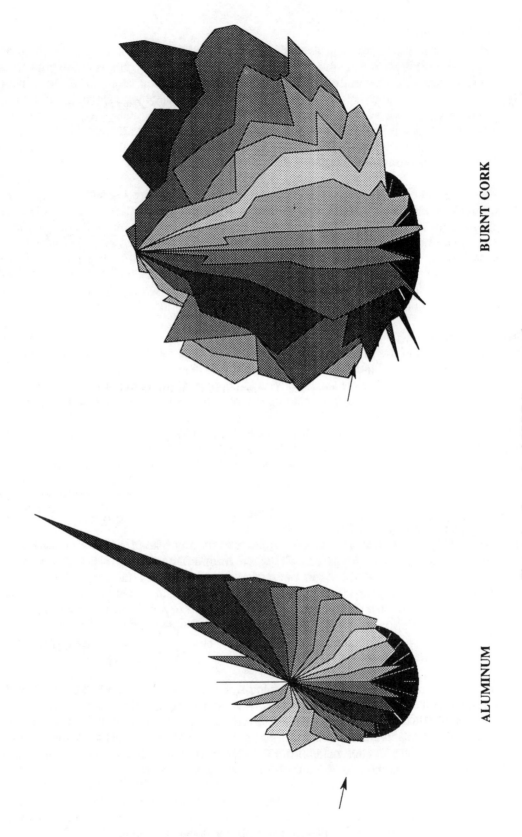

BURNT CORK

ALUMINUM

Figure 1. Example of BRDF Representation.

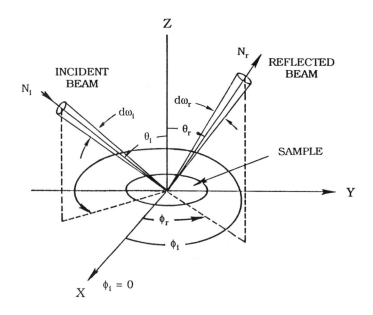

Figure 2.   Goniometric Coordinate System.

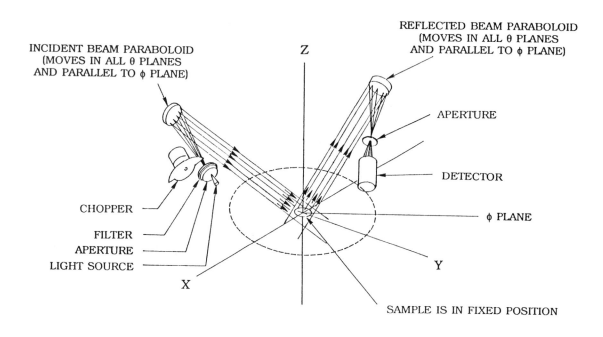

Figure 3.   True Goniometric System.

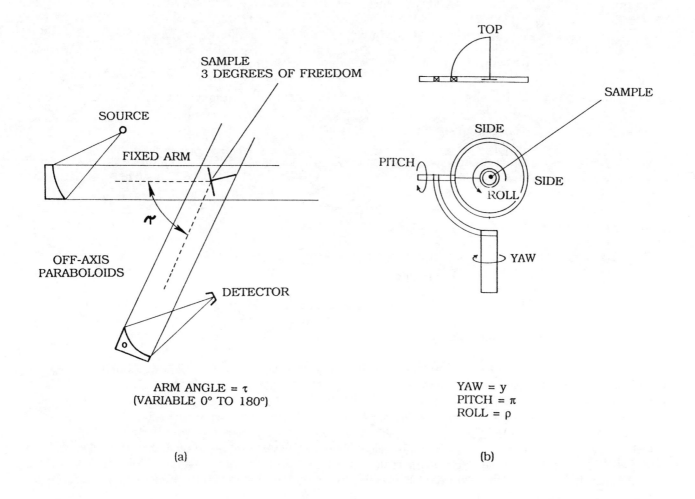

SAMPLE
3 DEGREES OF FREEDOM

SOURCE

FIXED ARM

TOP

SAMPLE

OFF-AXIS
PARABOLOIDS

PITCH

SIDE

SIDE

ROLL

DETECTOR

YAW

ARM ANGLE = τ
(VARIABLE 0° TO 180°)

YAW = y
PITCH = π
ROLL = ρ

(a)

(b)

Figure 4. Fixed Source Configuration.

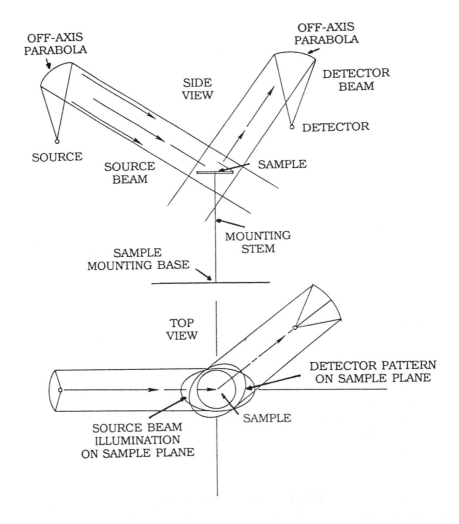

Figure 5. Mode 1, Source Detector Beams Slightly Larger than Sample.

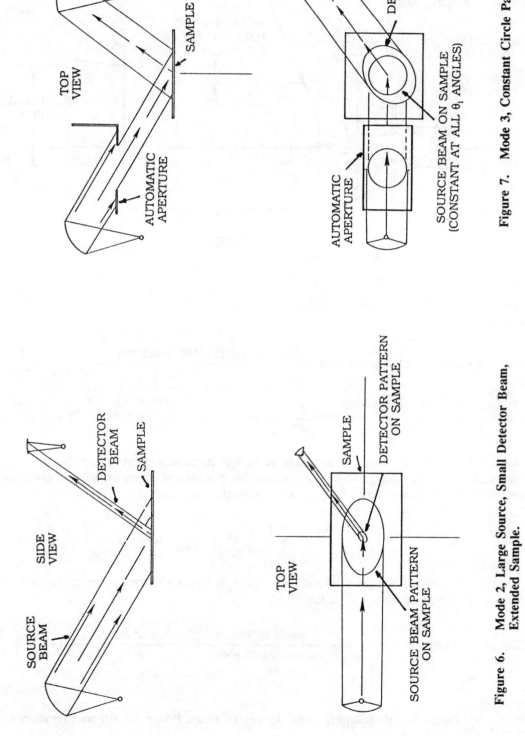

Figure 7. Mode 3, Constant Circle Pattern on Sample.

Figure 6. Mode 2, Large Source, Small Detector Beam, Extended Sample.

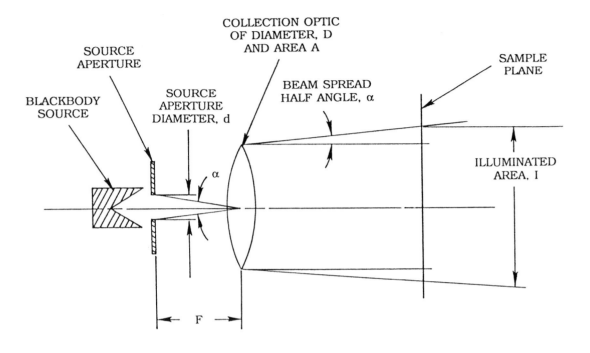

$$\text{BEAM SPREAD HALF ANGLE } \alpha = \text{ARC TAN} \left[\left(\frac{\text{SOURCE APERTURE DIAMETER, d}}{2} \right) \left(\frac{1}{\text{FOCAL LENGTH, F}} \right) \right]$$

$$= \text{ARC TAN} \frac{d}{2F}$$

THE POWER IN THE BEAM, P, WATTS, IS EQUAL TO N, THE RADIANCE OF THE SOURCE,, W/cm^2-STER, TIMES THE SOURCE APERTURE AREA $\pi/4$ d^2, cm^2, TIMES THE STERADIAN PORTION OF THE EMITTED RADIATION INTERCEPTED BY THE COLLECTING OPTICS - APPROXIMATELY A/F^2, STER.

$$P \approx N \left(\frac{\pi}{4} d^2 \right) \left(\frac{A}{F^2} \right) \approx K \frac{d^2}{F^2} A$$

THE ILLUMINATED AREA, I, IS EQUAL TO THE OPTIC AREA, A, TIMES A FACTOR G TO ACCOUNT FOR THE ANGLE OF INCIDENCE AND BEAM SPREAD: I = GA

$$\text{THE IRRADIANCE OF THE SAMPLE, H} = \frac{\text{BEAM POWER, P}}{\text{ILLUMINATED AREA, I}} \approx \left(\frac{K d^2 A}{F^2} \right) \left(\frac{1}{GA} \right) = \frac{K}{G} \left(\frac{d^2}{F^2} \right)$$

Figure 8.   Illuminating Beam Spread vs. Beam Power Density and Irradiance.

Helium-neon laser optics:
scattered light measurements and process control

B.E. Perilloux

Coherent, Inc., Auburn Group
2301 Lindbergh Street, Auburn, California 95603

ABSTRACT

Output coupler (OC) and high reflector (HR) thin-film coatings and substrates, that are employed in 632.8nm helium-neon (HeNe) lasers, are investigated for optical scatter. The measurement of scatter in this paper is in terms of bidirectional reflectance distribution function and calculated total integrated scatter, or BRDF and CTIS, respectively. Laser output power will be briefly reviewed as a function of total scatter loss from the OC and HR. Increasing amounts of loss (scatter), in individual sets of OC's and HR's, reduces the output power of each laser tube from a theoretical optimum output, for the same amount of loss. Sample optics were measured for BRDF and compared with visual microscope inspection. Statistical analysis of many thousands of coated optics provides insight to controlling causes of scatter in the manufacturing process. Here, the average and standard deviation of BRDF scatter data for many optics provide important data for process control. Various scatter data is presented from OC and HR production runs.

1   Introduction

Red HeNe lasers are employed in a large variety of optical systems such as bar code scanners, interferometers and hand-held pointers. Each application requires a different performance from the HeNe laser. Potential scatter loss from HR's and OC's significantly affects the output laser power, as well as other laser performance attributes.

Our measurement of optical scatter was previously accomplished by visual inspection with a microscope. In this paper, the basic theory and measurement of scatter is presented for manufacturing HeNe optics. First, the model of scatter and the units are reviewed, namely, BRDF and CTIS. Laser output power is then investigated as a function of loss. Next, actual BRDF and CTIS measurements are presented for sample optics. Scatter data from manufacturing will show process capability as a function of scatter.

2   BRDF and CTIS Calculations

The measurement of scatter in this paper is based on BRDF and subsequent CTIS values. As was previously noted by others[1], measured BRDF values are not a function of specular irradiance and scattered radiance, due to a gaussian beam profile and a non-isotropic surface. However, the accepted definition of BRDF as[1]

$$BRDF = (P_s/\Omega_s) / (P_i \cos \theta_s) \qquad (1)$$

still provides meaningful comparative measurements. P_s is the scattered power measured at the detector, P_i is the incident power, Ω_s is the solid angle of the detector and θ_s is the scatter angle. These BRDF measurements, either at discrete scatter angles, θ_s, or a range of angles, provides scatter data for optics that can be compared on an absolute basis. Also, because measured BRDF scatter curves from HeNe OC and HR optics are typically smooth and have minimal slope, a single scatter angle can be used to predict scatter.

For further interpretation of absolute scatter, BRDF curves can be employed to approximate CTIS, for a given scatter angle range. In this case the BRDF curve is generated by a scatterometer with either a continuously variable detector range, θ_s, or discrete detectors at different scatter angles. For our scatterometer, with discrete detectors, a cubic spline curve is fitted to the discrete BRDF data. For both cases, the resultant BRDF curve can be integrated over the scan range and multiplied by a constant to give CTIS as

$$CTIS = 1/R_s \int_{0}^{2\pi} \int_{\theta_1}^{\pi/2} BRDF \cos(\theta_s)d\theta_s \sin(\theta_s)d\phi_s \qquad (2)$$

which can be reduced to

$$CTIS = \pi /R_s \int_{\theta_1}^{\pi/2} BRDF \sin(2\theta_s)d\theta_s \qquad (3)$$

where R_s is the specular reflectance. Note that in Equation (3), BRDF is not cosine corrected, where the $\cos \theta_s$ term would have been dropped. Realizing that this definition of CTIS has specific limitations, it should be described as CTIS for a given scatter range, based on discrete or continuous BRDF data.

For the purposes of this paper, continuous BRDF curves are used to describe scatter from sample optics, along with corresponding CTIS values. For statistical analysis of production runs, one discrete scatter angle, or a single BRDF value, is used to predict scatter. For the theoretical laser output power in the next section, the terms CTIS and total scatter loss are assumed to be interchangeable.

3 Laser Performance and Scatter Loss

The optical gain of a HeNe laser, the beam mode and output power of the laser are in general, affected by the two cavity mirror surfaces of the HR and OC, and the gain medium or HeNe gas tube. Assuming an ideal or constant gain medium, the specular reflectivity of the HR and OR determine the cavity gain and output power. Conversely, optical losses from the HR and OC optics reduce cavity gain and output power. This optical loss occurs mostly from scatter and absorption. The HR may also leak or transmit some of the specular beam, but is not considered here.

As an added requirement for optimum output power, the specular reflectance of the OC must be coupled to the overall gain of the cavity. The equation for laser gain for a CW, low-gain laser is[2]

$$P_0/P_s = T/2 (g/(L+T) -1) \qquad (4)$$

where P_0/P_s is the relative output power, g is the laser gain, L is the total cavity loss and T is the OC transmission. Also, the surface radii of the HR and OC determine the resonator cavity[3], but for the purpose of this paper are not considered.

The optimum transmission is usually based on empirical data and estimated a priori from Equation (5) as[4]

$$T_0 = -L + \sqrt{g\ L} \qquad (5)$$

where L and g are assumed. Because the transmission of an OC is a fixed parameter after manufacturing, the variations in cavity loss (from scatter, for example) for the assumed transmission value will uncouple the transmission and lower the laser output power. Figure 1 shows the graph of relative output power as a function of total cavity loss. The optimum transmission of 1.236% was calculated from Equation (5), with assumed values of g=.05 and L=.01. The second curve in Figure 1 shows the loss of relative output power by not having the output transmission adjusted for the actual cavity loss. Here, lesser or greater amounts of loss will cause small reductions in relative output power. This graph illustrates two points: lowering the cavity loss (less scatter, e.g.) will provide substantial increases in power; and, cavity losses other than the assumed cavity loss will not significantly affect the optimum transmission or lower the output power.

4 BRDF/CTIS Measurements

Continuous BRDF measurements were taken for six samples of HR's and OC's. The samples chosen were at extremes for scattering surfaces, either very low or high. The samples are labeled as 'scatter' or 'residue' based on microscope inspection only. This BRDF data was measured by TMA Technologies, Inc. on a separate scatterometer as an independent verification of the BRDF data measured with our scatterometer. Using this BRDF data and Equation 2, CTIS was determined for each sample.

Figures 2 and 3 show the BRDF curves for the six samples. The parts were scanned from zero through ninety degrees. Using Equation 3, the CTIS values were determined from 2.0 degree to 90 degrees. Table 1 shows the CTIS values for the six samples.

TABLE 1

Sample	Description	CTIS(%)
#1	HR-clear surface	.0142
#2	HR-high residue	.1630
#3	HR-high scatter	.0320
#4	OC-clear surface	.0156
#5	OC-concentrated residue	.1020
#6	OC high scatter	.2110

For all of the samples measured, the BRDF curves are relatively smooth and have minimal slope. Therefore, as mentioned above, this indicates that scatter measured at a single angle will accurately predict overall scatter performance for similar optics.

Microscope pictures were taken of each of the surfaces of the six sample optics, and used as a visual comparison for the numerical CTIS values. Figure 4 show the surface of sample number 6. The other pictures were partially obscured and are not presented.

5 Manufacturing Process Control of Scatter

Substrates are manufactured with two objectives: minimal scatter (surface damage) and correct radius. After fabrication, multi-layer thin-film coatings are deposited by means of vacuum depositon to produce the desired specular reflectance or transmittance. During the deposition of the dielectric coatings, the process must be precisely controlled in order to produce the desired reflectance or transmittance, and minimize the absorption and scatter losses in the coatings.

After coating, BRDF scatter measurements are made on various optics. Based on this BRDF data, parts are accepted or rejected. The coated optics are measured with a custom scatterometer that was manufactured by TMA Technologies, Inc. To date, this scatterometer has measured many thousands of optics. Previously, individual optics were visually inspected using a microscope.

The scatterometer software also provides the average and standard deviation of the BRDF values, for each coating batch. This information is used to help determine causes of scatter in the manufacturing process.

Table 2 shows the of average and standard deviation of BRDF scatter from a typical lot of HR and OC optics. The BRDF measurements of the optics samples were taken immediately after the coatings were deposited.

TABLE 2

Sample	Avg. BRDF	Std. Dev. BRDF
HR	1.56E-4	1.04E-4
OC	2.41E-4	2.00E-4

In general, large average values of BRDF, for a given coating lot, suggest a process that is not adjusted for optimal performance. For example, a substrate polishing process may not be adjusted to provide the lowest scattering surface; hence, all of the surfaces would exhibit increased scatter.

Large standard deviations in BRDF scatter indicate a process that is out of statistical control limits. This potential random variation indicates the manufacturing process will, in general, have a significant number of highly scattering optics. In this case many of the parts may be rejected for excess scatter. Process improvements require the determination of one or more causes of scatter. This is usually accomplished empirically by adjusting individual process parameters.

6   Summary

The measurement of scatter was reviewed for the definition of BRDF. Also, an explicit expression was presented for calculated total integrated scatter as a function of a BRDF curve.

HeNe laser output power was reviewed as a function of total cavity loss and output coupler transmittance that is not optimally coupled for a given scatter loss.

Optical scatter measurements of HeNe optics has been utilized to measure thousands of optics. Typical samples of various levels of scatter on coated surfaces, and the BRDF curves shows that a single BRDF measurement can be used to predict scatter.

With an objective inspection method for BRDF scatter of HeNe optics, average and standard deviation of BRDF provides data for the control and improvement of the manufacturing process.

7   Acknowledgements

Thanks are due to Dr. John Stover of TMA Technologies, Inc. for helpful discussions and to Kim Woodward and Charles Langhorn, Coherent Auburn Group, for assistance with manuscript preparation and editing, respectively.

8   References

1.   Stover, J., Optical Scattering: Measurements and Techniques, ppg. 16-17, McGraw-Hill, 1990.

2.   Verdeyen, J., Laser Electronics, pg. 193., Prentice-Hall, 1981.

3.   Kogelnik, H. and Li, T., "Laser beams and resonators", Applied Optics, 5, ppg. 1550-1567, 1966.

4.   See Reference 1, pg. 191.

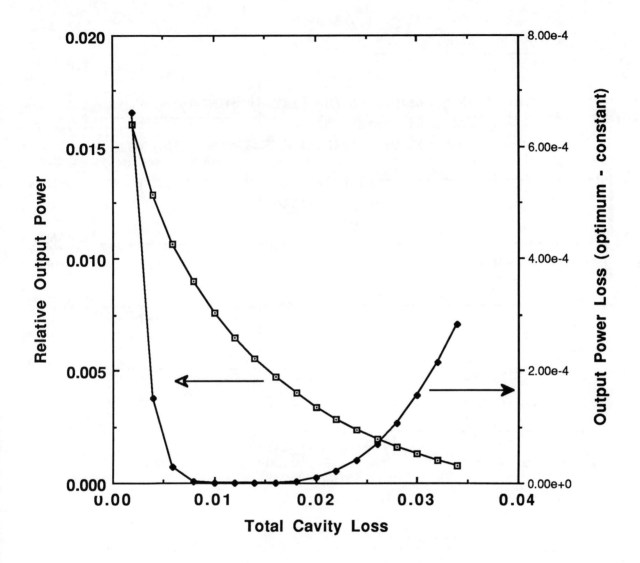

Figure 1 Output power and Power Loss versus
Total Cavity Loss

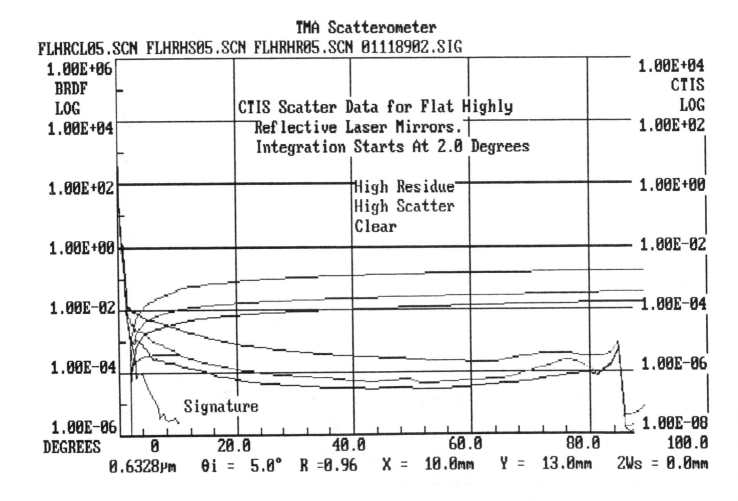

Figure 2. BRDF curves (lower three) for coated HR's with residue, scatter and clear surfaces. The CTIS integration start at 2.0 degrees.

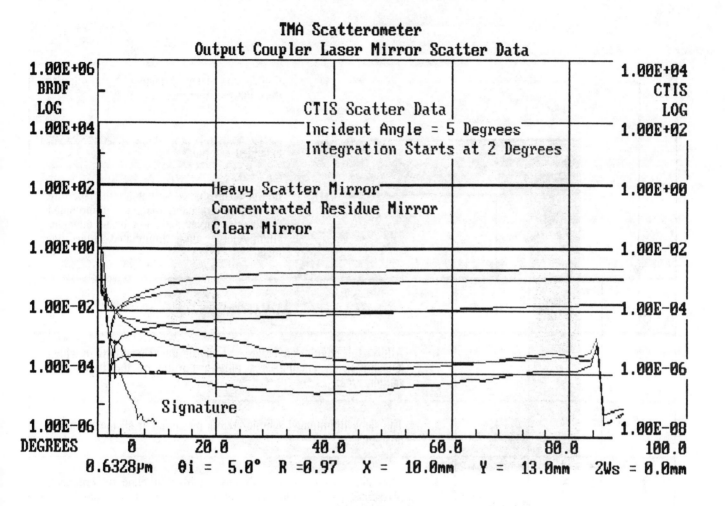

Figure 3. BRDF curves (lower three) for coated OC's with scatter, residue and clear surfaces. The CTIS integration start at 2.0 degrees.

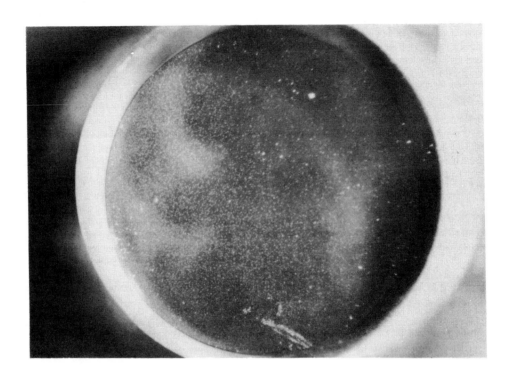

Figure 4. Microscope picture of an output coupler with large amounts of scatter.

Bidirectional reflectance distribution function raster scan technique for curved samples

M. B. McIntosh and J. R. McNeely

Oak Ridge K-25 Site*
Martin Marietta Energy Systems, Inc.
P. O. Box 2003
Oak Ridge, Tennessee 37831-7271

ABSTRACT

A measurement technique has been developed for mapping the bidirectional reflectance distribution function (BRDF) over the entire surface of a curved sample at fixed angles of incidence and scatter. The instrument used was the Toomay, Mathis & Associates, Inc. (TMA) Complete Angle Scatter Instrument at the Optical Characterization Laboratory (OCL) in Oak Ridge, Tennessee. Raster scans and maps of the BRDF of flat samples are relatively straightforward with this instrument, however, similar measurements of curved samples are more complex. The inherent problems involved in performing BRDF mapping on a non-flat optic will be discussed and one solution to those problems will be described for an Optics Manufacturing Operations Development and Integration Laboratories (MODIL) parabolic Assessment Mirror.

2. INTRODUCTION

X-Y raster scatter measurements of figured optics are difficult with some of the existing commercially available scatterometers. This is due to the inability of the scatterometer to correct for the continuously changing angle of incidence (θi) of the incoming laser beam as the optic is X-Y raster scanned. This results in difficulties in interpreting the raster scan BRDF data and limits the usefulness of the raster scan.

In this paper we explain one technique for obtaining useful X-Y raster scan data for figured optics. As a demonstration, we used a wide field-of-view off-axis parabola produced for the Optics MODIL at the Oak Ridge National Laboratory (ORNL). This mirror was 88 cm in width and 198 cm in length. It was manufactured by United Technologies Optical Systems out of Ceraform™ and coated with aluminum. The mirror was figured to have a focal length of 300 mm. The uniformity of the surface finish was a key evaluation. A raster scan scatter map allows for an assessment of the finish uniformity. Therefore, the following technique was developed to demonstrate the ability of the instrument to make such a measurement, for a curved sample.

3. DESCRIPTION OF TECHNIQUE

3.1 Scatterometer

The instrument used was the TMA Complete Angle Scan Instrument (CASI™) developed for the OCL at ORNL. The measurement was done at a wavelength of 10.6 microns with a CO_2 laser. The instrument is uniquely suited for advanced scatter measurement techniques due to its ability to be easily modified. The flexibility of the TMA Scatter

---

*The Oak Ridge K-25 Site is managed by Martin Marietta Energy Systems, Inc. for the U.S. Department of Energy under contract no. DE-AC05-84OR21400.

analysis software also makes this possible due to a built-in ability to halt the program temporarily. With the scatterometer at the OCL, a BRDF raster scan is performed by moving the optic along the X and Y axes defined in Figure 1.[1] This exposes the entire area of interest of the optic surface to the incident laser beam. The raster scan is made up of data is taken at discrete points determined by the step size used.

3.2 Modified BRDF raster scan technique

The key to this measurement is to be able to correct for any changes in the incidence angle caused by the rastering of the figured optic through the incident beam. This was accomplished by momentarily halting the raster scan program and readjusting the incidence angle. A target was positioned at the location where the reflected specular beam would terminate when the desired angle of incidence was achieved. A point of the raster scan was measured by the instrument. The computer would then initiated a movement to the next raster point to be measured. While this was under way a pause command was given to stop the data acquisition process as the optic was rastered to a new measurement position. It was necessary to block the 10.6 micron laser beam while the optic was being rastered to prevent damage to the detector. After the computer repositioned the optic, the data acquisition program halted and an operator manually readjusted the incidence angle. This was done by adjusting the Z axis, the tilt and the rotation of the optic as shown in Figure 1. The Z axis was adjusted first. This was done by rotating the optic around the center of rotation of the detector arm assembly. When the Z axis was properly adjusted, the front surface of the mirror was directly over the axis of rotation of the detector arm. This position could be located within an acceptable precision by observing the spot of the 0.6328 HeNe alignment laser on the optic. If the Z axis was adjusted correctly the spot remained at one position on the mirror as the optic was rotated. If not adjusted correctly the laser spot moved over the mirror surface as the optic is rotated. The tilt and rotation were adjusted next by reflecting the beam into the center of the target stationed at the appropriate location to provide the desired incidence angle.

For measurement of the Optics MODIL Assessment mirror, a large 10 mm spot size was used with a step size of 10 mm. This caused some loss of detail on the uniformity of the mirror surface but allowed for a reasonable assessment of the optic and reduced the amount of manual operator adjustment. The Detector assembly was positioned at 5° from the specular reflection with an incidence angle of 10°. Two raster scans were required to completely scan the clear aperture. For each scan there were 77 required readjustments of the incidence angle. The large spot at the sample ensured that the spot size at the receiver aperture was small enough to allow for near angle measurements. Normal TMA raster scan procedures were used to setup and initiate the data acquisition program. It should be noted that it was necessary to measure the power of the beam incident on the optic at a focus which will ensure all the light from the beam was falling onto the detector.

4. EXPERIMENTAL RESULTS

4.1 BRDF measurements

The two plots in Figures 2 and 3 show data produced by the using the modified X-Y raster technique. Figure 2 represents the left half of the Optics MODIL mirror front surface and Figure 3 represents the right. When these were compared to other partial scans performed on the mirror using the standard X-Y raster scan technique (Figures 4, 5, and 6), it could be seen that there was sharp discontinuities in BRDF values at the location where the partial scans overlap. This was caused by the slowly changing incidence angle as the optic was scanned. As the incidence angle changed, the detector position relative to the specular beam changed causing the BRDF discontinuity in the raster scan. This discontinuity was not present when the above technique is used. The modified X-Y raster technique therefore yields a better BRDF map which gives more information on the surface uniformity.

4.2 Technique pitfalls and problems

The modified raster scan technique depends heavily on the ability, patience, and precision of the operator. The most serious pitfall was allowing the beam to be steered into the detector. This was prevented by practice scans with the alignment laser and careful attention to the operation at hand. A potential source of errors was misalignment of the incidence angle. In the above measurements it was found that the error in the angle of incidence was no more than

$\pm 0.1°$. This can be improved in many ways. One way to improve the accuracy of the incidence angle adjustment would be to use a detector of some kind which would be positioned to give a maximum reading at the desired incidence angle. The error in the detector sweep radius due to Z axis misalignment was no more than $\pm 0.1\%$ if one assumes a possible error of ± 0.5 mm in adjustment.

4.3 Technique improvement

Elimination of operator interactions by totally automating the modified X-Y raster scan technique for curved samples would greatly improve the technique. Automation would also make smaller step sizes more practical. One way to do this would be by using the prescription for the optic and a large aperture detector. The computer could use the prescription of the optic to grossly adjust the incidence angle and then fine tune the adjustment by use of the detector. The Z axis adjustment could be made totally by prescription because small errors in this parameter do not lead to large errors in the BRDF measurement.

5. CONCLUSIONS

The ability to determine the uniformity of an optical surface by BRDF raster scanning is extremely useful. The technique outlined above extends this ability to figured optics. Though there are many pitfalls to making this measurement, it is possible to do so with a minimum of operator induced errors. Hopefully this technique, or variations of it, will be useful with most commercially available scatterometers. The technique can be automated and improved substantially.

6. ACKNOWLEDGEMENTS

Work at the OCL in Oak Ridge, Tennessee is sponsored by the Optics MODIL, Oak Ridge, Tennessee. The authors would especially like to thank U. C. Fulmer for his assistance in making BRDF measurements.

7. REFERENCES

1. D. R. Cheever, F. M. Cady, K. A. Klicker, and J. C. Stover, "Design Review of a Unique Complete Angle Scatter Instrument (CASI)," *Proc. SPIE*, Vol. 818, p. 13 (1987).

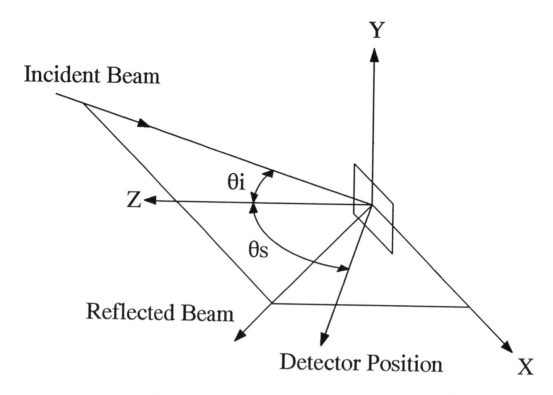

Fig. 1. Definition of the X, Y, Z axes and associated angles.[1]

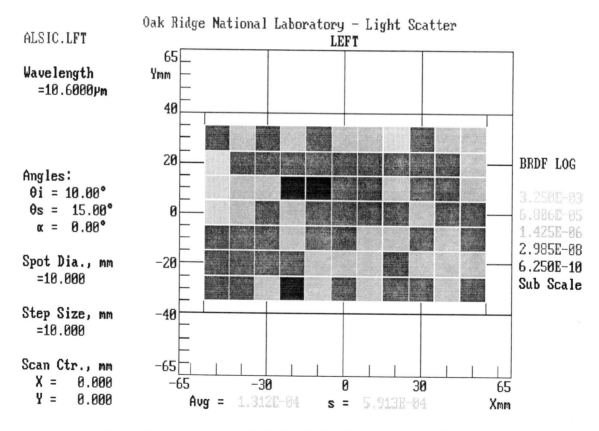

Fig. 2. Raster scan of the left side of the aluminum-coated Ceraform™
Assessment Mirror using the modified raster technique.

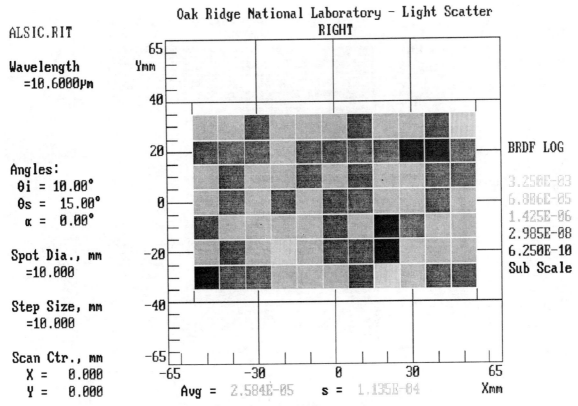

ALSIC.RIT

Oak Ridge National Laboratory - Light Scatter
RIGHT

Wavelength
=10.6000μm

Angles:
 θi = 10.00°
 θs = 15.00°
 α = 0.00°

Spot Dia., mm
=10.000

Step Size, mm
=10.000

Scan Ctr., mm
 X = 0.000
 Y = 0.000

BRDF LOG

3.250E-03
6.806E-05
1.425E-06
2.985E-08
6.250E-10
Sub Scale

Avg = 2.584E-05   s = 1.135E-04

Fig. 3. Raster scan of the right side of the aluminum-coated Ceraform™
Assessment Mirror using the modified raster technique.

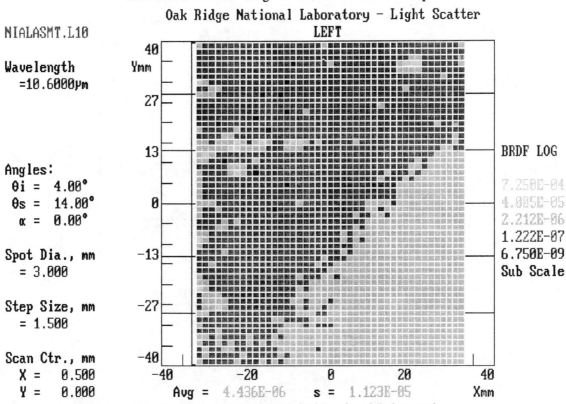

NIALASMT.L10

Oak Ridge National Laboratory - Light Scatter
LEFT

Wavelength
=10.6000μm

Angles:
 θi = 4.00°
 θs = 14.00°
 α = 0.00°

Spot Dia., mm
= 3.000

Step Size, mm
= 1.500

Scan Ctr., mm
 X = 0.500
 Y = 0.000

BRDF LOG

7.250E-04
4.085E-05
2.212E-06
1.222E-07
6.750E-09
Sub Scale

Avg = 4.436E-06   s = 1.123E-05

Fig. 4. Raster scan of the left side of a nickel-coated
aluminum Assessment Mirror using the standard raster technique.

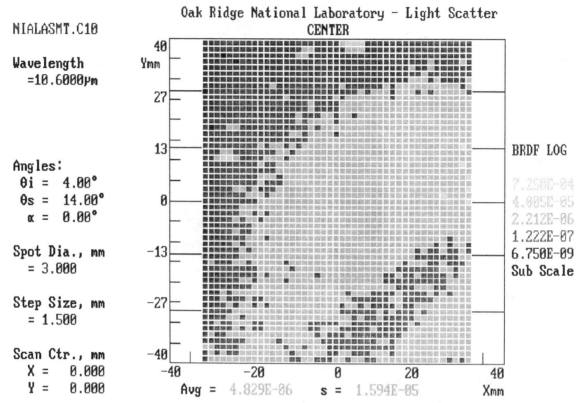

Fig. 5. Raster scan of the center of a nickel-coated
aluminum Assessment Mirror using the standard raster technique.

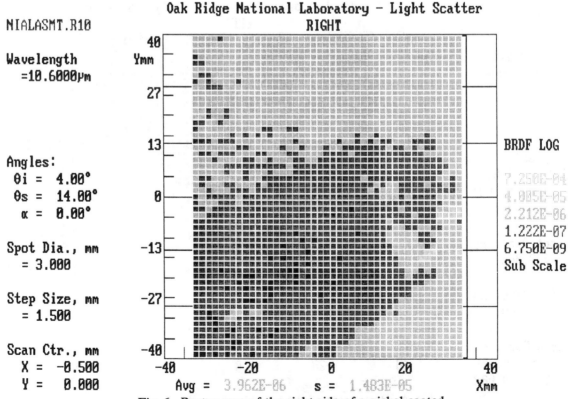

Fig. 6. Raster scan of the right side of a nickel-coated
aluminum Assessment Mirror using the standard raster technique.

Description and calibration of a fully automated infrared scatterometer

Stéphane Mainguy, Michel Olivier, Michel Josse,and Michel Guidon.

Commissariat à l'Energie Atomique
Centre d'Etudes Scientifiques et Techniques d'Aquitaine
Département Technique
Service Physique Expérimentale
Section Interaction Rayonnements Matériaux
B.P. n°2; 33114 LE BARP/FRANCE

ABSTRACT

A fully automated scatterometer, designed for BRDF measurements in the IR at about 10 μm, is described. Basically it works around a reflecting parabola (464 mm diameter, f/0.25) and permits measurements in and out of the plane of incidence. Optical properties of the parabolic mirror are emphasized by a ray-tracing technique which permits to determine correct illumination on the sample and detection conditions of scattered light. Advantages and drawbacks of such an instrument are discussed, as well as calibration procedures. As a conclusion, we present experimental results to illustrate the instrument capabilities.

INTRODUCTION

Bidirectional reflectivity measurements in the mid-infrared region ($\lambda \approx 10$ μm) are interesting for several reasons:

- first, they allow CO_2 laser optical components to be characterized from the point of view of scattering specifications[1,2].

- second, it is possible with this technique to study interaction between an infrared optical wave and a rough metallic or dielectric surface at frequencies where surface electromagnetic waves can be excited[3],

- third, bidirectional reflectivity measurement at wavelengths around 10 μm is a convenient way to obtain information about radiative properties of solid surfaces at room temperature. This last point is our main objective at the moment.

The relation between reflective and emissive properties is expressed by Kirchhoff's second law which states that for opaque materials the sum of the directional hemispheric reflectivity (i.e. the integral over an hemisphere of the cosine corrected BRDF) plus the directional emissivity equals unity[4]. Emissivity values for opaque materials near room temperature are often difficult to obtain, therefore, directional reflectivity measurements offer an alternative to perform this task.

Since we are studying very rough materials, light scattering occurs in the whole hemisphere above the sample. Consequently, to obtain reliable results it is necessary to measure scattered light outside the plane of incidence. It is well-known that out-of-plane measurement are rather tedious when performed with a conventional scatterometer[1]. At the lab we have designed and built a scatterometer which allows BRDF measurements in a very simple and straightforward way in and out of the plane of incidence. This scatterometer is an improvement of the Brimmer, Griffiths and Harrick design[5].

In this paper a presentation of the instrument is given. The following points will be emphasized:
1. principle of the scatterometer,
2. description of the instrument,
3. presentation of the parabolic mirror properties by a ray-tracing technique
4. advantages and limitations of the scatterometer
5. calibration and measurement procedures
6. experimental results

1. PRINCIPLE OF THE SCATTEROMETER

Basically the scatterometer uses a large parabolic mirror to steer the incoming beam on the sample at focus and to collect scattered light (fig.1).

A laser beam (another light source could be used) is directed onto the parabola , parallely to its axis, by means of a movable flat mirror (M1) oriented at 45° angle. After reflection on the parabola the beam hits the sample at an incidence angle which depends on the position of the moving mirror M1. By rotating the sample about the parabola axis various azimuthal angles can be achieved. Light scattered by the sample is collected in a similar way by means of another movable flat mirror (M2) oriented at 45° which can also move on a rotating stage about the parabola axis. Thus, in principle, any direction of scattering in the hemisphere above the sample can be studied. See figure 1 for angular notations.

It must be mentioned here that the idea of using a parabolic mirror for directional reflectivity experimental studies has been already proposed elsewhere. Brimmer and coworkers[5] performed diffuse reflectance FTIR spectroscopy on solid samples. More recently, in a FTIR spectrometer, Forrister and al.[6] measured directional reflectance of textured surfaces by using an accessory based on a parabolic mirror.

In these works no attempt has been made to study in details the optical properties of the parabolic mirror. Since this is important in our case, using a laser beam as a light source, we decided to study the properties of our set-up in order to perform reliable BRDF measurements. This point will be discussed in paragraph 3.

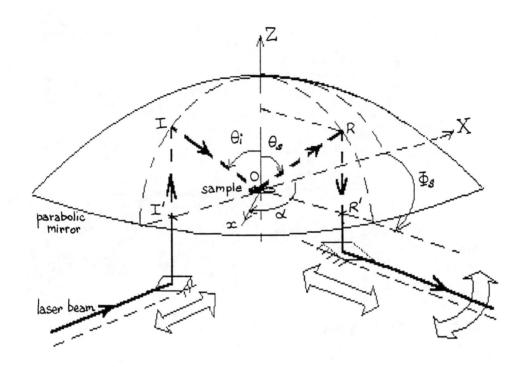

FIG.1 – PRINCIPLE OF THE SCATTEROMETER

ANGULAR NOTATIONS - in accordance with STOVER

    OX : horizontal reference axis
    OZ : vertical reference axis
    Ox : axis bound to the sample
      (axis of the parabolic mirror - normal to the sample)

$\theta i=(\overrightarrow{OZ},\overrightarrow{OI})$; $\theta s=(\overrightarrow{OZ},\overrightarrow{OR})$;
$\Phi i=(\overrightarrow{OI'},\overrightarrow{OX})=180$ deg ; $\Phi s=(\overrightarrow{OX},\overrightarrow{OR'})$

$\alpha=(\overrightarrow{OX},\overrightarrow{Ox})$: angular position of the sample

2.DESCRIPTION OF THE INSTRUMENT

Schematic views of the set-up are shown on fig. 2 and fig.3.

The light source is a waveguide CO_2 laser which offers a choice of 60 lines in the 9 to 11 μm range. Until now, only the main line at 10.6 μm has been used. It offers an output power of 4 W. A mechanical chopper rotating at about 220 Hz is placed at the output of the laser.

After the chopper there is a beam expander which allows, in connection with a focusing lens L1, to adjust the beam diameter and its divergence (see paragraph 3).

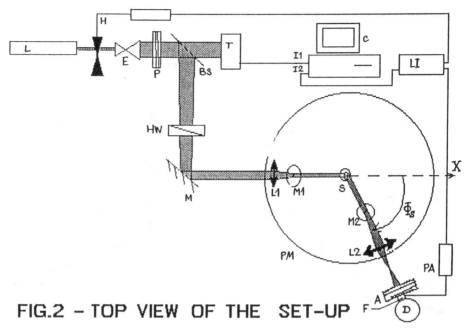

FIG.2 – TOP VIEW OF THE SET-UP

L : CO_2-laser ; H : chopper ; E : beam expander ;
P : polarizer; BS : beam splitter ; T : control thermopile ;
HW : half-wave plate; M : alignment mirror ; L1 : input lens ;
M1 : input mirror; PM : parabolic mirror ; S : sample ;
M2 : output mirror ; L2 : output lens ; A : analyzer ;
F : interferential filter at 10.6 μm ; D : MCT detector ;
PA : pre-amplifier ; LI : lock-in amplifier; I1 : control
signal input ; I2 : scatter signal input ; C : computer ;

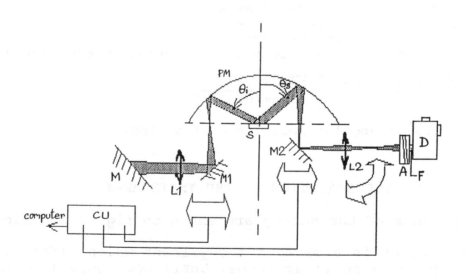

FIG.3 – FRONT VIEW OF THE SET-UP

M : alignment mirror ; L1 : input lens ; M1 : input mirror;
PM : parabolic mirror ; S : sample ; M2 : output mirror ;
L2 : output lens ; F : interferential filter at 10.6 μm ;
A : analyzer ; D : MCT detector ; CU : stepper motors control
unit ;

Then the beam goes through a germanium grid polarizer in order
to get a 99% vertical polarization. A ZnSe beam splitter oriented
at 45° and designed for vertical polarization shares the incoming
beam in two parts. A thermopile checks the laser stability on the
transmitted beam. The reflected beam goes through a half-wave plate
which determines the orientation of the beam polarization when it
reaches the sample.

Then, after reflection off an alignment metallic mirror, the
beam enters the main part of the scatterometer, i.e. the parabolic
mirror. It is made of solid aluminium and has been carefully
polished. The focal length is 116 mm, and opened outside 464 mm
(f/0.25). Before being reflected by the mobile input mirror M1, the
beam is focused by lens L1. L1 and M1 move together by means of
stepper-motor driven translation stage (travel: 240 mm). Each
position of this stage corresponds to a well-known incidence angle
Θ_i.

The sample is located on a stage with its upper surface at the
focus of the parabolic mirror. Sample height adjustments, up and
down translations and rotation about vertical axis are permitted.

Scattered light in a given direction is reflected off downwards
by the parabolic mirror and then is collected by a mobile optical
bench which includes the output mirror M2, an imaging lens L2, a
germanium grid analyser, an interference filter at 10.6 μm and a MCT
detector. The sensitive area (2 mm in diameter) of the detector is
situated in a plane which is the conjugate of the focal plane of
the parabolic mirror through lens L2.

This movable bench can be translated back and forth and rotated
about the parabola axis in order to collect light scattered at
different angles Θ_s and Φ_s. The angular range covered by the
scatterometer is summerized in table 1.

angle	min.	max.
Θ_i	5°	85°
Φ_i	180°	
Θ_s	5°	85°
Φ_s	-90°	160°
α	0°	360°

Table 1 - Angular range covered by the scatterometer.

The MCT output signal is processed with a lock-in amplifier in order to get a DC signal proportional to the scattered flux received by the detector. This signal together with the signal coming from the thermopile enters the computer via an A/D converter. In addition to data acquisition, the computer is used to drive the stepper-motors, to process the data, to display and store the results.

So in its present form our scatterometer is a fully automatic instrument that permits to make BRDF measurements.

3. PRESENTATION OF THE OPTICAL PROPERTIES OF THE PARABOLIC MIRROR BY RAY-TRACING TECHNIQUE

Reliable BRDF measurements need several experimental conditions:

- the sample must be illuminated by a parallel or slow diverging beam in order to make BRDF measurements at a definite incidence angle,

- the spot on the sample must be large enough to average the local surface inhomogeneities,

- on the detector, clear discrimination between different scattered directions must be achieved.

These conditions are easily fullfilled on simple goniometers set ups[1,7]. This is not the case when using a parabolic mirror. The main reasons are:

- after reflection on the parabola, a quasi parallel incoming beam is focused on the sample. Thus no definite incidence angle can be settled;

- focusing on the sample does not ensure the required averaging condition of a reasonably large spot;

- light scattered by the sample at a given angle is spread in a wide angular range after reflection on the parabola;

- solid detection angle depends on the parabola region used to collect scattered light.

However, it is possible to overcome these difficulties by two different ways:
   1 - by focusing the incoming beam properly,
   2 - by collecting the exit light with a field lens.

3.1 Sample illumination

Let's consider a slow diverging beam centered on vertical axis parallel to the paraboloid axis (see fig.4).

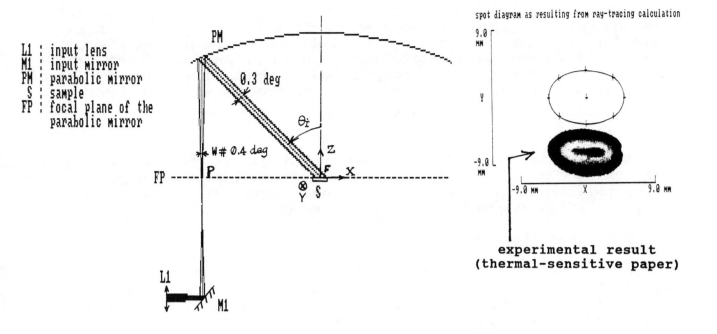

L1 : input lens
M1 : input mirror
PM : parabolic mirror
S : sample
FP : focal plane of the
     parabolic mirror

**experimental result
(thermal-sensitive paper)**

FIG.4 - CONDITIONS OF ILLUMINATION AT θ_i= 45 DEG (RAY-TRACING RESULTS)

This beam is focused at P on the focal plane of the parabola and its divergence w is controled by a beam expander at the laser output.

First, if the distance between P and the parabola focus F is small, the beam can be reflected in a parallel beam directed to the focal plane. The intersection will be a cicular spot since the angular incidence is small and paraxial approximation can be applied. Spot diameter d is :

$$d = f.w \qquad\qquad (1)$$

with f : focal length of the parabola
     w : beam divergence.

Second, if the distance between P and F is such that gaussian approximation is no longer valid, reflected rays off the parabola reach the focal plane of the parabola at high angle of incidence. Therefore the intersection between the beam and the sample becomes an elliptic spot.

A ray-tracing technique has been used in order to know precizely the size of the spot on the sample. It turns out that, for every position of P in the focal plane between F and parabola edge, a parallel beam can be directed to the sample, and spot size can be calculated. When P moves away from F, spot shape becomes more elliptic on the sample and input beam divergence must be corrected.

As an example, fig.4 presents a ray-tracing calculation for θi=45°.

Experimental tests have been carried out to measure the spot size using thermal sensitive paper and thermoluminescent plates. A very good agreement has been reached, even with small surface irregularities at the surface of the paraboloid.

3.2 Scattered light collection

Let's consider here light scattered in a particular direction θs by every illuminated point of the sample. After reflexion on the parabola, scattered light is concentrated on a small surface in the focal plane. When paraxial approximation is valid (i.e. for small θ_s), this surface is reduced to a secondary focal spot.

Under these conditions, the parabola works as a Fourier optic and a far field light distribution is obtained in its focal plane, which represents the BRDF of the sample.

A scattered beam off the sample at a particular angle is related to a small surface in the focal plane of the parabola. Using the same ray-tracing program as above, the shape and size of these surfaces can be calculated for different scatter angles and assuming an illuminated spot on the sample as shown on fig.4 (see fig.5).

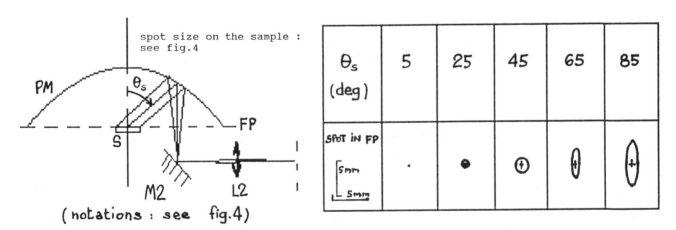

θ_s (deg)	5	25	45	65	85
SPOT IN FP ⌐5mm ⌞5mm	.	●	⊕	⦶	⦶

FIG.5 – SCATTERED LIGHT COLLECTION BY THE PARABOLIC MIRROR

In order to measure light scattered at a selected angle, it is necessary to collect all the flux in the corresponding focal plane surface. Lens L2 is used to relay that surface onto the detector sensitive area. The distance between L2 and parabola focal plane remains constant as well as the distance between L2 and detector when M2 moves to collect light at a different scatter angle.

According to the sample under test, lens L2 can be replaced with a different lens (having a longer or shorter focal length). This modifies the magnification of the image on the detector and therefore the angular resolution is changed. In any case, it is important that the image of the spot on the detector is smaller than its sensitive surface. For that reason large detector was chosen for the experiment (sensitive size diameter 2 mm).

The solid angle of detection can be defined as $\triangle \Omega_s$:

$$\triangle \Omega_s = \frac{\pi . D^2 . \sin^4 \theta_s}{16 \ f^2 . (1 - \cos \theta_s)^2} \tag{2}$$

θ_s : scatter angle ,
f : focal length of the parabola ,
D : diameter of the image of the detector sensitive area through L2 in the focal plane of the parabola.

4. ADVANTAGES AND LIMITATIONS OF THE SCATTEROMETER

This experimental set up was chosen because its implementation is relatively simple and it can be operated automatically under computer control.

Moreover, the sample is held in a fixed position, with respect to the incident beam, independently of the scattered light measurements. To our point of view, this is one of the main advantages of that experiment, as opposed to other set ups where the sample must be moved at different angles during data acquisition. This condition is favorable in our case when very inhomogeneous or anisotropic materials are considered.

In addition, the steadiness of the sample permits to implement complementary experiments in parallel to BRDF measurements (electrical or magnetical fields applied to the sample, temperature variation,...).

And finally, light scattering study of liquids at their surface is permitted with this set up because the surface under study remains horizontal during the experiment.

However, this design presents a couple of drawbacks:

- there is a considerable number of reflections for both the incident and the scattered beam between laser output and detector (45° mirrors and parabola). These miscellaneous mirrors together with other optical supports and mechanical mounts situated below the parabola give rise to stray light that may reach the detector and contribute to undesired signal. These effects have been reduced as much as possible. High quality mirrors have been selected, every mechanical support was black anodized and a reduced field of view is used on the detector;

- the reflection coefficients of the incoming beam and outgoing scattered light onto the parabola vary according to the different positions of the mirrors M1 and M2. The valuation of these reflections must be known accurately for each incident/reflected angle and each beam polarization used. The ratio between the incoming and outgoing light after reflections on different mirrors and parabola and considering a well-known sample is the so-called transfer function of the system. We calculated this ratio and obtained a good agreement with experimental measurements (see paragrah 5).

5. CALIBRATION AND MEASUREMENT PROCEDURES OF THE SCATTEROMETER

The calibration of the scatterometer requires the measurement of the incident power, an absolute calibration of the detectors and the evaluation of the transfer function of the scatterometer.

To understand these different operations for calibration, we shall focus our attention on the theory of the scatterometer.

The BRDF formula (in accordance with Stover[1]) is :

$$BRDF = \frac{P_s}{P_i \cdot \cos\theta_s \cdot \Delta\Omega_s} \tag{3}$$

θ_s : scatter angle,
P_i : incident power,
P_s : scattered power in the θ_s, Φ_s direction,
$\Delta\Omega_s$: solid angle of detection.

The scatterometer measurement uses a derivative expression which can be formally written as :

$$BRDF = \frac{1}{A \cdot B \cdot K \cdot \cos\theta_s \cdot \Delta\Omega_s} \cdot \frac{V^{out}_{LI}}{V^{out}_T} \tag{4}$$

with

A : calibration factor which relates the output power (i.e. measured after L2) to the lock-in output signal V^{out}_{LI} ,

B : transfer function of the scatterometer,

K : calibration factor which gives the input power (i.e. measured before L1) knowing the thermopile output signal V^{out}_{T} .

Coefficients A and K are obtained from preliminary experiments using a well calibrated power meter[8] (K depends on the polarization due to transmission through the half-wave plate).

Coefficient B is the product of the amplitudes of the two Fresnel reflections on the aluminium parabolic mirror and also multiplied by the two Fresnel reflection amplitudes of golden mirrors M1 and M2, and by transmission coefficients of lenses L1 and L2. B is named the "transfer function of the scatterometer". It is calculated using Fresnel formula and complex index of refraction for aluminium. A series of tests has been carried out, with the scatterometer working in the plane of incidence and considering a well-known silicon sample at various angles. Very good agreement has been obtained between theory and experiment.

Currently, the accuracy of our BRDF measurements is estimated around 10%, but we intend to improve it after a better evaluation of the calibration coefficients.

6. EXPERIMENTAL RESULTS

In this section we present experimental results of BRDF measurements obtained with our scatterometer.

We used a ZnSe lens L2 of focal length 127 mm, located at twice its focal length from the focal plane of the parabolic mirror and from the detection plane, so that: D=2 mm. According to (2), the solid angle of detection goes from 2.10^{-4} sr at $\theta_s=5°$, to 7.10^{-5} sr at $\theta_s=85°$.

Figure 6 shows a BRDF diagram in the incident plane ($\Phi_s=0°$) for an aluminium sample with very diffuse reflexion at 10.6 μm considering a s-polarized incident light. This diagram uses polar coordinates (θs, BRDF). Data were taken for each degree between 5 and 85 degrees.

Figure 7 presents three BRDF diagrams of a highly absorbing and rough dielectric material (r.m.s. roughness # 8 μm), for s-polarized incident light, in planes $\Phi_s=-5°$, $\Phi_s=0°$ (incident plane) and $\Phi_s=5°$. The scatterometer permits to measure a weak specular peak superimposed on a diffuse reflexion lobe.

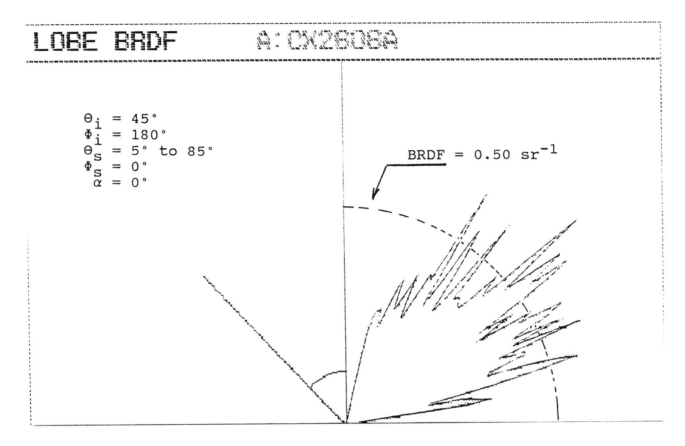

FIG. **6** - BRDF DIAGRAM FOR A ROUGH ALUMINIUM SAMPLE
(incident plane)

CONCLUSION

This scatterometer based on an original design has just started to give experimental results.

BRDF measurements on rough samples have given reliable data.

Calibration tests will be improved in the near future in order to get more accurate BRDF data.

ACKNOWLEDGEMENTS

The authors would like to thank Dr. J.-J. Greffet (Laboratoire EM2C du CNRS et de l'Ecole Centrale Paris) for useful discussions and assistance, Mr. B. Horbette (CEA - CESTA) for his constant support and Mr. J.-M. Lubicz for help in the ray-tracing calculations.

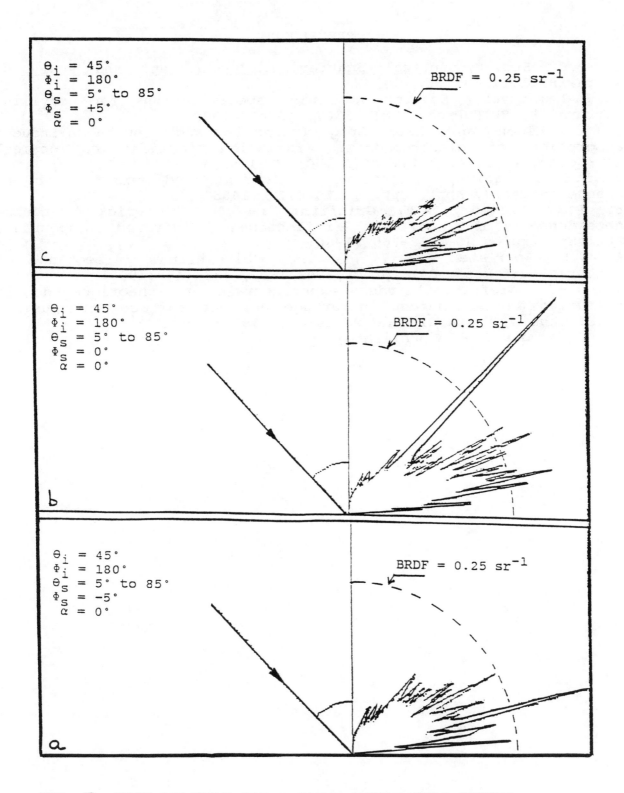

FIG. 7 BRDF DIAGRAMS FOR A ROUGH DIELECTRIC SAMPLE

a : Φ_S = -5°
b : Φ_S = 0° (incident plane)
c : Φ_S = +5°

REFERENCES

1. J.C. Stover, <u>Optical Scattering: Measurement and Analysis</u>, McGraw-Hill (1990)

2. J.C. Stover, "Scatter or finish specifications for beryllium mirrors ?", SPIE Proc., Vol. 1307, 40-45 (1990)

3. J.M.Elson and C.C. Sung,"Intrinsic and roughness-induced absorption of electromagnetic radiation incident on optical surfaces", Appl. Opt. <u>21</u>, 1496-1501 (1982).

4. F.E. Nicodemus, "Directional reflectance and emissivity of an opaque surface", Appl. Opt. <u>4</u>, 767-773 (1965)

5. P.J. Brimmer, P.R. Griffiths and N.J. Harrick, "Angular dependence of diffuse reflectance infrared spectra", Appl.Spectrosc. <u>40</u>, 258-265 (1986)

6. W.B. Forrister and T.L. Starr, "Directional reflectance of textured surfaces", SPIE proc., Vol.1307, 438-445 (1990)

7. J.-J. Greffet, "Etudes experimentale et théorique de la diffusion du rayonnement infrarouge par des surfaces rugueuses", Ph.D. Thesis, Université de Paris-Sud (1988)

8. S. Mainguy (to be published)

Determination of thin film roughness and volume structure parameters from light scattering investigations

Angela Duparré and Samer Kassam

Friedrich Schiller University Jena, Physics Department
Max-Wien-Platz 1, O-6900 Jena, Federal Republic of Germany

ABSTRACT

Light scattering measurements can be used for determining roughness as well as volume structure parameters of optical thin films. A method of structural analysis is outlined, which allows a quantitative estimation of roughness parameters, mean columnar diameter, packing density, and the evolutionary exponent, respectively.

1. MOTIVATION

Since high quality optical coatings spread a wide range of applications, light scattering from thin films mostly has been considered as a source of losses limiting the performance of optical devices. On the other hand, light scattering can be used as a sensitive tool for a non-contact and non-destructive analysis of optical thin film morphology. A number of publications in the last decade witness the growing interest in this aspect. Until now most of the scattering theories and measurements have been confined to roughness scattering[1-3], whereas a treatment of volume scattering can hardly be found[4,5].

For films with a columnar structure, which is very frequently met in optical thin film coatings, a method of quantitative determination of both roughness and volume structure parameters is presented.

2. STRATEGY

In order to derive the morphology parameters from scattering measurement data, it becomes necessary to separate roughness scattering from scattering caused by volume inhomogeneities. Fig. 1 demonstrates that pure surface roughness scattering can be obtained by evaporating an additional thin aluminium film which preserves the surface profile[6]. If, on the other hand, the columnar structured film is deposited on a highly reflecting substrate, pure volume scattering measurement can be achieved by choosing the appropriate optical film thickness (multiples of $\lambda/2$ for normal incidence, λ is the wavelength)[7]. This configuration ensures that the electrical field strength at the interfaces nearly vanishes, and, thus, interface scattering is suppressed. After a successful separation of roughness and volume scattering the measured data can easily be fitted to the appropriate theories. In the following two paragraphs we outline how the desired structure parameters can be extracted from ARS (angle resolved scattering) and TIS (total integrated scattering) measurements.

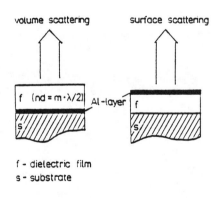

Fig. 1. Configuration for a separation of roughness from volume scattering.

3. ANGLE RESOLVED SCATTERING

3.1 Theory

In the case of a stratified columnar structured medium the expressions for the normalized intensity of surface roughness scattering[1-3] and volume scattering[7] have, in principle, the same form:

$$\frac{1}{P_0}\left\langle \frac{dP}{d\Omega} \right\rangle = F(\Theta_0, \varphi, \Theta, pol., d, \lambda, \varepsilon_i)\, g(\vec{k} - \vec{k}_0)\ , \tag{1}$$

where F is the optical factor which is independent of statistical properties, and which can be calculated from the corresponding theories. The second factor $g(\vec{k} - \vec{k}_0)$ represents the average power spectral density of surface roughness and volume fluctuations, respectively. g is the Fourier transform of the corresponding autocorrelation function $G(\vec{\tau})$ and, hence, contains the structural information of interest.

3.2 Equipment

The ARS-measurements were performed at the He-Ne laser wavelength $\lambda = 632.8$nm using the scatterometer schematically shown in Fig. 2. The goniometer arm automatically rotates about the sample surface. Mostly scattering was measured at normal incidence $\Theta_0 = 0^0$, in each case for s-polarized and p-polarized scattered light.

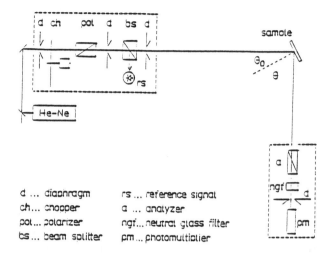

d ... diaphragm   rs ... reference signal
ch... chopper   a ... analyzer
pol... polarizer   ngf...neutral glass filter
bs... beam splitter   pm ...photomultiplier

Fig. 2. Experimental set-up used for ARS-measurements at $\lambda = 632.8$nm.

3.3 Example: Volume scattering from PbF$_2$-films

If the distribution of the volume fluctuations can be assumed to be homogeneous and isotropic in lateral direction, and for normal incidence, $g(\vec{k})$ can easily be derived from the Gaussian formed function $G(\vec{\tau})$:

$$G(\tau) = \xi^2 \exp\left(-\frac{\tau^2}{\tau_c^2}\right) \quad \overset{FT}{\Longleftrightarrow} \quad g(k) = \pi \xi^2 \tau_c^2 \exp\left(-\frac{[k\tau_c]^2}{4}\right), \quad k = \frac{2\pi}{\lambda}\sin\Theta, \quad (2)$$

where Θ is the angle of scattering, and ξ is the rms-fluctuation strength which is related to the packing density[5] p:

$$\xi^2 = (\varepsilon_{void} - \varepsilon)^2\, p\,(1-p). \quad (3)$$

$\varepsilon, \varepsilon_{void}$ are the dielectric constants of the film material and the voids, respectively. The correlation length τ_c represents the mean columnar diameter.

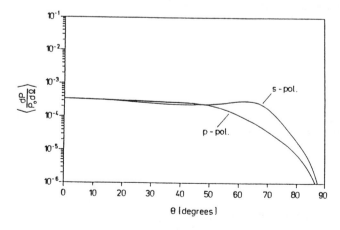

Fig. 3a. Volume scattering of a PbF$_2$-film (optical thickness $nd = 2\lambda$) with arbitrarily chosen statistical parameters at $\lambda = 632.8$nm and normal incidence.

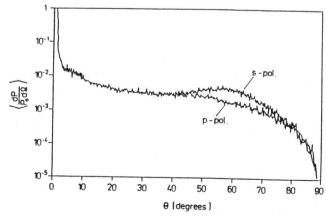

Fig. 3b. Experimental results for the same geometrical layer configuration as in Fig. 3a.

The ARS-experiments have been performed on PbF_2-films exhibiting a pronounced columnar structure. Fig. 3b shows the measured data, Fig. 3a demonstrates the theoretical behaviour[7] of volume scattering, assuming $G(\vec{\tau})$ to be of a Gaussian type with, at first, arbitrary parameters ξ and τ_c.

Owing to $\tau_c/\lambda \ll 1$, the comparison of theory[7] with experiment only yields the product $\tau_c \xi$. But from the estimation of the roughness correlation length, which is assumed to be quite similar to τ_c of the volume, it became possible to determine the packing density of the film as $p \approx 98.5\%$.

4. TOTAL INTEGRATED SCATTERING

4.1 Equipment

The TIS-measurements were made at both the He-Ne laser wavelength $\lambda = 632.8$nm and the He-Cd laser wavelength $\lambda = 325$nm at nearly normal incidence. The backscattered light was collected by a Coblentz sphere within an angular range from 2^0 to 84^0. A schematic illustration is given in Fig. 4.

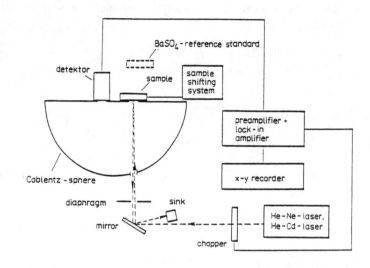

Fig. 4. Experimental set-up used for TIS-measurements at $\lambda = 632.8$nm and $\lambda = 325$nm.

4.2 Theory

In general, TIS-values can be theoretically obtained by integrating Eq.1 over the upper hemisphere. For the limiting cases where the autocorrelation length becomes very large or small as compared to the wavelength, analytic expressions can be given:
Surface scattering[3]:

$$\tau_c \gg \lambda \qquad TIS = \left(\frac{4\pi\sigma}{\lambda}\right)^2 \qquad \tau_c \ll \lambda \qquad TIS = \frac{64}{3}\frac{\pi^4\tau_c^2\sigma^2}{\lambda^4} \ , \qquad (4)$$

where σ is the rms-roughness.
Volume scattering (columnar structure)[7]:

$$\tau_c \ll \lambda \qquad TIS = \tilde{F}_{opt}\frac{\xi^2\tau_c^2}{\lambda^2} \ . \qquad (5)$$

$\tilde{F}_{opt}$ is the optical factor of total volume scattering.

4.3 Example: Surface scattering of PbF_2-films

Total integrated surface scattering was measured on PbF_2-films for different film thicknesses. From Fig. 5 the enhancement of TIS intensity with increasing film thickness becomes obvious. As has been outlined in more detail in Ref. 6, this scattering curve can be directly used for determining the evolution law of the columnar film structure, which means a quantitative description of the development of the mean columnar diameter $\bar{D}$ (assumed to be equal to τ_c) with film thickness[8]:

$$\bar{D} = D_0 \left(\frac{d}{d_0}\right)^\beta , \tag{6}$$

where β is the evolutionary exponent, d is the film thickness, d_0 a reference thickness and D_0 is the mean columnar diameter at thickness d_0. From the scattering data in Fig. 5 the exponent $\beta = 0.53$ was obtained[6].

Another application of TIS-experiments was the independent determination of the rms-roughness σ and the mean columnar diameter $\bar{D}$ by measuring surface scattering at two different wavelengths. From such experiments we obtained for a 1μm thick PbF_2-film:

$$\left.\begin{array}{ll}\lambda = 632.8\,\text{nm} & TIS = 0.54\% \\ \lambda = 325\quad\text{nm} & TIS = 5.2\%\end{array}\right\}\quad\begin{array}{l}\sigma = 9\,\text{nm} \\ \bar{D} = 72\,\text{nm}\end{array}\quad. \tag{7}$$

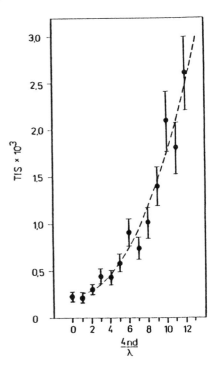

Fig. 5. Enhancement of surface scattering ($\lambda = 632.8$nm) with increasing film thickness.

5. ACKNOWLEDGEMENTS

The authors wish to thank J. Neubert for performing the ARS-measurements and R. Dohle for the evaporation of the PbF_2-films.

6. REFERENCES

1. J. M. Elson, "Infrared light scattering from surfaces covered with multiple dielectric overlayers," Appl. Optics 16, 2872-2881 (1977)

2. P. Bousquet, F. Flory, and P. Roche, "Scattering from multilayer thin films: theory and experiment," J. Opt. Soc. Am. 71, 1115-1123 (1981)

3. J. M. Elson, J. P. Rahn, and J. M. Bennett, "Relationship of the Total Integrated Scattering From Multilayer-Coated Optics to Angle of Incidence, Polarization, Correlation Length, and Roughness Cross Correlation Properties," Appl. Optics 22, 3207-3219 (1983)

4. C. K. Carniglia, "Scalar scattering theory for multilayer optical coatings," Opt. Eng. 18, 104-114 (1979)

5. J. M. Elson, "Theory of light scattering from a rough surface in an inhomogeneous dielectric permittivity," Phys. Rev. B30, 5460-5480 (1984)

6. A. Duparré, R. Dohle, and H. Müller, "Relation between light scattering and morphology of columnar structured optical thin films," J. Mod. Optics 37, 1383-1390 (1990)

7. S. Kassam, A. Duparré, P. Bussemer, K. Hehl, and J. Neubert, "Light Scattering from the Volume of Optical Thin Films: Theory and Experiment," Appl. Optics (1991) (in the press)

8. R. Messier, "Toward quantification of thin film morphology," J. Vac. Sci. Technol. A4, 496-499 (1986)

OPTICAL SCATTER:
APPLICATIONS, MEASUREMENT, AND THEORY

Volume 1530

SESSION 5

Scatter Measurements

Chair
Jeffrey W. Garrett
Lockheed Missiles & Space Company, Inc.

Scatter and contamination of a low-scatter mirror

J. R. McNeely, M. B. McIntosh, and M.A. Akerman

Oak Ridge K-25 Site*
Martin Marietta Energy Systems, Inc.
P. O. Box 2003
Oak Ridge, Tennessee 37831-7271

ABSTRACT

The contamination of high-quality, space-borne optics by particles originating from baffle systems could significantly alter the performance of the optics. To assess this potential problem, the bidirectional reflectance distribution function (BRDF) of a low-scatter beryllium mirror was measured at the 10.6 μm wavelength with the mirror in the "cleaned" state and after controlled contaminations with aluminum oxide powder up to 100 μm in size. The aluminum oxide powder was used to simulate particles which could be released from a typical baffle material. After contamination, the particle size distribution on the mirror surface was statistically sampled using a scanning electron microscope (SEM) image analysis technique. The BRDF measurements of the contaminated mirror were compared to Mie scattering theory calculations for sub-wavelength, wavelength, and super-wavelength particles sizes.

2. INTRODUCTION

The effect of particulate contamination on the optical scatter produced by mirrors has been studied by several authors.[1-5] These studies show that particles on a low-scatter mirror surface can increase the optical scatter orders of magnitude. These results have been modelled using Mie scattering theory with varying degrees of success.[1,2,4,6] Mie scattering theory originally derived for diffraction by a single sphere can apply to diffraction by multiple spheres of the same size and composition. A modified Mie theory developed by Nahm[2,12] works reasonably well at a visible wavelength (0.6328 μm) for polystyrene spheres of uniform size (1 μm) on a mirror surface. In practice, real contamination particles can be irregular in shape and can vary in size from sub-wavelength to many wavelengths. These factors combine to complicate the analysis of scattering from contaminated mirror surfaces.[1]

Baffle materials containing aluminum oxide such as Martin Black (anodized aluminum) have been used in a number of infrared space telescope applications including the Infrared Astronomy Satellite (IRAS).[7] Optical baffle vanes incorporated in these telescopes have roughened surfaces designed to absorb electromagnetic radiation at the operating wavelength. Such a surface can also become a source of particulate or out-gassing contamination under certain conditions leading to degradation of optical system performance.

In this paper, the scatter and contamination of a low-scatter beryllium mirror with particles similar in size, shape, and composition to particles that could be released from a typical baffle material such as Martin Black are examined. The approach was to measure the optical scatter from the mirror at the 10.6 μm wavelength with the mirror in the "cleaned" state and after carefully controlled contamination in a clean room with aluminum oxide powder. The powder varied in size from sub-wavelength to tens of wavelengths. Particle densities on the contaminated mirror surface were statistically sampled using an SEM image analysis technique. Calculations of expected optical scatter based on Mie scattering theory were made using the measured particle densities and assuming spherical particles. The mirror BRDF measurements, the controlled contaminations, and the particle size measurements were carried out in a Class 10 clean room to minimize extraneous effects due to laboratory dust.

---

*The Oak Ridge K-25 Site is managed by Martin Marietta Energy Systems, Inc. for the U.S. Department of Energy under contract no. DE-AC05-84OR21400.

The next section describes the experimental approach in more detail. The following two sections give the experimental results and the Mie scattering theory calculations. Finally, some conclusions are drawn in the last section.

3. EXPERIMENTAL DESCRIPTION

3.1 BRDF measurement technique

The angular distribution of optical scatter produced by a contaminated, low-scatter, flat, beryllium mirror was measured using an automated TMA Complete Angle Scatter Instrument (CASI™) operating at the 10.6 μm wavelength using a CO_2 laser source. The instrument was previously described by Stover and Cheever.[8,9] The instrument shown in Figure 1 is located in a clean room at the Optical Characterization Laboratory.[10] The BRDF measurements were made near the center of the mirror. The illuminated spot on the mirror was approximately 5 mm in diameter. The incident polarization was perpendicular to the plane of incidence and the angle of incidence was 10°. Angular scans of the scatter distribution were taken in the plane-of-incidence from -80° to +80° relative to the specular laser beam.

3.2 Aluminum oxide powder characterization

The beryllium mirror was contaminated with aluminum oxide powder having three previously measured size distributions: 1 to 20 μm (small size) , 3 to 60 μm (medium size), and 4 to 100 μm (large size). The average particle sizes of the three distributions were 7 μm, 28 μm, and 45 μm, respectively. These particle size distributions for the powder were determined with a Sharples Micromerograph instrument.[11] The three particle size distributions are consistent in size with microscopic features on typical baffle material surfaces which could be released as particulates causing contamination of optical surfaces.

To contaminate the mirror, the powder was entrained in a stream of dry nitrogen gas directed toward the mirror surface. The mirror surface was held in a vertical position during the contamination.

3.3 Particle size distribution sampling technique

The contaminated mirror surface was examined in the clean room with an ISI-CL8 SEM using a Link AN-10000 x-ray spectrometer and image analyzer. Particle size distributions were sampled at multiple locations on the mirror surface uniformly spread over the area of the illuminated spot. Particles were counted at a magnification of 200x. At this magnification, approximately 7.5% of the area of the illuminated spot was covered in the nine locations examined. The minimum detectable particle size at 200x was about 3 μm. Each particle was sized by averaging six linear dimension measurements taken 60° apart. Chemical typing of the particles by x-ray analysis made it possible to identify, count, and size only the contamination particles. The cleaned mirror surface was also examined by this technique at 200x and 1000x. No particles larger than 2 μm were observed on the cleaned mirror surface. The SEM image size measurement was calibrated using an NIST dimensional standard SRM 484.

3.4 Experimental procedure

A five-step procedure was used to examine the optical scattering characteristics of the cleaned and the contaminated beryllium mirror. First, the BRDF was measured with the mirror in the "cleaned" state. Second, the mirror was contaminated and the BRDF was remeasured. Third, the mirror was examined with the SEM to determine the particle size distribution on the surface. Fourth, the BRDF was remeasured after the SEM examination to ensure that the scattering and, thus, the particle size distribution was not altered. Finally, the mirror was cleaned and the BRDF was remeasured. The process was repeated for the three particle size distributions. The results of the experiment are described in the next section.

4. EXPERIMENTAL RESULTS

4.1 BRDF measurements

Angular BRDF scans of the cleaned and the contaminated beryllium mirror are shown in Figure 2. The BRDF of the cleaned, low-scatter mirror was about 5×10^{-6} sr^{-1} at 5° from the specular laser beam. The BRDF of the mirror contaminated with the large particles was 3.5×10^{-2} sr^{-1} at 5° or about four orders of magnitude higher than for the cleaned mirror. The scattering with the large particles was strongly dependent on angle and peaked at small angles less than 10° as shown in Figure 2. At medium angles ($\approx 30°$), there was little difference in scatter between the large- and medium-size contaminating particles. At larger angles greater than 60°, there was a strong upward trend in the BRDF with the large particles.

The scattering with the medium-size particles was about three orders of magnitude higher than for the cleaned mirror and less dependent on angle than for the larger-size particles. With the small particles, the BRDF was about an order of magnitude higher than the cleaned mirror at 5°. The results in Figure 2 clearly illustrate the effect of particulate contamination on the scatter performance of a low-scatter beryllium mirror at infrared wavelengths. Table 1 summarizes the cleaned and the contaminated mirror BRDF measurements at 10.6 μm as a function of particle size range on the mirror surface.

Table 1. BRDF of the Cleaned and the Contaminated Beryllium Mirror

Mirror Condition	Measured $A\ell_2O_3$ Particle Size Range (μm)	BRDF (Sr^{-1}) 5° from specular	30° from specular
Cleaned	--	5.5×10^{-6}	1.3×10^{-6}
Contaminated	3-8	6.9×10^{-6}	1.2×10^{-6}
Contaminated	3-14	1.0×10^{-4}	4.6×10^{-5}
Contaminated	3-40	7.2×10^{-3}	6.1×10^{-4}
Contaminated	3-110	3.5×10^{-2}	6.6×10^{-4}

The small-size (3 to 14 μm) particle distribution was modified by gently blowing the mirror surface with dry nitrogen gas. This removed the 9 to 14 μm particles and some of the smaller-size (<9 μm) particles from the mirror surface. The scattering with the modified particle size distribution (3 to 8 μm) was essentially the same as that of the cleaned mirror as shown in Table 1 and Figure 3. Visible particles as large as 8 μm were still present on the mirror surface (See Section 4.2). A similar result was reported earlier by Young[3] in which a contaminated mirror showed little change in scatter at 10.6 μm. This result shows that scattering by sub-wavelength (<10.6 μm) size aluminum oxide particles can be negligible depending on the size and quantity of particles present on the mirror surface.

Figure 3 also illustrates the reproducibility of the measured BRDF of the contaminated mirror before and after SEM examination. The plot shows a small decrease in BRDF after the SEM examination. This result indicates that the SEM examination did not significantly alter the particle size distribution on the mirror surface and that the aluminum oxide particles were well-adhered to the surface.

The BRDF measurements of the cleaned mirror at various stages in the experiment are shown in Figure 4. The BRDF of the cleaned mirror averaged $5.3 \pm 0.2 \times 10^{-6}$ sr^{-1} through four contamination and cleaning cycles. These results indicate $\pm 3\%$ reproducibility for both the scatterometer and the cleaning procedure used. Note that the instrument signature for the scatterometer is also shown in Figure 4.

4.2 Particle size distribution sampling

The mirror surface contaminated with the small-size particles is shown in Figure 5 at a magnification of 200x. The aluminum oxide particles are irregular in shape and vary in size from 3 to 14 μm. The total density of particles averaged about 12,000/cm^2 over the 3 to 14 μm size range. After removal of the larger-size particles by blow-off with nitrogen gas, the total density of particles dropped to about 3,600/cm^2 in the 3 to 8 μm range. For the medium- and large-size particle distributions, the total densities were about 5,000/cm^2 and 1,500/cm^2, respectively. The particle distributions averaged over the multiple sampling areas are listed in Table 2. Comparison of the SEM and Micromerograph-derived size ranges shows that some of the larger particles in the original powder distributions were not present on the mirror surface. They evidently did not stick to the mirror surface during the contamination process. The measured particle densities in Table 2, together with statistical 95% confidence limits on the densities, were used as input for the Mie scattering theory calculations described in the next section.

Table 2. Aluminum Oxide Particle Size Distributions on the Beryllium Mirror Surface versus Powder Size Range

Particle Size Range (μm)	Number of Al$_2$O$_3$ Particles (cm^{-2})			
	1-20 μm Powder	1-20 μm Powder[a]	3-60 μm Powder	4-100 μm Powder
3-4	4514	2430	763	659
4-6	4583	972	903	347
6-8	1875	208	556	139
8-10	486	0	208	35
10-12	69	0	486	0
12-14	69	0	277	0
14-16	0	0	277	0
16-18	0	0	347	0
18-20	0	0	208	0
20-30	0	0	625	0
30-40	0	0	208	104
40-50	0	0	0	0
50-60	0	0	0	69
60-70	0	0	0	0
70-80	0	0	0	0
80-90	0	0	0	35
90-100	0	0	0	35
100-110	0	0	0	69

[a]After blowing mirror surface with dry nitrogen gas.

5. MIE SCATTERING THEORY CALCULATIONS AND COMPARISONS

Figures 6, 7, 8, and 9 are comparisons of Mie scattering theory calculations based on a computer code BHMIE due to Bohren and Huffman[12] and the angular resolved scatter plots shown in the previous section. The Mie theory calculations were done for a polarization normal to the scatterometer detector scan plane. The detector was sensitive to all polarizations of the scatter. The index of refraction of aluminum oxide used in the Mie theory calculations was $0.54 - 0.07i$ at 10.6 μm.[13,14] Maximum, minimum, and mean BRDF predictions are plotted to reflect a range based on the 95% confidence limits for the particle size distribution sampling. The Mie predictions tend to be higher than the BRDF measurements for the sub-wavelength size aluminum oxide particles as shown in Figure 6. In fact, the measured BRDF with the sub-wavelength (3 to 8 μm) size particles was essentially the same as for the clean mirror even though 30% of the particles remained on the mirror after blowing with dry nitrogen. This should be expected since the total scattering cross-section is proportional to particle area and the particles remaining on the mirror after blow-off had a cross-section approximately 10 times lower than for the 3 to 14 μm size particles. The Mie predictions are more than an order of magnitude higher than the measured BRDF for the sub-wavelength size particles.

The Mie predictions are lower than the BRDF measurements for the super-wavelength size particles as shown in Figures 8 and 9. This result is consistent with previous results.[1,6] Zerull[6] suggests that this effect may be due to a flattening of the diffraction peak for irregularly-shaped particles. The difference between the prediction and the measurement in the case of the large particles (Figure 9) is particularly exaggerated at small and large angles. It is anticipated that these effects are also due to particle shape effects.

The best agreement between the Mie predictions and the measurements is for the wavelength size (3 to 14 μm) aluminum oxide particles as shown in Figure 7. In this case, the BRDF measurement is between the minimum and mean Mie predictions.

6. CONCLUSIONS

The addition of dielectric, baffle-like aluminum oxide particles to a low-scatter beryllium mirror surface increases the infrared scatter by several orders of magnitude particularly at small angles for super-wavelength size particles. There is reasonably good agreement between the Mie theory calculations and the BRDF measurements at the 10.6 μm wavelength for particles comparable in size with the scattering wavelength. For particles of this size, the calculations faithfully reproduce the shape of the scatter distribution as a function of angle. The Mie theory calculations underestimate the scattering for real super-wavelength size particle distributions on the mirror surface and overestimate the scattering for a sub-wavelength size particle distribution. For the sub-wavelength size particle distribution, the increase in scatter in this experiment was negligible at the 10.6 μm wavelength.

Further refinements of this work are expected to provide improved predictions of scatter for different types of particles and to provide a figure of merit for a particular baffle material based on the distribution of particle sizes as well as the total number of particles ejected in various tests. Modified Mie[2] or other scattering theory calculations combined with an improved SEM particle size distribution measurement technique will be tested to achieve better agreement between BRDF measurements and calculations for particulate contaminated mirror surfaces.

7. ACKNOWLEDGEMENTS

Work at the Optical Characterization Laboratory in Oak Ridge, Tennessee is sponsored by the U.S. Army Strategic Defense Command, Huntsville, Alabama. The authors would especially like to thank U. C. Fulmer for the BRDF measurements and W. H. Davis for the SEM work. The authors also wish to acknowledge D. E. White for providing the aluminum oxide powder and the Micromerograph particle size data.

8. REFERENCES

1. Kie Nahm, Paul R. Spyak, and William L. Wolfe, "Scattering from Contaminated Surfaces," *Proc. SPIE*, Vol. 1165, Scatter from Optical Components (1989).

2. Kie Nahm and William L. Wolfe, "Light Scattering by Polystyrene Spheres on a Mirror," *Proc. SPIE*, Vol. 675, Stray Radiation V (1986).

3. R. P. Young, *Degradation of Low Scatter Mirrors By Particle Contamination*, Report No. AD/A-004 103, Arnold Engineering Development Center (AEDC), Arnold Air Force Station, Tennessee 37389 (1975).

4. R. P. Young, "Low-Scatter Mirror Degradation by Particle Contamination," *Opt. Eng.* **15**, 516-520 (1976).

5. Y. Wang, "Mathematical Model for Scattering from Mirrors," Report No. NPO-17050/6567, Jet Propulsion Laboratory (JPL), California Institute of Technology, Pasadena, California (1988).

6. R. H. Zerull, R. H. Geise, and K. Weiss, "Scattering Functions of Nonspherical Dielectric and Absorbing Particles vs. Mie Theory," *Appl. Opt.* **16**(4), pp. 777-778 (1977).

7. S. M. Pompea, D. F. Shepard, and S. Anderson, "BRDF Measurements at 6328 Angstroms and 10.6 Micrometers of Optical Black Surfaces for Space Telescopes," *Proc. SPIE*, Vol. 967, Stray Light and Contamination in Optical Systems (1988).

8. John C. Stover, "Optical Scattering Measurement and Analysis," p. 109, McGraw-Hill, Inc., New York, 1990.

9. D. R. Cheever, F. M. Cady, K. A. Klicker, and J. C. Stover, "Design Review of a Unique Complete Angle-Scatter Instrument (CASI)," *Proc. SPIE*, Vol. 818, p. 13 (1987).

10. The Optical Characterization Laboratory is operated under the auspices of the Department of Energy to characterize optics for the Department of Defense.

11. "Particle Size Distribution Analysis," Sharples Corporation Research Laboratories, Bulletin 101.

12. C. E. Bohren and D. R. Huffman, "Absorption and Scattering of Light by Small Particles," Wiley, New York, 1983.

13. F. Gervais in "Handbook of Optical Constants of Solids, Vol. II," ed. E. D. Palik, p. 761, Academic Press (1991).

14. C. A. Worrell, "Infrared Optical Constants for CO_2 Laser Waveguide Materials," *Journal of Material Science*, **21**, pp. 781-787 (1986).

Fig. 1. Complete angle scatter instrument at the
Optical Characterization Laboratory.

Oak Ridge National Laboratory - Light Scatter

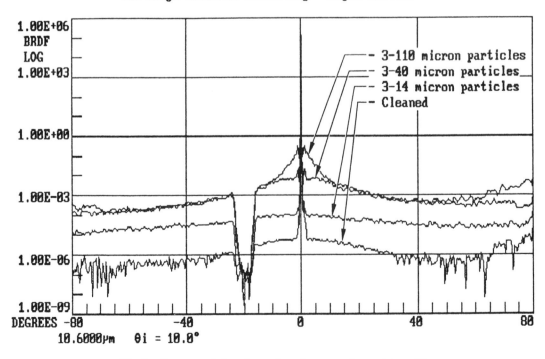

Fig 2. Scatter of the cleaned and contaminated beryllium
mirror versus contaminating particle size range.

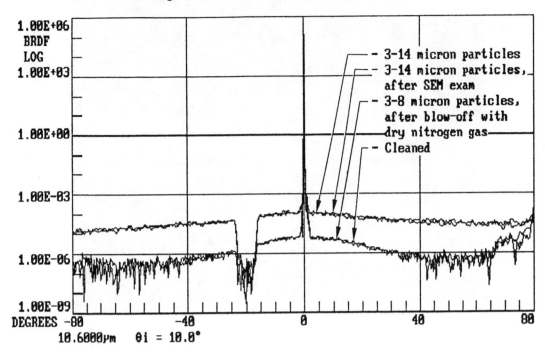

Fig. 3. Scatter of the cleaned and contaminated beryllium
mirror before and after SEM examination and
after blow-off with dry nitrogen gas.

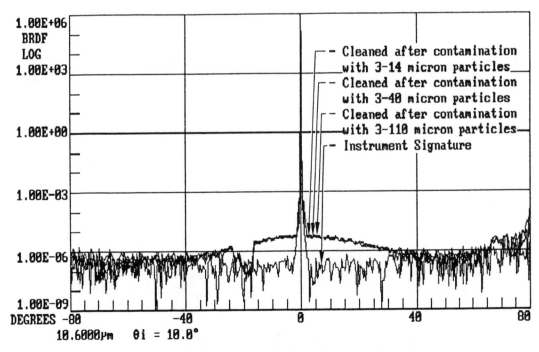

Fig. 4. Scatter of the cleaned beryllium mirror after
contamination with three particle size ranges.

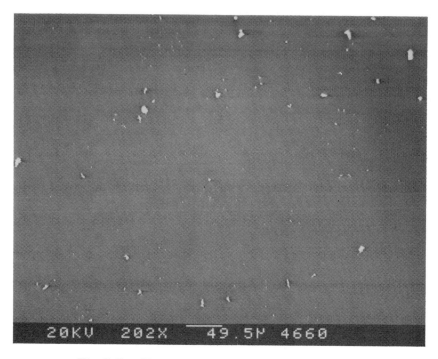

Fig. 5. Beryllium mirror surface contaminated with
3 to 14 μm size aluminum oxide particles.

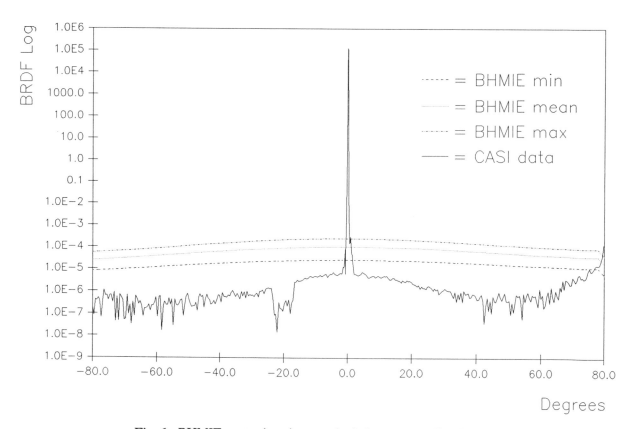

Fig. 6. BHMIE scattering theory calculations versus CASI BRDF
data at the 10.6 μm wavelength for a low-scatter beryllium
mirror contaminated with 3 to 8 μm size particles.

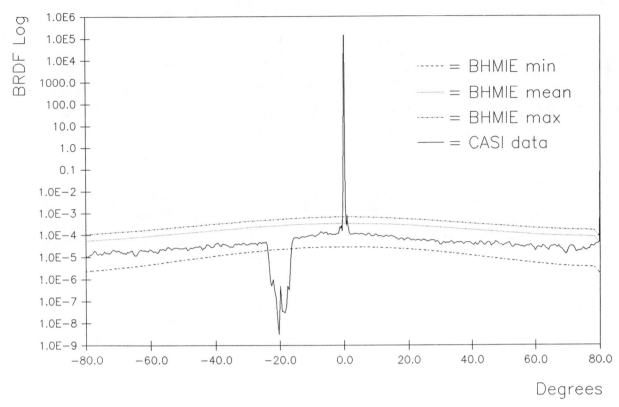

Fig. 7. BHMIE scattering theory calculations versus CASI BRDF data at the 10.6 μm wavelength for a low-scatter beryllium mirror contaminated with 3 to 14 μm size particles.

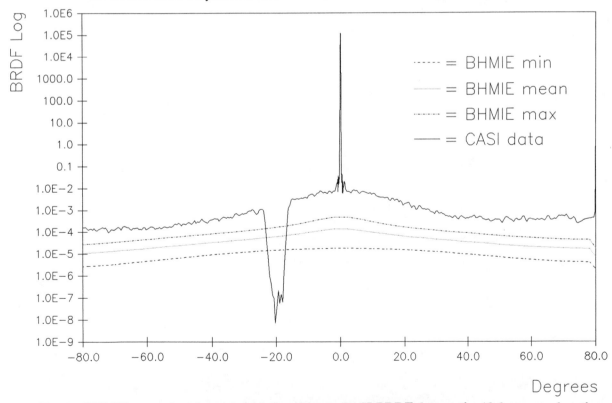

Fig. 8. BHMIE scattering theory calculations versus CASI BRDF data at the 10.6 μm wavelength for a low-scatter beryllium mirror contaminated with 3 to 40 μm size particles.

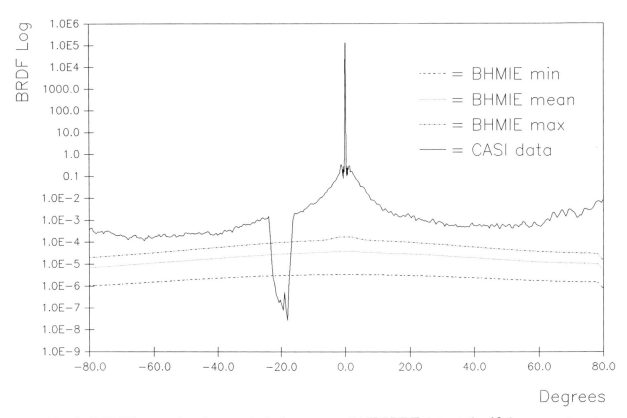

Fig. 9. BHMIE scattering theory calculations versus CASI BRDF data at the 10.6 μm wavelength for a low-scatter beryllium mirror contaminated with 3 to 110 μm size particles.

Light scatter variations with respect to wafer orientation in GaAs

Jeff L. Brown

USAF Solid State Electronics Directorate
Wright Laboratory
WL/ELOT
Wright-Patterson AFB, OH 45433

ABSTRACT

Attempting to determine the quality and uniformity of a smooth surface from examination of BRDF raster scans can be a risky business as evidenced by studies of commercially grown and polished GaAs wafers. To demonstrate this, raster scan variations with respect to wafer orientation are presented along with a method for analyzing these variations without the need for making multiple raster scans. Some possible sources of the orientational variations are indicated by the data.

1. INTRODUCTION

Optical scattering has come to be recognized as a sensitive, non-destructive evaluation (NDE) technique for evaluating smooth surfaces. It has been applied to a large number of material systems and is an active research area.[1] However, choosing the appropriate subset of the bidirectional scatter distribution function (BSDF) to measure for a particular material is rarely a trivial matter.

For instance, when it is desirable to measure the polish uniformity over a large area of a sample, one may choose to make raster scans of the surface. Quite often, after determining the appropriate angles of incidence and scatter, raster step sizes, and polarization states, the sample is mounted with little consideration given to the rotary orientation of the sample with respect to, say, the plane of incidence of the laser beam. Since the BSDF can, in general, be expected to change whenever any of the directional variables are changed,[2] rotary orientation must be considered.

This paper will show that, for a scatterometer restricted to measuring scatter in the plane of incidence, rotary orientation can have a large effect on the results of raster scans, at least when measuring GaAs wafers. A graphical extension of a technique developed by Silva, Orazio, and Stowell[3] shows how preferential scatter directions vary across a sample and gives some clues as to the source of these variations. The techniques and effects shown here are most likely applicable to other materials as well.

2. BACKGROUND

Silva, et al., previously presented data showing some effects of changing the rotary orientation of a sample.[3] (Rotary orientation is here defined as the angle of clockwise rotation about an axis normal to the sample surface and measured between the plane of incidence of the laser beam and some registration mark on the sample, as in Fig. 1.) The scatterometer used in their work was designed to measure bidirectional reflectance distribution function (BRDF) in the plane of incidence, except that it also employed a rotation stage behind the X and Y translation stages used for scanning a sample.[4] Properly aligned, this type of scatterometer has the ability to rotate the sample about any point on the sample surface, since the rotary axis, denoted R-axis, is normal to the sample surface and intersects the laser beam and detector viewing axis at the same point on the sample. See Fig. 2. It is important to note that the X and Y translation stages used for raster scans, and thus the (x,y) coordinate system, rotate with the rotation stage. Silva and his collaborators used their scatterometer to generate polar plots showing how the scatter varied when the sample was rotated through 360°. Several of these polar plots were averaged in order to determine the rotary orientation most likely to yield the minimum and maximum average scatter for raster scans of a particular sample.

SPIE Vol. 1530 Optical Scatter: Applications, Measurement, and Theory (1991) / 299

Rotary Orientation

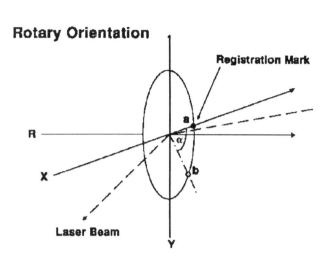

Fig. 1. Rotary orientation of a sample is defined as the angle of clockwise rotation about the R-axis measured from the plane of incidence of the laser beam. (a) R=0° orientation, (b) R=α orientation (registration mark rotated as shown.)

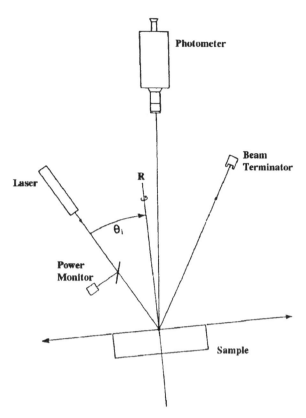

Fig. 2. Top view of scatterometer layout.[4]

A scatterometer with the same attributes as described above has been used in the present work to measure optical scatter of a large number of commercially produced, 3" diameter GaAs wafers. In the beginning, it was not clear what rotary orientation to use for the purpose of comparing different wafers from the same manufacturer and from different manufacturers. To see if there was any crystallographic effect on scattering, two orientations were chosen: the first one with the plane of incidence parallel to the [011] direction, and the second perpendicular to the [011] direction.[5] Certain symmetric patterns, such as the four-fold, cloverleaf pattern of dislocation defects, appear in the raster scan data of other measurements,[6] so this was a logical choice. Fig. 3 is a typical example of the resulting raster scans.

Fig. 3a is the raster scan and associated histogram of a wafer with a rotary orientation of 0°, denoted R = 0°. Fig. 3b is the raster scan for the same sample with a rotary orientation of R = 90° (clockwise rotation.) A third orientation, R = 60°, is also shown in Fig 3c. The patterns in these raster scans bear no resemblance to any of the typical defect patterns seen by other techniques. More striking, however, is the fact that the high scatter zone seems to have rotated in a counterclockwise direction in going from the R = 0° orientation to the R = 90° orientation. Keeping in mind that the raster scan coordinate system is the same for both scans, one must assume that different areas on the wafer have different directional scattering preferences. The darkest (highest scatter) area for different orientations does not occur in the same location on the wafer.

After observing several wafers with similar varying patterns as in Fig. 3, it soon became apparent that crystal structure and the well known defects had little to do with the resulting raster scan patterns. Since all wafers were destined for destructive testing, a method was developed to gather as much information as possible about the scattering characteristics of these wafers in an efficient manner.

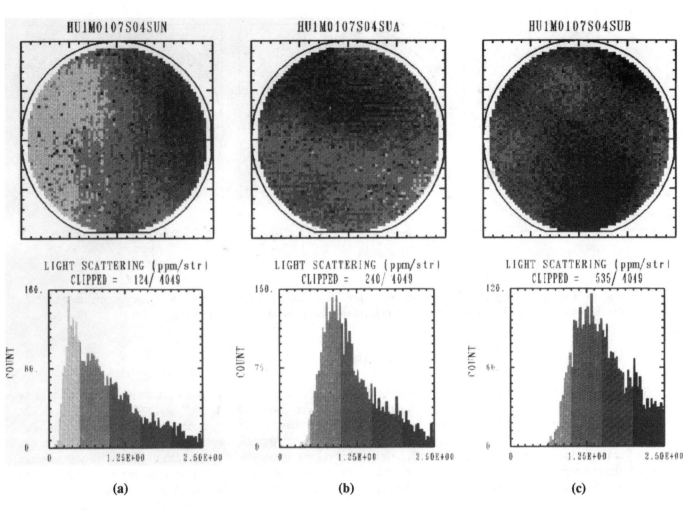

Fig. 3. Raster scans and associated histograms for a GaAs wafer in 3 different rotary orientations: (a) R = 0°, (b) R = 90°, and (c) R = 60°. The data is BRDF in units of ppm/str. Some points have been clipped from the histograms as shown.

3. METHOD

Following Silva, et al., the method involves measuring the BRDF while a wafer is spun about several points across the wafer. The resultant polar plots show orientational scattering preferences, and these polar plots can be displayed in such a way that raster scan variations can be predicted.

Fig. 4 shows a typical polar plot of the scatter for a single rotary scan, five rotary scans from different points on the sample, and the envelope of 13 rotary scans from a typical data set. For these polar plots, even though the sample is spun clockwise, the data is plotted counterclockwise as if the plane of incidence is rotating and the sample is fixed. The angles associated with the minimum and maximum scatter of the envelope are useful if the scans comprising the envelope are similar, implying uniformity of directional scattering preferences. However, it is more useful to display the polar plots in a graphical way that pinpoints the data location of the individual scans as in Fig. 5.

In this figure, the large circle represents a wafer in its R = 0° orientation. Superimposed on this wafer are the individual polar plots, where the centers of the plots indicate the position on the wafer at which the rotary scans were taken. This graphical representation forms a concise picture of how directional scattering preferences vary across the wafer.

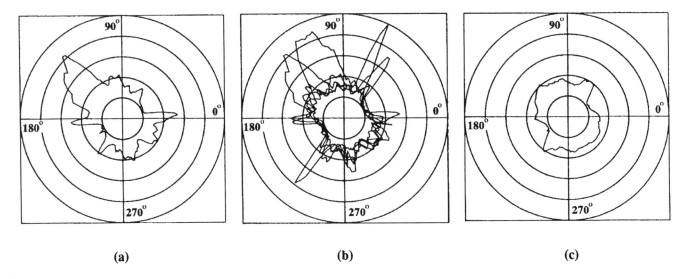

(a)   (b)   (c)

Fig. 4. Typical polar plots of BRDF vs. angle for (a) a single rotary scan, (b) five rotary scans, and (c) the envelope of 13 rotary scans. The maximum value for these plots is 7.3 ppm/str.

4. DISCUSSION

To demonstrate the usefulness of the graphical technique of Fig. 5, consider the three raster scans and corresponding polar plot analysis of Fig 6. The three raster scans of Fig. 6a, 6b, and 6c were taken at R = 0°, 90°, and 196° respectively. In Fig. 6d, observe the magnitude of each polar plot in the direction corresponding to the rotary orientation for each raster scan. Fig. 6e shows pointers to aid in reading the polar plots. For the case of R = 0°, the polar plots show a general trend of increasing scatter from right to left across the wafer. This is exactly what occurs in the R = 0° raster scan of Fig. 6a. For the case of R = 90°, the polar plots show a nearly constant magnitude, and this is what is seen in the corresponding R = 90° raster scan of Fig. 6b. The same analysis can be applied to the third case, Fig. 6c, to see how the polar plots correlate with the pattern of this raster scan. For any other rotary angle, one should be able to construct the patterns of the raster scan that would result for that particular rotary orientation directly from the polar plot analysis, at least to the extent that the polar plot data density allows.

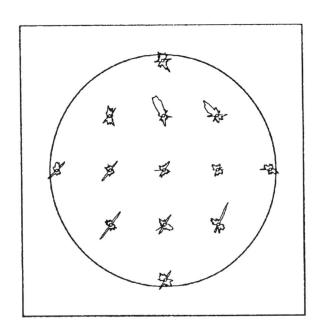

Fig. 5. Polar plot analysis.

Not all wafers show variations with respect to rotary orientation; but for those that do, the variation can be large. Fig. 7 is a population (percentile) chart with one curve for each raster scan in Fig. 6. Each curve is a plot of BRDF against each data point's normalized rank order for the sorted data set making up a raster scan. This is a handy way of viewing the statistics of a raster scan data set for easy comparison. The 50th percentile (median) and the spread of the data (proportional to the standard deviation) are two quick comparisons one can make. For the three curves in Fig. 7, two of them have similar statistics, but the third, the R = 90° orientation, has a lower median by about a factor of two and a tighter distribution. This can also be seen to a certain extent in the histograms of Fig. 6.

A more drastic example of directional scattering preferences and magnitude variations is shown in Fig. 8. (This data is presented without the benefit of raster scans because the gray

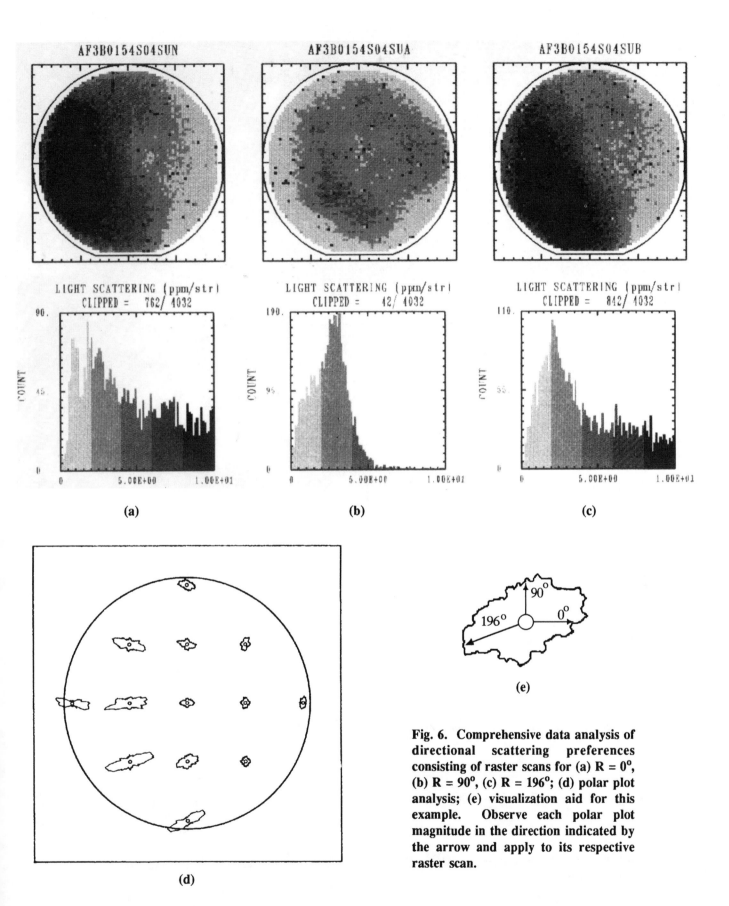

AF3B0154S04SUN

LIGHT SCATTERING (ppm/str)
CLIPPED = 762/ 4032

(a)

AF3B0154S04SUA

LIGHT SCATTERING (ppm/str)
CLIPPED = 42/ 4032

(b)

AF3B0154S04SUB

LIGHT SCATTERING (ppm/str)
CLIPPED = 842/ 4032

(c)

(d)

(e)

Fig. 6. Comprehensive data analysis of directional scattering preferences consisting of raster scans for (a) R = 0°, (b) R = 90°, (c) R = 196°; (d) polar plot analysis; (e) visualization aid for this example. Observe each polar plot magnitude in the direction indicated by the arrow and apply to its respective raster scan.

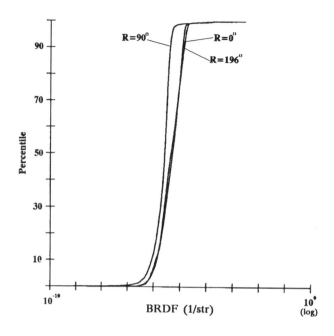

Fig. 7. Population chart for the raster scan data of Fig. 6.

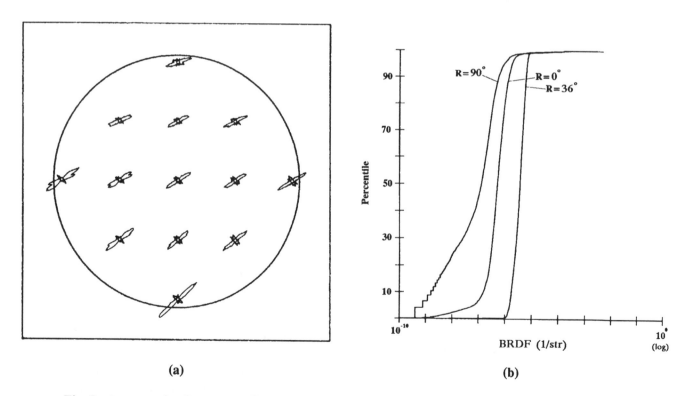

(a)

(b)

Fig. 8. An example of strong, uniform directional scattering preference. (a) Polar plot analysis and (b) population chart.

scale plotting routines are presently limited to linear scaling.) The directional preference is nearly uniform across the surface, and the population chart shows a large variation in median scatter, by about a factor of 30 between lowest and highest.

It isn't clear at this time what causes these orientational scattering preferences, but one possible explanation for this is subsurface damage with enough order to produce diffraction lobes in the BRDF which rotate past the detector during a rotary scan. Residual polishing scratches below the surface might be expected to cause such a diffraction effect, but the data of Fig. 8 implies scratches running in one direction and extending clear across a 3" diameter wafer. Perhaps the cause is residual subsurface sawing damage.

Although the actual source of these variations has not been determined, there are some indications based on trends of the data for over 100 wafers. The same orientational scattering preferences are almost never repeated on adjacent wafers cut from the same crystal, suggesting that the variations are due to differences in cutting and polishing. As mentioned earlier, not all wafers exhibit variations with respect to rotation. The trait, however, is specific to particular manufacturers across several crystals. This also would implicate polishing practices specific to a particular manufacturer. These possibilities bear further investigation.

5. CONCLUSIONS

The examples shown here demonstrate that sample orientation needs to be considered when interpreting raster scan data from in-plane scatterometers. GaAs is a particularly interesting example which can exhibit large variations with respect to rotary orientation. The polar plot analysis developed for studying these variations has proven useful in an extensive GaAs materials study. The end result of this analysis is a set of polar plots which provides more information about the scattering behavior of a wafer than can be obtained with a small number of raster scans. This information becomes important when interpreting raster scans over large areas. A small change in rotary orientation of the wafer can make a large difference in the raster scan patterns and change the median BRDF for the raster scan by an order of magnitude.

6. ACKNOWLEDGMENTS

The author is indebted to John Hoeft for scatterometer software development and many technical discussions. He also wishes to thank Chris Blouch, for his help in preparing gray scale wafer maps, Jim Sewell, principal architect of the software which generates the wafer maps, John Scheihing for his expertise in desk top publishing, and Dave Look for many long discussions on semiconductor materials metrology.

7. REFERENCES

1. J. C. Stover, "Scatter from Optical Components: An Overview," **Scatter from Optical Components**, John C. Stover, Editor, Proc. SPIE 1165, pp. 2-9, (1989).

2. J. C. Stover, **Optical Scattering: Measurement and Analysis**, p. 16, McGraw-Hill, Inc., New York, 1990.

3. Robert M. Silva, Fred D. Orazio, Jr., and W. Kent Stowell, "Scatter Evaluation of Supersmooth Surfaces," **Scattering in Optical Materials**, Soloman Musikant, Editor, Proc. SPIE 362, pp. 71-71 (1982).

4. Fred D. Orazio, Jr., W. Kent Stowell, and Robert M. Silva, "Instrumentation of a Variable Angle Scatterometer (VAS)," **Scattering in Optical Materials**, Soloman Musikant, Editor, Proc. SPIE 362, pp. 165-171 (1982).

5. **1990 SEMI International Standards**, Vol. 3, pp 125-147, Semiconductor Equipment and Materials International, Mountain View, CA, 1990.

6. See, for example, D. C. Look, et al., "Wafer-Level Correlations of EL2, Dislocation Density, and FET Saturation Current at Various Processing Stages," **Journal of Electronic Materials**, Vol. 18, No. 4, pp. 487-492, 1989.

Stray Light Reduction in a WFOV Star Tracker Lens*

by Isabella T. Lewis, Arno G. Ledebuhr , Timothy S. Axelrod,
and Scott A. Ruddell (Lawrence Livermore National Laboratory)

Lawrence Livermore National Laboratory (LLNL) has recently developed a wide-field-of-view (28° x 44°) camera for use as a star tracker navigational sensor. As for all sensors, stray light rejection performance is critical. Due to the baffle dimensions dictated by the large field angles, the 2-part sunshade/baffle configuration commonly seen on space-born telescopes is impractical. Meeting the required stray light rejection performance (of 10^{-7} Point Source Transmittance, (PST)) with a 1-part baffle required iterative APART modeling (APART is an industry standard stray light evaluation program), hardware testing, and mechanical design correction.

This paper presents a chronology of lens and baffle improvements that resulted in the meeting of the stray light rejection goal outside the solar exclusion angle of the baffle stage. Comparisons with APART analyses are given, and future improvements in mechanical design are discussed. Stray light testing methods and associated experimental difficulties are presented.

Introduction.

In order to provide navigational updates, the LLNL star tracker must be able to detect and centroid stars as dim as a 4.5 visual magnitude. This requirement is derived by stepping the viewing field of view (FOV), which is 28° x 44°, through all possible sky viewing angles, and counting the dimmest star that must be seen to ensure that at least 5 stars will be visible in all viewing directions. Five stars is the minumum number required to ensure a unique match in all cases. The addition of stray light into the problem affects image processing both by increasing the noise (through the shot noise and fixed pattern noise associated with the stray light signal) and by creating a non-uniform background. Although it is not the limiting factor in this situation, system performance can also be limited by the dynamic range allowed in the analog to digital (A/D) converter in conjunction with the stray light signal variation across the focal plane.

To determine how the star tracker is affected by stray light, a first order understanding of the signal processing and of the noise of the focal plane is required. After a working knowledge is attained, the stray light signal (magnitude and pattern) can be evaluated in terms of impact to system performance.

This paper is divided into four basic sections. The first is a calculation of the allowable stray light through the understanding of the signal processing in the star tracker. The second is a description of the stray light testing methods, and the problems with measurements of especially dim PST levels. The third is a description of the mechanical hardware and a recap of the stray light levels seen in the prototype configuration, along with a comparison with APART results. The last section details the experimental steps used to isolate and correct the worst causes of stray light, leading up to the present configuration.

1. Allowable Stray Light

* Work performed under the auspices of the U.S. Department of Energy by the Lawrence Livermore National Laboratory under contract No. W-7405-ENG-48.

0-8194-0658-9/91/$4.00

Figure 1 shows the optical schematic and light gathering properties of the star tracker optics. Note that the concentric optical design form necessarily results in a spherical (concentric) focal surface, which implies the need for a fiber optic faceplate (FOFP) field flattener. Optical properties are given in figure 2. Note that the FOFP creates strong transmission losses on-axis, and even stronger losses in the off-axis portion of the lens. The selection of this design form does, however, allow a much faster optics system than would flat field optical designs, more than compensating for the FOFPs transmission loss through greater light gathering capability. This light gathering capability is essential in detecting the dimmer required stars. A 200 msec exposure of a magnitude 4.5 star generates a total signal of 4000 photo-electrons at the center of the focal plane, falling to less than 3000 photo-electrons at the edge of the field of view (FOV). Due to the controlled blur circle of the optics, this signal is spread over a minimum of 4 pixels, with less than 35% of the total signal residing on a single pixel when the blur is centered on a pixel. For a star image centroid located at the intersection of four pixels, the four brightest pixels capture just 20% (each) of the total image signal.

Stars are identified by meeting two requirements. The first (possible star detection) is that the signal levels in adjacent pixels scanned in a row by row left to right produce a jump of at least N counts. N is adjusted from a large number (such as 10) progressively smaller, allowing more candidate stars until at least 10 stars pass this initial thresholding test plus the following verification test. The blur verification test requires that the pixels around the peak pixel (found in the area of the jump in signal seen in the first test) are above the average background signal, consistent with the nominal image blur shape and the intensity seen in the peak pixel. The blur verification algorithm adds the four pixels adjacent to the peak pixel then subtracts the four diagonally adjacent pixels for comparison with the peak pixel less the average background. Local background for the peak pixel is found by averaging the pixels in the periphery of the 5 x 5 pixel area around the peak pixel. The addition/subtraction of the adjacent pixels automatically eliminates the average background from the +/- sum. With background subtraction on a local scale, and initial peak detection done between adjacent pixels, a slow variation in stray light signal can be permitted without harming the ability of the sensor. A uniform background degrades performance only by the increase in net noise due to background shot noise and fixed pattern response non-uniformity.

In a case where there is essentially no stray light, the noise seen in the focal plane is limited by fixed pattern dark current non-uniformity and by readout noise. In all hardware versions through the present, the fixed pattern noise is not corrected, and thus must be included in analyses of false stars. Present hardware versions have the A/D sensitivity set such that 1 count is equal to the 1-sigma rms focal plane noise from a dark frame. At the operational integration times, this sets 1 count of noise equal to a signal generating 115 photo-electrons. For a typical m_v 4.5 star image, the expected signal rise between adjacent pixels can be less than 550 photo-electrons, depending on image plane location and image centroiding with respect to pixel center. A background signal increases noise through shot noise and through response non-uniformity. With response non-uniformity of 1%, a uniform background signal of 5,000 photo-electrons implies 50 photo-electrons rms additional noise (in the absence of fixed pattern noise compensation) plus 71 photo-electrons rms from shot noise, increasing the net noise from 115 electrons rms to 144 electrons rms.

To determine navigational orientation, the star tracker must match detected stars against its star catalog. A minimum of 5 stars that can be found in the catalog must be seen in each viewing position to achieve a positive match. Since the star tracker's algorithm uses only the brightest 10 stars in each frame for processing, only stars brighter than instrument magnitude[1] 4.5 are selected for the star tracker catalog. This avoids wasting memory and processing time and reduces the chance that random focal plane noise that is classified as a star can contaminate the star match.

Under nominal conditions (no stray light), the rms noise level in the focal plane is 1 count (115 electrons rms). For a threshold value of $N = 4$ in the first test, the probability of detecting a rise of 4 counts between adjacent pixels is roughly 10^{-4} per pixel pair. Multiplying by the number of pixels in the focal plane (2×10^5), roughly 20 possible stars would be identified from noise sources. Of these, roughly 1/2 would be eliminated by the blur verification algorithm. This leaves about 10 false stars in the image files, which is the tolerable limit in weak star fields. At this $N = 4$ counts threshold level, an $m_v = 4.5$ star has roughly an 80% probability of being detected. At a threshold level of 5, the probability of detecting the $m_v = 4.5$ star drops to below 50%. This is the minimum allowable threshold limit (based on limiting false star identification) when the rms noise increases from 1 count (115 electrons) to 1.25 counts (144 electrons). This occurs with a stray light signal of 5000 photo-electrons. With the sun as the stray light source, an attenuation factor from aperture flux levels (the PST) of at least 10^{-7} is needed to bring the stray light signal to the desired 5000 photo-electron level.

Note that the 5000 signal criteria is equivalent to roughly 44 counts, which is 1/6 of the total A/D range available. In this instance, the stray light signal level is limited by noise before A/D saturation with variation across the focal plane is a concern.

2. Stray Light Testing

Although bidirectional reflectance distribution function (BRDF) testing of individual surfaces is now a common practice, with more than one model of commercial testing unit available, stray light testing of telescope assemblies is still difficult. This is due to both the lower levels under consideration (10^{-5} to 10^{-14} PSTs in extreme cases, compared to 10^{-4} to 10^{-7} for single surface BRDFs), and to the difficulty of locating a probe in the focal plane. Any probe other than the actual detector at or near the focal plane disturbs the stray light fluxes (through reflections off the probe back into the lens system then back to the probe again).

The low PST design requirements necessarily imply quite small fluxes in the focal plane, which are difficult to detect. Generally, the stray light flux is specified and designed to have little impact on the system performance. Hence, it is generally barely detectable at the focal plane for the stray light source in operation. For visible systems, the stray light source of concern is often the sun. For testing purposes, one would ideally want to use a source brighter than the sun to raise the stray light signal to easily detectable levels. Unfortunately, this isn't possible without using a fairly divergent source (i.e. a source subtending 10° to the sensor, rather than the 1/2° subtended by the sun), which "blurs" the available PST vs q information. Our test design minimized the source angle (which reduces the irradiance at the aperture) while constraining the observed flux to be "easily detectable".

Detection of lens system scattering (PST) is complicated by the imaging of increased scene radiance, which will be present to some degree when the stray light source is on. Background radiance can come from stray light source illumination of any structures within the FOV of the lens or from scattering of the

---

[1] Instrument magnitude is equal to visual magnitude at G0 (5900K) star color temperature.

atmosphere (dust, etc.) in front of the camera.

Figure 3 illustrates the scattering difficulties associated with in-laboratory tests with relatively small rooms. The illumination is scattered by the test unit (and by chamber walls behind the test unit). Some of the scattered light is directed towards the walls in front of the test lens within the lens FOV and is imaged as a background source. This is added to the stray light signal caused by scattering of teh illumination flux at the lens aperture. As soon as the scene radiance signal becomes an appreciable fraction of the stray light lens scattered signal, the PST measurement is ambiguous. Limiting the scene radiance signal to 10% of the stray light signal at 10^{-7} PST implies a laboratory greater than 18 feet across. This is in the best of conditions, assuming the stray light source can be controlled and no stray light escapes directly to the rear wall, etc.

Since we did not have the luxury of a large lab, we moved outside where the "walls" were at infinity. Figure 4 shows the test set up after several rounds of refinement. We adopted a 650 W quartz tungsten halogen lamp. Detection was accomplished by the focal plane of the sensor, with integration times boosted slightly to exaggerate the stray light signal.

Since PST varies with angle, we would ideally want a collimated source. Increasing the distance between the source and the baffle decreases the subtended source angle, but at the same time this reduced the flux at the baffle aperture. We desired a signal increase of at least 3 counts from a 10^{-7} PST using the test configuration. (The sun at 10^{-7} PST would yield 64 counts.) This allowed the stray light source flux at the aperture to be as weak as 3/64 the sun's radiance. After accounting for the "redder" color temperature of the QTH lamp, which better matches the Thomson CCD response, the required spectral radiance at 550 μm is permitted to be as low as 1/40 the sun's spectral flux, or 5 mW/cm^2-μm. At the nominal spacing chosen for the experiments, the QTH lamp flux registered 10 to 20 mW/cm^2-μm. Since flux density varies rapidly with distance at the nominal range, radiometer flux measurements were taken at the plane of the baffle for each stray light exposure.

Tests were run near Livermore, California. The evenings were frequently hazy, and the air fairly damp. We encountered two problems associated with this climate. Firstly, background levels were considerably higher than expected from a dark frame[2] due to sky radiance and imaging of faint clouds and haze. Since cloud motion between exposures could create apparent differences in background level, we developed a routine of taking a "light-off" frame (i.e. stray light source unplugged), followed in rapid succession by a "light-on" frame. Elapsed time between comparison light-off/ light-on frames is less than 1 minute. Still, we took the precaution of experimenting only on relatively cloudless nights, and occasionally exposed a light-off/light-on/light-off sequence, comparing light-off frames, to verify that the background was stable over the period of the exposures. Through this verification procedure, operation was found to be permissible in lightly opaque cloud conditions.

In order to quantify the amount of light scattered by the atmosphere above the lens, the direct illumination of the baffle was blocked, while the scattering from the air column above the lens was not. This is illustrated in figure 5. Typical illumination beam diameters at the lens were several feet, thus the

---

[2] Here, "dark frame" refers to in-laboratory images taken with the aperture of the lens blocked, permitting no background flux to the detector. Under these circumstances, a finite signal is still present due to dark current.

background blockage of a small aerosol section needed to completely shade the baffle does not significantly alter the aerosol measurement. Typical aerosol measurements were less than or equal to 10^{-7} PST equivalent, and were sampled throughout each evening of experiments. At this magnitude, they had to be subtracted from the total light-on values, but after this was done, the resulting baffle PST measurement was not considered to be compromised by the atmospheric background. Larger aerosol contents would have been considered too large for accurate 10^{-7} PST experiments.

Data reduction from the three types of images was done by subtracting the averages of a 20 x 20 pixel area of the background from the same portion of the sky image on the adjacent light-off, light-on, and aerosol scattering images. PST was calculated based on the difference in counts attributable to baffle scattering and the theoretical unattenuated signal expected from the stray light source flux at the aperture. Data is thought to be accurate to 25% down to PSTs as low as 10^{-7}, limited mainly by calibration.

3. Prototype Mechanical Hardware

Figure 6 shows the prototype mechanical hardware configuration. Figure 7 shows the surfaces modeled in the original APART analysis, which was performed after the prototype hardware was fabricated, but before the testing program had begun.

The baffle shown in the figure 7 was designed as a "2-part" baffle. The portion with vane tips modeled as scattering surface 3 (the inner section) shades the lens surfaces from direct illumination at angles greater than 52°. The outer baffle section (vane tips modeled as surface 33) is designed so that the inner baffle section blocks further direct illumination from the outer vanes tips to the lens surfaces, and acts to block illumination of the inner vane tips at angles greater than 74°. For stray light illumination angles greater than 52°, the most direct scattering path is off the inner baffle section to an optical lens or housing surface to the focal surface: 2 bounces. For stray light illumination angles greater than 74°, the most direct scattering path is off the outer baffle section to the inner baffle section to an optical lens or housing surface to the focal surface: 3 bounces. Modeling and measurements showed the PST to be fairly uniform over the focal surface and relatively constant from 52° to 74°, falling 2 to 3 orders of magnitude at stray light angles greater than 74° which required the the additional bounce to reach the focal plane.

The tests of the original hardware prototype showed PSTs on the order of 10^{-4} and 10^{-6} in the 52° to 74° and 74° to 90° angular regions, respectively. At the time, predictions based on APART models[3] led to 3 orders of magnitude lower PST. The discrepancy in measurements and models was quickly traced to the evident (in hindsight) but unmodeled scattering paths through the FOFP and from the rear housing walls. These were identified by the preferred method of all stray light sleuths: looking from the perspective of the focal plane at various stray light source angles.

4. Stray Light Reduction Steps

The FOFP (the worst of the two offending surfaces) was not included in the preliminary APART models. Ideally, light entering the front of a fiber optic faceplate is transmitted straight through to the rear of the faceplate, which is butted directly to the detector. At the fast F/number and large off-axis angle (30°) seen in the star tracker design, the FOFP signals hitting off-axis cross talked severely. At angles just greater

---

[3] Modeling was done by 2 independent stray light groups, neither of which was BREAULT Research, the designers of the APART code.

than 28° (the design exclusion angle for the inner baffle vane tips), light scattered from a single bounce from the vane tips entered the FOFP just outside the FOV of the detector, but immediately spread, resulting in a large stray light signal in the focal plane. Figure 8 illustrates this #1 offending path. Blocking was accomplished simply by coating the FOFP with an opaque black paint outside the clear aperture required for the active format. After this correction, stray light levels were reduced from 10^{-4} to 10^{-5} in the 52° to 74° region.

The rear housing walls were the second most serious stray light path. The light scattered from the inner baffle vane tips could reach this housing wall and directly reflect onto the focal surface. While the APART models included a rear housing wall and identified this wall as a "critical object"[4], it was inadvertently omitted from the list of scattering path combinations for analysis. Experiments with a temporary vane that blocked the housing wall from being illuminated, schematically shown in figure 9, reduced the PST from on the order of 10^{-5} to 10^{-6}.

At this point, stray light levels were still a factor of 10 larger than predicted by stray light analyses. The major scattering paths remaining were vane tips to aperture stop to focal plane. An auxiliary APART analysis showed that diffuse scatter off the finite width of the vane tips dominated diffractive scattering for the radii used in our manufactured assembly. Hence, scattering was controlled by the radius of the edge and the blackening material. The prototype hardware specified 0.0025 inch maximum vane tip edge radii, and the aperture stop was created by notching a groove into the central optical element (which is a sphere) with a glass saw blade and painting the notch with FLOQUIL' blackening agent. Neither the vane tips or the aperture stop in the preliminary configuration can be said to have been optimized.

Various alternate methods were chosen for the aperture stop and the vane edges, along with a general study of the reflectance of black paints. Since the paint thickness varies with application as does the reflectance, a combined measure of the scattering off a processed vane was indicated. The scattering off a variety of painted sharp edges was measured directly using a scatterometer at TMA, curiously showing little difference between paints even though the base reflectance from the paints varies by a factor of 2.

Since little improvement was easily achievable in the vane tips, attention was turned to the second of the two scattering surfaces, the aperture stop. Several alternate techniques were examined. The first used a split ball central element with a curve polished to a sharp edge on one of the two hemispheres, using OCA Applied Optics' special index matching blackening formulation[5] to create the opaque shield. This worked quite well, and brought the measured PST down to the 10^{-7} value predicted by the APART models in the 52° to 74° stray light angle region. However, this technique requires more numerous and more complex fabrication steps than most other methods, and the search for another alternate aperture stop technique was continued. None of the other methods were found to be as effective. Among the alternates were split ball techniques with a lithographic application of an opaque chromium-based shield and with the inclusion of a very thin (0.0005 inches thick) blackened metal shim. Neither of these was shown to be as effective as the index-matching black on the polished edge. Each of the 4 techniques (the prototype, index matched black, plus the two other alternate techniques) is illustrated in figure 10.

---

[4] A critical object in APART is one that is in direct view of the focal plane.

[5] Reference Lewis et. al., Low scattering edge-blackening compounds for refractive optical elements, August, 1989 SPIE proceedings, Vol. 1165, p. 227+

4. Present Configuration

Maintaining the stray light goal of 10^{-7} PST to allow optimum star tracker performance required painting of the out-of-format portion of the FOFP, enlargement of the rear housing wall with the inclusion of a rear wall-blocking vane, and a change in the aperture stop manufacturing. A schematic of the improved star tracker verison is given in figure 11.

Future designs may have the option to increase the length of the inner baffle portion, and possibly exclude the outer baffle section. At the present baffling efficiency, the additional reduction afforded by the outer section at angles greater than 74° provide an insignificant margin of improvement on the system: the 10^{-7} PST achieved with a single inner baffle portion is within 25% of the nominal net system performance, limiting the total noise level increase to 25% of the level seen in a dark frame. With the omission of the outer baffle (which does little to improve performance at stray light angles less than 74°), the length of the inner baffle skirting the 30° FOV can be increased as shown in figure 12 which has 2 benefits. Firstly, the angle at which the first optical element is shielded from direct illumination is decreased, allowing operation at nearer sun angles. Secondly, the PST of the overall system showed a slight tendency to decrease (inversely with the baffle length) as the front lip of the baffle was extended.

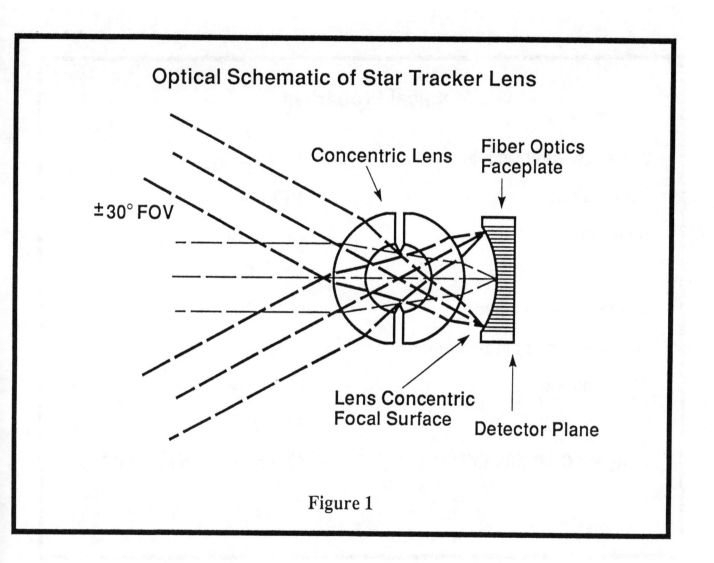

Figure 1

Optical Properties

Entrance Pupil Diameter	1.4 cm
Focal Length	1.77 cm
Response Waveband	0.4 μm to 1.1 μm
Field of View (FOV)	28° x 42°
On-Axis Transmission	50%
Transmission at Maximum FOV	30%
Blur Diameter	50 μm
Detector IFOV	1.3 mrad
Detector Quantum Efficiency	40% peak from 0.6 to 0.9 μm

Figure 2

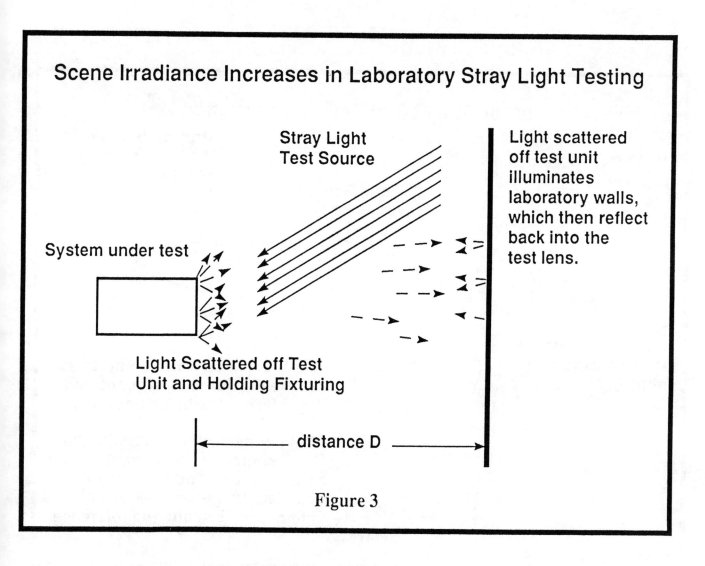

Figure 3

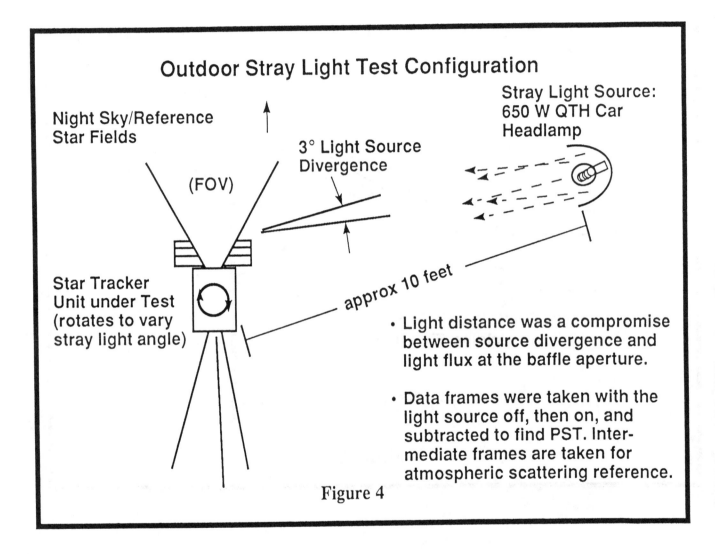

Outdoor Stray Light Test Configuration

Night Sky/Reference
Star Fields

(FOV)

3° Light Source
Divergence

Stray Light Source:
650 W QTH Car
Headlamp

Star Tracker
Unit under Test
(rotates to vary
stray light angle)

approx 10 feet

• Light distance was a compromise
between source divergence and
light flux at the baffle aperture.

• Data frames were taken with the
light source off, then on, and
subtracted to find PST. Intermediate frames are taken for
atmospheric scattering reference.

Figure 4

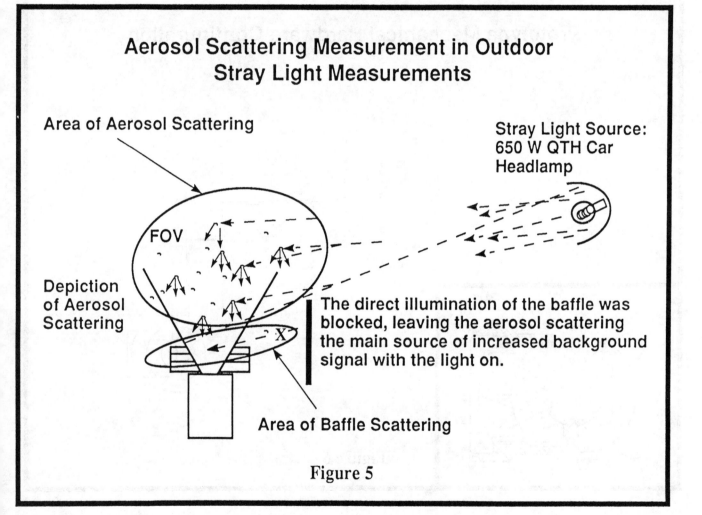

Aerosol Scattering Measurement in Outdoor Stray Light Measurements

Area of Aerosol Scattering

Stray Light Source: 650 W QTH Car Headlamp

FOV

Depiction of Aerosol Scattering

The direct illumination of the baffle was blocked, leaving the aerosol scattering the main source of increased background signal with the light on.

Area of Baffle Scattering

Figure 5

Prototype Mechanical Hardware Configuration

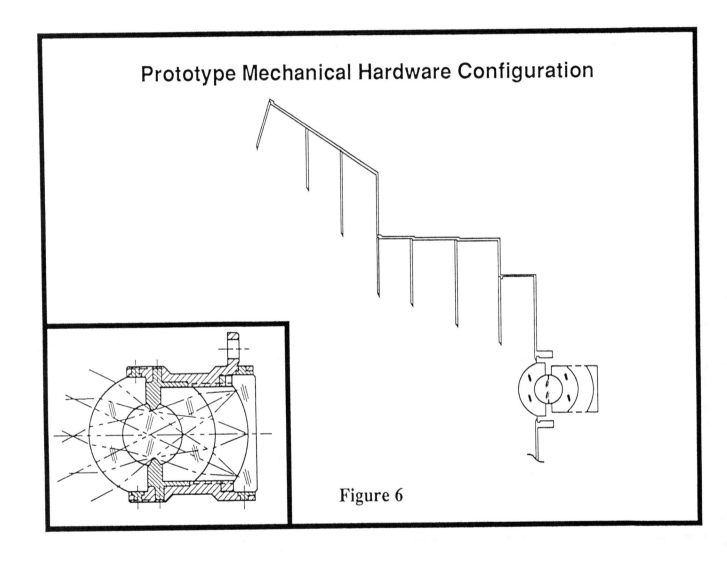

Figure 6

APART Model Depiction

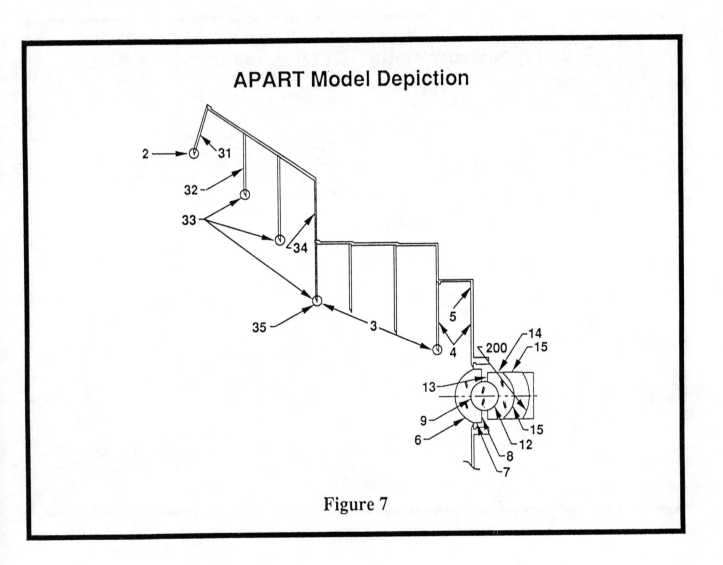

Figure 7

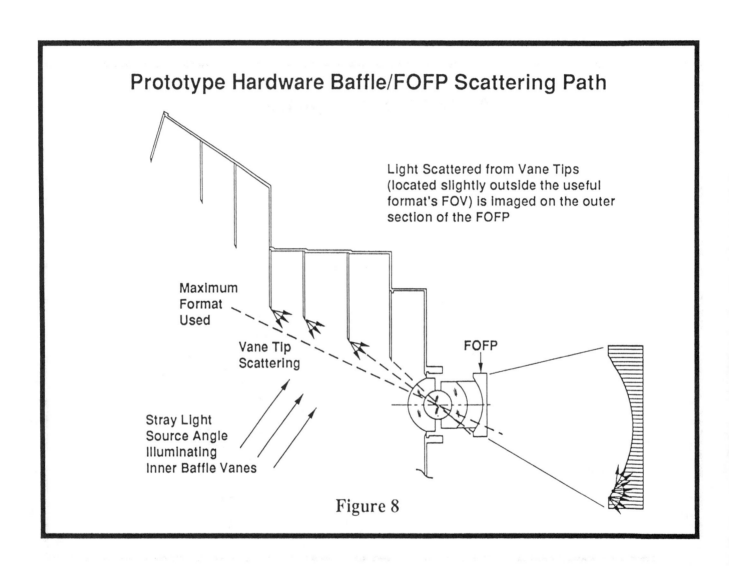

Prototype Hardware Baffle/FOFP Scattering Path

Light Scattered from Vane Tips (located slightly outside the useful format's FOV) is imaged on the outer section of the FOFP

Maximum
Format
Used

Vane Tip
Scattering

FOFP

Stray Light
Source Angle
Illuminating
Inner Baffle Vanes

Figure 8

Prototype Mechanical Hardware Baffle/Rear Wall Scattering Path with Blocking Vane Illustration

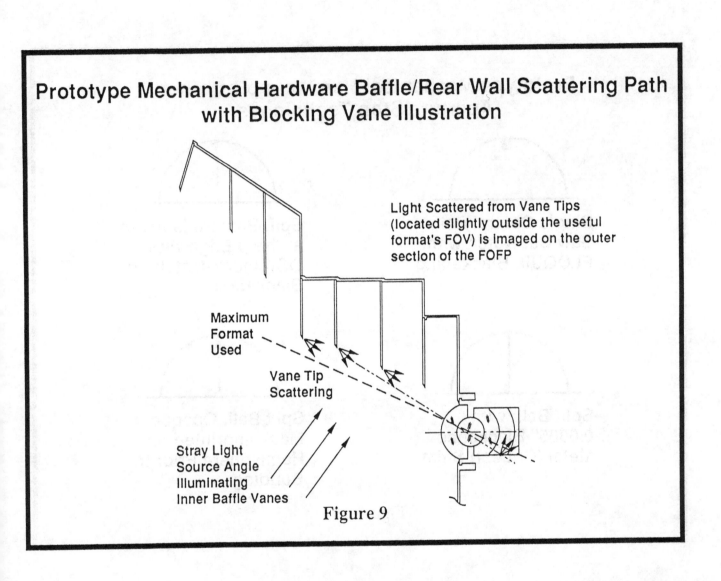

Light Scattered from Vane Tips (located slightly outside the useful format's FOV) is imaged on the outer section of the FOFP

Maximum Format Used

Vane Tip Scattering

Stray Light Source Angle Illuminating Inner Baffle Vanes

Figure 9

Prototype Mechanical Hardware plus other Aperture Stop Techniques

Single Central Ball:
Saw Cut Notch with
FLOQUIL Blackening

Split Ball: Polished to
a Sharp Edge with
OCA Index-Matching
Blackening

Split Ball: with
0.0005'' Blackened
Metal (Cr best) Shim

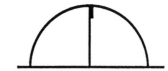

Split Ball: Opaque
Mask Imprinted on
Hemisphere Prior to
Bonding

Figure 10

Refined Hardware with Stray Light Improvements

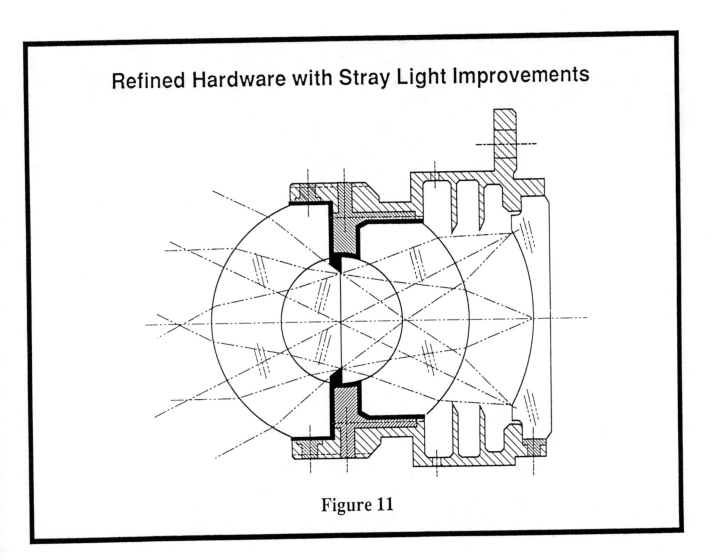

Figure 11

Suggested Alternate Baffling for Minimum Exclusion Angle

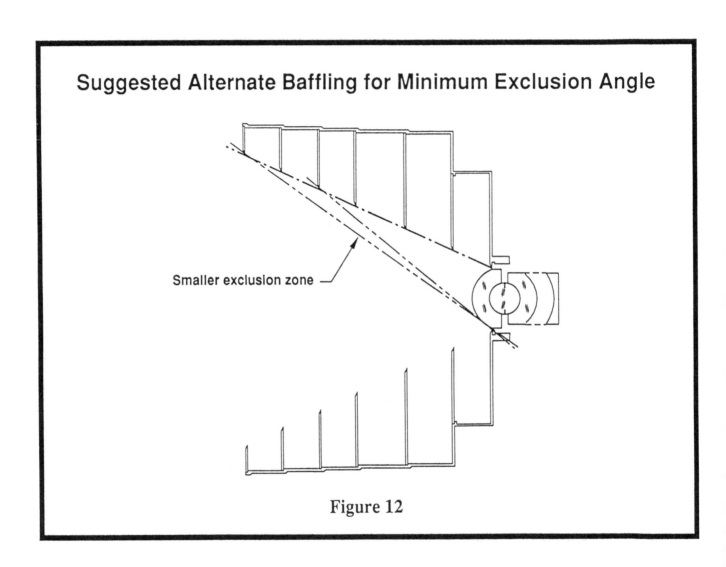

Smaller exclusion zone

Figure 12

Scattering in paper coatings

Timo Hyvärinen and Juha Sumen

Technical Research Centre of Finland, Electronics Laboratory
P.O.Box 200, SF-90571 Oulu, Finland

ABSTRACT

Paper coatings are porous structures of pigments, binder and air. This research aims at developing a model for the prediction of the bulk optical properties of coating from its structure and material characteristics. Two theoretical approaches have been studied and compared. In both cases the coating is treated as a structured layer consisting of individual scatterers in a medium, and the pores are regarded as scatterers. In the first approach they are assumed to be independent, the second approach takes approximately into account that due to their close positions the pores are in fact correlated scatterers. The approximation used limits the consideration to pair correlation, only. Comparison of the theories to reflectance measurements on polystyrene pigment coatings having uniform pore sizes in the range from 0.13 to 0.95 μm showed that for pore size parameters less than 5 the dependent scattering model provides a good prediction for the reflectance, but for larger pore size parameters the reflection behaviour approaches that predicted by the independent scattering theory.

1. INTRODUCTION

Printing papers are often coated to reach a finer pore structure and a smoother surface which improve the quality of the print and the opacity of the paper. Coatings are applied on paper as an aqueous suspension of a pigment and a binder in soluble or particulate form. Kaolin and calcium carbonate are conventional pigment materials, recently also different types of plastic pigments have been developed. Water is removed by drying, and at the end of the coating process a porous structure consisting of pigments, binder and air has formed. In paper coatings the void fraction ranges from 0.2 to 0.4 and that of the pigments from 0.75 to 0.5, the rest being occupied by the binder. The pore size may vary from 0.05 to about 1 μm, and the thickness of the coating from 5 μm to several tens of μm.

For coating research and development purposes it is essential to have an understanding on how the structure of the coating affects the optical properties. In this work the main interest is in coating reflectance and opacity. The bulk optical properties of paper and coating have mainly been investigated using the Kubelka-Munk (KM) or similar theories.[1,2,3] KM theory requires two experimental parameters (KM scattering and absorption coefficient) which can not directly be connected to the structure of the coating. To overcome this problem we have studied two more sophisticated theoretical approaches which enable the calculation of the optical properties from the real physical structure characteristics and material properties. Both of the approaches start from the properties of single scatterers, and use the radiative transfer theory (RTT)[4,5] for modelling multiple scattering in the coating. This implies that the coating is treated as a structured layer, not as a surface, only. The pores are more apart from each other than the pigments which, in addition, are usually bound to each other with a binder material having the refractive index close to that of the pigment material. It is therefore more feasible to consider the pores as single scatterers. In the first approach the pores are assumed to be independent, and their optical properties are calculated in a straightforward way from the Mie theory.[6] However, even the pores are situated so close to each other that their positions are correlated, and from the scattering theory point of view, paper coating is a dense distribution of scatterers for which the assumption of independent scattering is not valid. In this case the material's scattering and absorption coefficients per unit length are nonlinear functions of the scatterer's concentration. This phenomenon, studied in the second approach of this work, is often called dependent scattering. The theoretical and mathematical treatment of dependent scattering is very complicated. We have used the approach developed by Varadan et al [7] and Tsang et al [8].

The experimental verification of the theories is made difficult by the problems both in characterizing the coating structure reliably and in manufacturing coatings with a known pore structure. Also, only rough information is available on the imaginary refractive indices of pigment and binder materials. Previous experiments in which coatings have been characterized for both optical and structure properties are limited to small pore sizes[9,10] or to concise optical measurements.[11] To obtain more experimental data for the verification of the theoretical models we have prepared a set of coatings from polystyrene (PS) spheres, and measured their spectral reflectances. The coatings have uniform pore sizes covering the pore size range encountered in practical coatings.

2. THEORY

2.1 Mie theory + radiative transfer theory

The calculation of the optical properties of a single independent scatterer requires information on the size and shape of the scatterer as well as on its imaginary refractive index relative to that of the surrounding medium. Throughout this work we assume the pores to be spherical. This is expected to be a good approximation for all coatings except for those made of platelike kaolin pigments. For them a spheroid model might be a better choice. The imaginary refractive index of the pigment/binder mixture is directly calculated from the individual values as a volume weighted average. With these simplifications, the extinction cross section C_e and asymmetry parameter g for a pore are calculated from the Mie theory. The total extinction coefficient β_t for the coating, including absorption by the medium, is

$$\beta_t = \beta_e + \beta_m = 3fC_e/4\pi a^3 + \beta_m \,, \tag{1}$$

where n_o is the number of pores per unit volume, f is the porosity of the coating, a is the radius of a pore, and β_m is the absorption coefficient of the medium. For a pore the extinction and scattering cross sections are the same, and therefore, the albedo w for a coating volume element is obtained by

$$w = \beta_e/\beta_t \,. \tag{2}$$

In order to calculate the reflectance of the coating layer, β_t, w and g, together with suitable boundary conditions, are used as inputs for the RTT. Paper reflectance and opacity are usually determined with an integrating sphere set-up giving diffuse illumination on the sample and measuring the hemispherically integrated reflectance. The area under measurement is much larger than the thickness of the coating and paper, and therefore, the radiative transfer equation (RTE) for a parallel plane geometry can be used as a good approximation. A further simplification is achieved by using the so-called Eddington and delta-Eddington approximations to solve the RTE.[12,13] These approximations are well suited for cases where parameters averaged over the angle distribution (such as the hemispherically integrated reflectance) are determined. The Eddington approximation is used for asymmetry values g < 0.6 and the delta-Eddington approximation for g ≥ 0.6.[14] With diffuse illumination the boundary reflections do not substantially depend on the actual shape and roughness of the surface, and therefore we have chosen the mathematically simplest case, a smooth surface, to describe the coating surfaces. With these assumptions, and after an azimuthal averaging, the RTE takes the following form in the Eddington approximation:

$$\mu\frac{d(I_o+\mu I_1)}{d\tau} = - (I_o + \mu I_1) + w(I_o + g\mu I_1) + wF_of_d(1 + 3g\mu\mu_o')\exp(-\tau/\mu_o')/4\pi + wF_of_u(1 - 3g\mu\mu_o')\exp(\tau/\mu_o')/4\pi \,, \tag{3}$$

where $I_o+\mu I_1=I_o(\tau)+\mu I_1(\tau)=I_d(\tau)$ is the diffuse intensity as divided to two components by the Eddington approximation, F_o is the incident irradiance, $\mu=\cos\theta$ is the cosine of the angle of observation, $\mu_o'=\cos\theta_o'$ is the cosine of the angle of the incident radiation in the coating, and τ is the optical distance, $\tau = \beta_t z$ (Fig. 1). Eq. 3 has an equivalent form in the delta-Eddington approximation, but transformations for the parameters τ, w and g are needed.[13] The last two terms in Eq. 3 take into account the multiple boundary reflections of the incident beam. They are significant with very thin coatings and for angles near to the normal of the boundary, only. For angles near to the normal the reflectivities for radiation coming from air to coating and vice versa are close to each other, and f_d and f_u can be approximated by

$$f_d \approx \{1 - r(\mu_o)\}/\{1 - r(\mu_o)^2\exp(-2\tau_2/\mu_o')\} \,, \text{ and} \tag{4}$$

$$f_u \approx \{1 - r(\mu_o)\}r(\mu_o)\exp(-2\tau_2/\mu_o')/\{1 - r(\mu_o)^2\exp(-2\tau_2/\mu_o')\} \,, \tag{5}$$

where τ_2 is the optical thickness of the coating, and $r(\mu_o)$ is the single boundary reflection for radiation from air to coating with an angle θ_o and is determined by n_1, n_2 and θ_o. n_2 is determined here from the refractive indices of air and pigment/binder matrix as a volume weighted average.

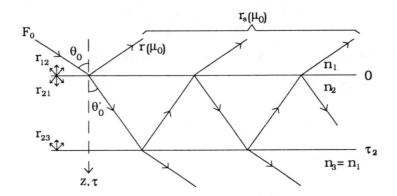

Figure 1. Reflectivities and angles for radiative transfer in coating.

Boundary conditions for the diffuse irradiances are:

$$F_-(0{:}\mu_o') = F_+(0{:}\mu_o')r_{21} \text{ , and} \tag{6}$$

$$F_+(\tau_2{:}\mu_o') = F_-(\tau_2{:}\mu_o')r_{23} \text{ ,} \tag{7}$$

where F_+ and F_- are upward and downward directed irradiances, and r_{21} and r_{23} are boundary reflections for diffuse illumination coming from coating side (Fig 1). In this case $r_{21} = r_{23}$, and r_{21} is obtained from r_{12} (the reflection for diffuse illumination from air to coating) by

$$r_{21} = 1 - (1-r_{12})/(n_2/n_1)^2 \text{ .} \tag{8}$$

The diffuse intensity components I_o and I_1 can now be solved using the procedure described by Shettle and Weinman,[12] and the total hemispherically integrated reflectance for incident radiation in the direction μ_o is

$$R_s(\mu_o) = \frac{(1-r_{21})\pi\{I_o(0{:}\mu_o) - 2I_1(0{:}\mu_o)/3\}}{\mu_o F_o} + r_s(\mu_o) \text{ ,} \tag{9}$$

where

$$r_s(\mu_o) = r(\mu_o) + \{1 - r(\mu_o)\}f_u \text{ .} \tag{10}$$

The reflectance for diffuse illumination is obtained by integrating over all incident directions μ_o,

$$R_{sd} = 2 \int_0^1 \mu_o R_s(\mu_o)d\mu_o \text{ .} \tag{11}$$

The extension of the model for a pore size distribution is straightforward; the distribution weighted single scattering properties are first calculated and then used in the RTT.

2.2 Dependent scattering + radiative transfer theory

The basis of the theoretical approach which we use for dependent scattering is described in detail by Tsang et al.[8] It starts with the configurational averaging (over the scatterer positions) of the multiple scattering equations. This leads to a hierarchy of equations which is truncated by using the quasicrystalline approximation (QCA), resulting in the consideration of pair correlation between the scatterers, only. By assuming spherical scatterers of one species embedded in a medium with a propagation constant k, a plane normally incident wave, and a plane wave solution for the coherent (average) field, and by using the T-matrix formalism for the scatterers, the following generalized Evald Oseen and Lorentz-Lorenz equations for the dependent (effective) propagation constant K for a dense scattering and absorbing medium can be derived [8]:

$$K - k = -\frac{\pi i n_0}{k^2} \sum_n (2n+1)\left[T_n^M X_{1n}^M + T_n^N X_{1n}^N\right], \text{ and} \tag{12}$$

$$X_{1\upsilon}^M = -2\pi n_0 \sum_n \sum_p (2n+1)\left[L_p(k,K|b) + M_p(k,K|b)\right]T_n^M X_{1n}^M a(1,n|-1,\upsilon|p) A(n,\upsilon,p) + \tag{13}$$
$$T_n^N X_{1n}^N a(1,n|-1,\upsilon|p,p-1) B(n,\upsilon,p),$$

$$X_{1\upsilon}^N = -2\pi n_0 \sum_n \sum_p (2n+1)\left[L_p(k,K|b) + M_p(k,K|b)\right]T_n^M X_{1n}^M a(1,n|-1,\upsilon|p,p-1) B(n,\upsilon,p) + \tag{14}$$
$$T_n^N X_{1n}^N a(1,n|-1,\upsilon|p) A(n,\upsilon,p)$$

where
$$L_p(k,K|b) = -\frac{b^2}{K^2 - k^2}\left[kh_p{}'(kb)j_p(Kb) - Kh_p(kb)j_p{}'(Kb)\right], \text{ and} \tag{15}$$

$$M_p(k,K|b) = \int_b^\infty r^2\left[g(r) - 1\right]h_p(kr)j_p(Kr)dr \tag{16}$$

and A, B and a are coefficients depending on the harmonic indices n, υ and p. T_n^N and T_n^M are T-matrix elements for a sphere, and are related to the complex conjugates of the Mie coefficients a_n and b_n by $T_n^M = -b_n{}^*$ and $T_n^N = -a_n{}^*$. b is the diameter of the sphere, h_p is Hankel and j_p spherical Bessel function, and g(r) is the pair distribution function describing the ratio of the number of scatterers per unit volume at a radial distance r from the centre of a reference particle to the average number density. Pair distribution functions have been studied in statistical mechanics, and based on comparisons to Monte Carlo simulation which enables the exact pair distribution function to be calculated, the simplest and most satisfactory approximation for the pair distribution function of nonpenetrable spherical particles is the Percus Yevick model.[8] It is used in this work. K and the unknown field coefficients X_{1n}^M and X_{1n}^N can be numerically solved from Eqs. 12 - 14. The real and imaginary parts of K are proportional to the dependent refractive index n_e and extinction coefficient β_{te} of the dense medium, respectively,

$$n_e = \lambda \text{Re}(K)/2\pi, \text{ and} \tag{17}$$

$$\beta_{te} = 2\text{Im}(K). \tag{18}$$

The dependent extinction coefficient β_{ee} and cross section C_{ee} for a pore and the dependent albedo w_e for a coating volume element in the dense medium can now be calculated by

$$\beta_{ee} = 2\text{Im}(K-k), \tag{19}$$

$$C_{ee} = 2\text{Im}(K-k)/n_0, \text{ and} \tag{20}$$

$$w_e = \beta_{ee}/\beta_{te}. \tag{21}$$

In order to obtain the angle distribution for dependent single scattering the incoherent field has to be calculated, but with the first order multiple scattering contribution, only. This is done by applying the distorted Born approximation (DBA) in deriving the equation for the incoherent field.[8] By assuming a randomly polarized incident wave, the following equation for the dependent scattering cross section C_{se} of a spherical scatterer can be derived:

$$C_{se} = \frac{\pi}{|k|^2} \int_0^\pi \left[\left|\sum_n \frac{2n+1}{n(n+1)}\left\{T_n^M X_{1n}^M \pi_n(\cos\theta_s) + T_n^N X_{1n}^N \tau_n(\cos\theta_s)\right\}\right|^2 + \right.$$
$$\left.\left|\sum_n \frac{2n+1}{n(n+1)}\left\{T_n^M X_{1n}^M \tau_n(\cos\theta_s) + T_n^N X_{1n}^N \pi_n(\cos\theta_s)\right\}\right|^2\right] W(\theta_s,\phi_s)\sin\theta_s d\theta_s, \tag{22}$$

where

$$W(\theta_s, \phi_s) = 1 + n_0 \mathrm{Re}\left[\int_{-\infty}^{\infty} [g(r) - 1] \exp[i(\mathrm{Re}K - k_s) \cdot r] \right] dr \tag{23}$$

and π_n and τ_n are the angle function notations used in Mie theory.[6] (θ_s, ϕ_s) is the scattering angle. The vector K goes downward and the vector k_s in the scattered direction. The effective asymmetry parameter g_e is calculated by multiplying the integrand in Eq. 22 by $\cos\theta_s$ and then dividing the result by C_{se}. It is interesting to compare Eq. 22 with the classical Mie theory. In Mie theory the coefficients X and W are absent. For a sparse concentration of scatterers the values of X and W approach 1, and the two cases reduce to each other.

The reflectance of the coating layer is calculated from the RTT as in Sect. 2.1, now using the dependent scattering parameters β_{te}, w_e, g_e and n_e instead of the independent ones. The extension of the dependent scattering theory for a scatterer size distribution is mathematically very complicated and has not yet been studied in this work.

2.3 Comparison between the theories

A computer program has been generated including a Mie scattering code for mono- and polydispersions, a dependent scattering code for monodispersions and a code for the approximate solution of the RTE. The Mie code is based on that developed by Wiscombe[15], and has been extended for an absorbing matrix. The calculation of the dependent propagation coefficient K has been generated using the procedure outlined by Tsang et al.[8] In maximum 25 equations (n = 25) in Eqs. 13 and 14 can be retained in the calculation in its present form. This enables dependent scattering calculations for pore size parameters up to about 20. The input parameters required by the computer program are summarized in Table 1. They all are real physical parameters.

The two theoretical approaches have been compared in the case of a porous PS coating. Values used in the calculations for the imaginary refractive index of PS are presented in Table 2. Values for the real part in the visible range have been taken from Vasalos[16], and the values in the near infrared region have been extrapolated from these. The imaginary part has been measured, but due to the low level of absorption, is a rough estimate only. Fig. 2 shows the extinction efficiency and asymmetry parameter for a pore in a PS matrix as calculated from Mie and the dependent scattering theories. Pore size ranges from 0.05 to 1 μm corresponding to a size parameter from 0.74 to 10.9 at 460 nm wavelength. For pore size parameters smaller than 4.5 the dependent extinction efficiency is smaller than the independent value, the relative difference increasing with smaller pore sizes. For pore size parameters larger than 4.5 the dependent extinction efficiency exceeds the independent value, the first maximum having a strong amplification with an increasing concentration. The dependent asymmetry parameter is smaller than the Mie value for all pore sizes. In Fig. 3(a) the calculated dependent and independent extinction coefficients for pores in polystyrene are compared as a function of concentration (porosity). In the independent theory the extinction coefficient increases linearly with porosity. According to the dependent scattering theory the extinction coefficient value saturates, for small size parameters even at 10% porosity, and there are very appreciable deviations from the linear behaviour. These deviations can be expected by physical intuition; if the volume fraction of scatterers is increased, the limit where the entire volume is occupied by the scatterers will be approached and the medium becomes homogeneous again with no scattering. The porosity value at which dependent extinction coefficient reaches its maximum increases with size parameter and is about 35% for a size parameter of 4.5.

The KM scattering coefficient S per square weight ω is commonly used in paper industry to describe the opacity or 'hiding power' of paper and coating. S relates the reflectance of a coating layer (R_0) to the reflectance of an infinitely thick coating (R_∞) in the following way:

$$S = \frac{1}{\omega} \frac{R_\infty}{1 - R_\infty^2} \ln\left[\frac{R_\infty(1 - R_0 R_\infty)}{R_\infty - R_0} \right] \tag{24}$$

Fig. 3(b) compares the KM scattering coefficient values calculated from the dependent and independent reflection values for a 30 μm thick PS coating. The dependent scattering theory predicts significantly smaller reflectance and KM scattering coefficient values for pore size parameters less than 1.5. The KM scattering coefficient reaches its first maximum at a slightly larger pore size in the dependent scattering theory, and after that the dependent scattering coefficient values are larger. With high porosities there is also a clear second maximum in the KM scattering coefficient at a size parameter of about 6 according to the dependent scattering theory.

Table 1. Input parameters in the theoretical models.

	indep. model	dep. model
Pore size	x	x
Pore size distribution	x	
Imaginary refractive index of pigment and binder	x	x
Porosity	x	x
Square weight or thickness	x	x
Illumination (directed or diffuse)	x	x
Wavelength	x	x
Pair distribution function for pores		x

Table 2. Imaginary refractive index of polystyrene.

$\lambda(\mu m)$	m
0.5	$1.602 + i10^{-6}$
0.6	$1.588 + i10^{-6}$
0.7	$1.583 + i10^{-6}$
0.8	$1.578 + i10^{-6}$
0.9	$1.575 + i10^{-6}$
1.0	$1.574 + i10^{-6}$
1.1	$1.572 + i10^{-6}$
1.2	$1.570 + i10^{-6}$

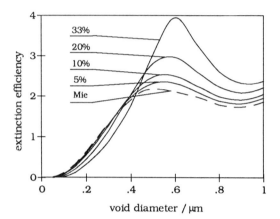
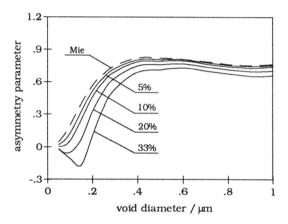

Figure 2. Calculated independent and dependent (for 5, 10, 20 and 33 % porosities) extinction efficiency and asymmetry parameter for a pore in a polystyrene matrix as a function of the pore size. λ = 460 nm.

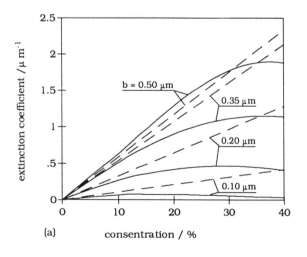
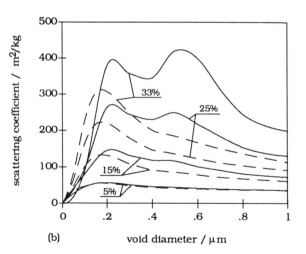

(a)   (b)

Figure 3. (a) Calculated extinction coefficient for pores in a polystyrene matrix as a function of porosity, b is pore diameter, λ = 540 nm. (b) Calculated independent and dependent (for 5, 15, 25 and 33 % porosities) KM scattering coefficient of a porous polystyrene coating as a function of the pore size, λ = 460 nm.

This comparison has revealed considerable differences between the two theories both in the single scattering properties of uniform pores and in the bulk optical properties of a coating having a uniform pore size. Practical coating structures always have some kind of a pore size distribution. Comparison for pore size distributions is left for future work because of the difficulties in extending the dependent scattering approach for size distributions. Our preliminary reasoning is, however, that the dependent scattering phenomena will appear less significantly with larger pore size distributions, and the dependent scattering properties approach those predicted by the independent scattering theory.

3. COMPARISON WITH EXPERIMENTS

Mie theory for independent single scattering and the RTT for multiple scattering in distributions of uncorrelated scatterers have been extensively studied experimentally, and are generally accepted to be valid in these cases. There exists much less experimental work with dependent scattering, the published investigations being limited to experiments with water suspensions of monosize PS spheres.[16,17,18,19,20] The results show, for size parameters up to about 7 and concentrations up to 40%, a good agreement between theory and experiment for the dependent extinction coefficient (coherent wave extinction). Extinction measurements with larger relative refractive index differences than between PS and water require very thin sample layers and are therefore difficult to arrange. The direct measurement of the dependent single scattering albedo and asymmetry parameter is also very complicated and has not been attempted. In the first phase, reported in this work, we have tried to verify the validity of the theoretical models for coating structures having a fairly uniform pore size. This has been carried out by measuring the spectral hemispherically integrated reflectance of binderless coatings prepared from PS spheres.

3.1 Measurement set-up

The reflectance measurement set-up is depicted in Fig. 4. Light from a monochromator is modulated and led to an integrating sphere in order to obtain diffuse illumination on the sample surface and a hemispherically integrated reflectance measurement. Diameter of the sample port is 20 mm and the diameter of the area seen by the detector on the sample surface is 10 mm. A substitution method in which the sample and reference are measured at the same port at different times is used. The system is calibrated by diffuse reflection standards having a direct traceability to NIST. In reflection measurement the accuracy of the set-up is estimated to be ± 1%.

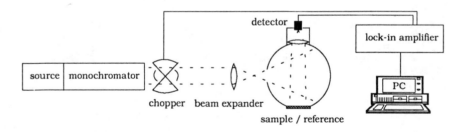

Figure 4. Reflectance measurement set-up.

3.2 Polystyrene sphere coatings

The coatings were applied manually on glass substrates as aqueous suspensions, and were dried in room temperature. 1% of carboxymethyl cellulose was added to prevent cracking during drying. Six different coatings were prepared using the sphere sizes d shown in Table 3. The analysis of the scanning electron micrographs of the upper and broken cross-sectional surfaces of the coatings showed the structure of the coatings to be quite homogeneous with a fairly uniform pore size (Fig. 5). The hydraulic pore diameter[21] $b_h = 2fd/3(1-f)$ was used as an approximation for an effective spherical pore diameter (Table 3). For porosity the value of 33% which is the average measured by Alince and Lepoutre for polystyrene coatings prepared in the same way. The thicknesses of the coatings were determined after the reflectance measurements by removing parts of the coating around the measured area and scanning several times across the surface by a mechanical profilometer.

Table 3. Coatings prepared from uniform PS spheres.

Sample	Sphere diam. (μm)	Effective pore diam. (μm)	Coating thickness (μm)
p039	0.39	0.13	44
p080	0.80	0.26	33
p127	1.27	0.42	62
p150	1.50	0.49	57
p236	2.36	0.77	60
p295	2.95	0.95	39

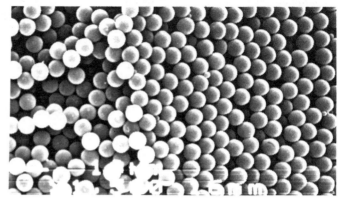

Figure 5. Scanning electron micrograph showing parts of an upper (right) and a broken cross-sectional surface (left) of the coating sample p236.

3.3 Measurements vs. theory

The hemispherically integrated reflectances of the prepared coatings were measured in the wavelength range 650-1100 nm, and are shown in Fig. 6 together with the calculated reflectance values. It is worth emphasising that the theoretical calculations are performed without any kind of parameter fitting. There is a good agreement between experiment and the dependent scattering model for all but the coating with the largest pore size. However, either of the theories is not capable of predicting the small periodical fluctuations appearing in the measured reflection spectra, most clearly on samples p080, p127, p150 and p236.

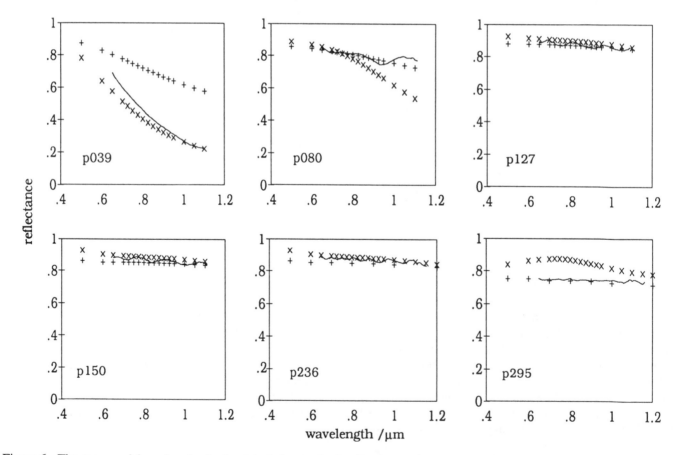

Figure 6. The measured (——) and calculated (+ indep., x dep.) reflection spectra of six polystyrene sphere coatings characterized in Table 3.

The KM scattering coefficient per square weight was calculated from the reflection values at 700 nm, and is depicted in Fig. 7 as a function of the pore size. Also these results show a good agreement between experiment and the dependent scattering theory for pore size parameters up to about 5. After that the measured reflection behaviour seems to approach that predicted by the independent theory. It is interesting to notice that the second maximum in KM scattering coefficient at the size parameter of about 6 (corresponding to a pore size of about 0.9 µm at λ = 700 nm) is much weaker than predicted by the dependent scattering theory.

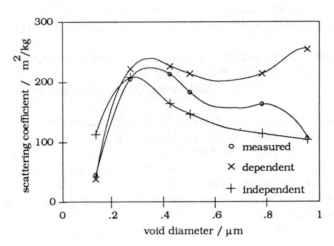

Figure 7. The measured and calculated KM scattering coefficient at λ = 700 nm for the polystyrene pigment coatings as a function of pore size.

4. DISCUSSION

Both of the introduced theoretical approaches are capable of predicting the trends in the bulk optical properties of paper coating very well without the need for experimental or fitted parameters. For the first time it is also possible to have reliable absolute values for optical properties of coating directly from the real structure and material characteristics. As could be expected, the dependent scattering theory provides a better absolute model for small (< 5) pore size parameters, whereas it remains to be investigated whether the reason for its failure at larger size parameters is in the theory itself, in the approximations made in deriving the equations for the dependent single scattering parameters, or in numerical solutions. This problem may be associated to discrepancies found between the calculated values of C_{ee} (Eq. 20) and C_{se} (Eq. 22) for larger size parameters, although the values should be the same for a pore which is a nonabsorbing scatterer. C_{se} is not used in the RTT, but the asymmetry parameter which is calculated from the same equation is required in RTT.

It was interesting to notice that also the independent scattering theory, being mathematically more simple and significantly less CPU time consuming, introduces a fairly good quantitative approximation at least for angle integrated optical properties even for coatings with a uniform pore size. We expect that with larger pore size distributions the dependent scattering phenomena will become less significant and the independent scattering theory will provide a very feasible model. This will be investigated in the second phase of the research by using real coatings. An interesting future opportunity is the combination of a rough surface scattering theory to this kind of bulk scattering theory in order to model quantitative angle resolved optical properties.

REFERENCES

1. G. Kortum, *Reflectance spectroscopy*, Springer Verlag, Berlin, 1969.
2. A.M. Scallan, and J. Borch, "An interpretation of paper reflectance based upon morphology: general applicability", *Tappi*, Vol. 57, 143-147, 1974
3. J.V. Robinson, and E.G. Linke, "Theory of the opacity of films of coating pigment and adhesive. A method for calculating the opacity of coating", *Tappi*, Vol. 46, 384-390, 1963.
4. S. Chandrasekhar, *Radiative transfer*, Dover Publications, Inc., New York, 1960.
5. A. Ishimaru, *Wave propagation and scattering in random media, Vol. 1 and Vol. 2*, Academic Press, New York, 1978
6. H.C. van de Hulst, *Light scattering by small particles*, Wiley, New York, 1957.

7. V.N. Bringi, V.V. Varadan, and V.K. Varadan, "Coherent wave attenuation by a random distribution of particles", *Radio Sci.*, Vol. 17, 946-952, 1982.

8. L. Tsang, J.A. Kong, and R.T. Shin, *Theory of microwave remote sensing*, Wiley-Interscience, New York, 1985.

9. B. Alince, and P. Lepoutre, "Light-scattering of coatings formed from polystyrene pigment particles", *J. Colloid & Interface Sci.*, Vol. 76, 182-187, 1980.

10. G. Lindblad, T. Iversen, H. Bergenblad and M. Rigdahl, "Light scattering ability of coatings as a tool to protect paper and board against photo-oxidation", *Nordic Pulp and Paper Res. J.*, No. 4, 253-257, 1989.

11. N.A. Climpson, and J.H. Taylor, "Pore size distributions and optical scattering coefficients of clay structures", *Tappi*, Vol. 59, 89-92, 1976.

12. E.P. Shettle, and J.A. Weinman, "The transfer of solar irradiance through inhomogeneous turbid atmospheres evaluated by Eddington's approximaton", *J. Atmos. Sci.*, Vol. 27, 1048-1055, 1970.

13. J.H. Joseph, W.J. Wiscombe and J.A. Weinman, "The delta-Eddington-approximation for radiative flux transfer", *J. Atmos. Sci.*, Vol. 33, 2452-2459, 1976.

14. W.J. Wiscombe, and J.H. Joseph, "The range of validity of the Eddington approximation", *Icarus*, Vol. 32, 362-377, 1977.

15. W.J. Wiscombe, *Mie scattering calculations: advances in technique and fast, vector-speed computer codes*, National Center for Atmospheric Research, Report no NCAR/TN-140+STR, 1979.

16. I.A. Vasalos, *Effect of separation distance on the optical properties of dense dielectric particle suspensions*, Doctoral Dissertation, Massachusetts Institute of Technology, 1969.

17. V.K. Varadan, V.N. Bringi, V.V. Varadan and A. Ishimaru, "Multiple scattering theory for waves in discrete random media and comparisons with experiments", *Radio Sci.*, Vol. 18, 321-327, 1983.

18. V.V. Varadan, Y. Ma and V.K. Varadan, "Propagator model including multipole fields for discrete random media", *J. Opt. Soc. Am. A*, Vol. 2, 2195-2201, 1985.

19. H.C. Hottel, A.F. Sarofim, W.H. Dalzell and I.A. Vasalos, "Optical properties of coatings. Effect of pigment concentration", *AIAA Journal*, Vol 9, 1895-1898, 1971.

20. T. Hyvärinen, and J. Sumen, "Optical properties of porous pigment coatings; dependent vs. independent scattering", to be published in Proc. ICO Topical meeting on atmospheric, volume, and surface scattering and propagation, Florence, 27-30 Aug., 1991.

21. L.C. Graton, and H.J. Fraser, "Systematic packing of spheres - with particular relation to porosity and permeability", *J. Geol.*, Vol. 43, 785-909, 1935.

IR BRDF measurements of space shuttle tiles*

R. P. Young, B. E. Wood, and P. L. Stewart

Calspan Corporation/AEDC Operations
Arnold Engineering Development Center
Arnold Air Force Base, Tennessee 37389

1. INTRODUCTION

Ground-based infrared (IR) cameras are used to support a series of experiments associated with studying the condition of space shuttle tiles while the shuttle is in flight. This includes studying the reflective properties of the shuttle tiles while the shuttle is in orbit. The IR cameras used in these experiments are sensitive in the 2.0- to 5.5-µm wavelength range; therefore, the shuttle tile optical properties are required for this same spectral range. An optical property of particular interest is the tile Bidirectional Reflectance Distribution Function (BRDF). The tile BRDF is input into radiation models to help predict the expected power levels observed by the IR cameras. The objective of the tests reported herein was to measure the tile BRDF over the range of 2.0 to 5.5 microns as a function of the reflectance angle. Both black and white as well as flown and unflown shuttle tiles were measured.

The radiation sources used for BRDF measurements are typically lasers,[1] thus restricting the availability of radiation wavelengths available. It was desired that the BRDF measurements of the shuttle tiles be made at nominally 0.5-µm intervals between 2.0 and 5.5-µm. Therefore, a broadband (blackbody) radiation source was selected for these measurements. Narrowband filters provided the wavelength selection. The test results are presented as plots of the spectral BRDF as a function of the angle measured from the surface normal.

2. TEST APPARATUS

The BRDF test equipment (Fig. 1) was mounted on a 5- by 10-ft optical table located in a large downflow clean room. The test setup included a source assembly, chopper, flat folding mirror, focusing mirror, test sample support assembly, rotary arm assembly, and an infrared (IR) detector. The source assembly was an Optronics Laboratories IR source attachment (Model 740-20IR) for a Model 746 spectrometer. It contained an IR (Nernst) glower and a 150-w quartz halogen visible source, either of which could be selected by rotation of the spherical imaging mirror. The IR glower was operated at nominally 1,000°C. The spectral

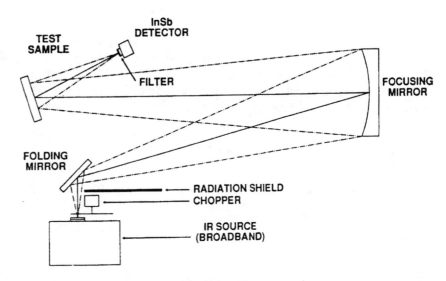

Fig. 1. Diagram of BRDF test equipment.

* The research reported herein was performed by the Arnold Engineering Development Center (AEDC), Air Force Systems Command. Work and analysis for this research were done by personnel of Calspan Corporation/AEDC Operations, operating contractor for the AEDC aerospace flight dynamics facilities. Further reproduction is authorized to satisfy needs of the U.S. Government.

radiance of a 1,000°C blackbody is presented in Fig. 2. The selected source lamp was imaged at the source assembly output aperture by a spherical mirror. The source assembly output aperture was rectangular (0.18 X 0.3 in.). A Stanford Research Systems Model SR540 variable speed optical chopper was mounted at the source assembly output aperture. The chopper was operated at 300 Hz.

The source output was directed toward an 8-in. diam, 40-in. focal length mirror by a flat folding mirror. The large mirror reimaged the source aperture at the detector. Both mirrors were mounted in 2-axis gimbaled mounts to facilitate optical alignment. The source beam incident on the test sample was slightly elliptical (nominally 1.1 in. diam.).

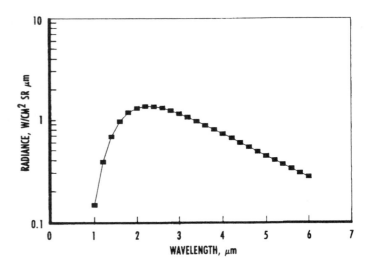

Fig. 2. Blackbody radiance.

The test samples were mounted on a mounting assembly that had 4 degrees of freedom, two in translation and two in rotation. The sample assembly could be translated along the optical axis to accommodate different sample thicknesses. Translation perpendicular to that axis was provided to center the sample mount relative to the optical beam. The rotational adjustments were used to align the test sample surface normal relative to the incident radiation. The sample mount was designed to accept 2-in.-diam test samples. A flame-sprayed aluminum test sample was used as a diffuse reflectance reference and a flat mirror was used for system alignment. The spectral reflectance of the diffuse aluminum reflector is shown in Fig. 3. A tile sample adaptor was used to mount the tile samples.

The scattered energy reflected from the test samples was measured using an InfaRed Associates photovoltaic Indium Antimonide (InSb) detector. The spectral detectivity of the detector is shown in Fig. 4. The detector was mounted within a dewar assembly with a sapphire window. The detector was 4 mm diam and operated at 77 K. The detector was operated in the current mode. Its output was measured using a low-noise Field Effects Transistor (FET) input and a Transimpedance Amplifier (TIA) designed and fabricated at AEDC. The TIA feedback resistance was 10 K. Bandpass filters for spectral measurements were mounted over the dewar window. Optical properties of the bandpass filters are provided in Table 1.

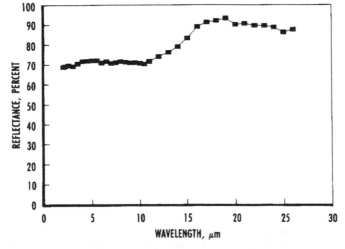

Fig. 3. Hemispherical relfectance of the diffuse aluminum reference surface.

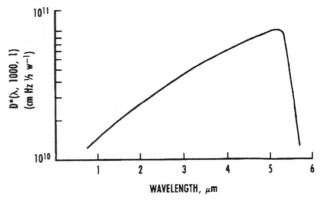

Fig. 4. InSb detector spectral dectivity.

Table 1. Filter Optical Properties

FILTER ID	WAVELENGTH, μm			
	L_1	L_2	L_c	DELTA L
4B8	2.05	2.49	2.27	0.44
4C1	2.45	2.52	2.49	0.07
4C8	2.70	2.91	2.81	0.21
6B2	3.42	3.50	3.46	0.08
6B1	4.14	4.35	4.25	0.21
6B5	4.56	4.79	4.68	0.23
8C2	5.09	5.43	5.26	0.34

L_1 - Filter short wavelength cutoff wavelength, μm
L_2 - Filter long wavelength cutoff wavelength, μm
L_c - Filter center wavelength wavelength, μm
DELTA L - Bandwidth, μm

The detector dewar was mounted on the rotary arm assembly. The assembly included an aluminum bar and an Aerotech, Inc. motor-driven rotary stage. The aluminum bar was supported on its free end with a low-friction slide. The rotational resolution was 120 motor steps per one degree of stage rotation. The detector-to-test sample distance was nominally 9.8 in.

The BRDF measurements system instrumentation included a chopper motor controller, a detector (TIA) voltage measurement instrument, and a rotary stage motor drive controller. The optical chopper drive unit provided the chopper motor power, chopper frequency readout, and a reference signal at the same frequency as the chopped optical signal. The TIA output voltage was measured using a Stanford Research phase-lock amplifier. The reference signal from the chopper drive controller was the reference signal for the phase-lock amplifier. The TIA was powered from a ±12-v power supply (four 6-v batteries). The rotary stage stepping motor was driven using an Aerotech stepping motor drive. The number of steps input to the rotary stage drive motor (for a desired drive increment) was selected by thumbwheel switches on the front of the controller.

3. TEST ARTICLES

The test articles were samples of NASA space shuttle orbiter thermal protection tiles (Fig. 5). Four tile samples were used for this measurements program. Two of the tiles were white and two were black. One each of the white and black tiles were flown on past shuttle orbiter missions and two were not flown. Tiles were identified as follows:

Sample 1: BLACK HRSI TILE
V070-391061-172
FLOWN ON CHALLENGER
AEDC IDENTIFICATION:
    FLOWN BLACK

Sample 2: BLACK HRSI TILE
V070-391061-1452
UNFLOWN
AEDC IDENTIFICATION:
    UNFLOWN BLACK

Sample 3: HITE LRSI TILE
V070-297309-084
FLOWN ON COLUMBIA
AEDC IDENTIFICATION:
    FLOWN WHITE

Sample 4: WHITE LRSI TILE
V070-394017-032
UNFLOWN
AEDC IDENTIFICATION:
    UNFLOWN WHITE

The white tiles were 8 in. square and 1/2 in. thick. The black samples were 6-in. square and 1.5-in. thick.

BLACK HRSI TILE
V070-391061-172
FLOWN ON CHALLENGER

BLACK HRSI TILE
V070-391061-145
UNFLOWN

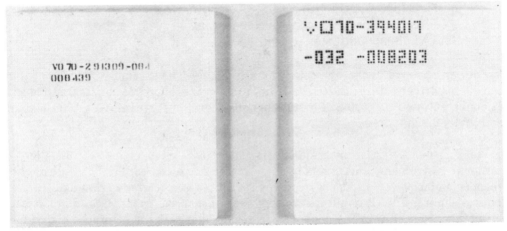

WHITE LRSI TILE
V070-291309-084
FLOWN ON COLUMBIA

WHITE LRSI TILE
V070-394017-032
UNFLOWN

Fig. 5. Shuttle tile test sample.

4. TEST DESCRIPTION

The tile BRDF measurements were taken in sets for each wavelength. After the system alignment was completed, test runs were made in the following order: (1) aluminum diffuser, (2) unflown white left, (3) unflown white right, (4) flown white left, (5) flown white right, (6) unflown black left, (7) flown black left, (8) flown black right, and (9) another aluminum diffuser. The left and right designations refer to the two different measurement locations on the tiles' surfaces. Data were not recorded for unflown black right because that area of the black tile was marked with a red dye.

The alignment procedure started by positioning the rotary arm at 5 deg from the test sample normal. The desired optical filter was installed on the detector dewar. The flat test mirror was installed in the sample holder and its surface was centered over the center of the rotary arm alignment pin. The visible source was selected and the visible beam was centered vertically on the test mirror using the large mirror vertical gimbal adjustments. The visible beam was centered horizontally over the rotary arm alignment pin using the large mirror horizontal gimbal adjustment. The test mirror gimbal was then used to position the source image at a marked reference position on the detector dewar. The angle of incidence of the optical beam relative to the mirror surface normal was nominally 9 deg. However, the angle of incidence for the tile samples was dependent on the parallelism of the tile front and back surfaces.

The visible source was turned off and the IR source was selected. The IR source was allowed nominally a 15-min warm-up. After the phase-lock amplifier was properly adjusted, the detector signal was recorded for that angular position. The rotary arm was moved to the next angular position and the detector output again recorded. The sequence of moving the rotary arm and recording the detector signal was repeated until the rotary arm was 34 deg from the test surface normal. That concluded the series of measurements for one test run.

The aluminum diffuser was removed and the rotary arm was returned to the start position. The tile adapter was installed in the test sample mount. The unflown white tile was installed in the sample adapter (identification number up). The left edge was positioned so that the illuminated spot on the tile was 2.25 in. from the left edge of the tile. The vertical spot position was 2.25 in. from the bottom of the tile for all tile measurements. Detector signal versus arm rotary position was recorded using the same procedure as was used to measure the scattered radiation from the diffuser sample. The tile sample was then translated in the tile adapter so that the illuminated spot was 2.13 in. from the right edge of the tile, and the measurements were repeated. This same procedure was used for the flown white tile and for both black tiles with the following exceptions. Only the left side of the unflown black tile was measured. The illuminated spot for the unflown black tile was 1.8 in. from the tile's left edge. The illuminated spots for the flown black sample were 2.1 in. from the left and right edges. After the tile measurements were completed, the aluminum diffuser run was repeated. This completed one test set of 9 runs. The entire test alignment and measurement procedure was performed for each optical filter and for the 0- to 5.5-µm (no filter) test runs.

5. DATA REDUCTION

The Bidirectional Reflectance Distribution Function (BRDF) for an opaque surface was defined in Ref. 2 for the general case of a surface illuminated by a distant source as:

$$BRDF = dN_r/dH_i \tag{1}$$

where N_r is the radiance reflected from the reflecting surface and H_i is the irradiance incident on the reflecting surface. The distance of the source of radiation to the reflecting surface was nominally nine times the diameter of the source aperture; thus, the distant source assumption was approximated. The area illuminated on the reflecting test sample was small in comparison to the distance between the detector and test surface; therefore, the differential notation can be dropped. Equation (1) then becomes:

$$BRDF = N_r/H_i \tag{2}$$

The incident irradiation on the test sample surface and the reflected radiance were not measured parameters, but were deduced from the voltage output of the system radiometer. The incident irradiance was much too high to be directly measured with the radiometer. Instead, it was calculated from the measurement of the radiometer signal obtained from the radiance reflected from a diffuse reflector. The following equation defines the measured BRDF in terms of the parameters measured by the scatterometer data acquisition system.

$$BRDF = \frac{\rho_D V_s \cos(\theta_r)_D}{\pi V_D \cos(\theta_r)_s} \tag{3}$$

where V_s and V_D are the radiometer signals when the mirror and diffuser surfaces are viewed, respectively. The subscript (D) refers to the diffuser and the subscript (s) refers to the tile test sample. The cosine term in the numerator is referenced to the scattering angle when the diffuser is viewed, and the cosine term in the denominator is referenced to the scattering angle when the test surface (tile) is viewed. Both angles are referenced to their respective surface normals. Note that $V_D/(\cos\theta_r)D$ is a single-valued parameter.

6. UNCERTAINTY/PRECISION OF MEASUREMENTS

The BRDF of a test sample was not a measured quantity, but was a calculated function dependent on the reference diffuser reflectivity and the output of an IR detector that viewed the illuminated diffuser and the test sample surfaces. Errors in the measured quantities propagate to the BRDF through the functional relationship as presented in Eq. (3). The detector output voltage from the diffuse and test samples was used as a voltage ratio, and the dynamic range of the voltage ratio was relatively low for these tests. Stray radiation contributions to BRDF measurements errors presented in Ref. 2 were not significant for the tile BRDF measurements reported herein because of the relatively high scatter from the test sample surfaces.

The uncertainty of the BRDF measurements resulting from the uncertainty of the measured parameters was calculated to be nominally 5 percent. The fact that the reference diffuse reflector was not perfectly Lambertian was not included in this uncertainty calculation. It was estimated that the non-Lambertian characteristic of the reference diffuser accounted for an additional 5-percent uncertainty in the BRDF measurements. The total uncertainty in the BRDF measurements is estimated to be less than 10 percent.

The angular position of the rotary arm, as measured relative to the test surface normal, was established using an alignment mirror. The uncertainty of this angular alignment was less ± 0.1 deg. However, the actual angle from the front surface of a test sample was dependent on the parallelism of the front and back surfaces. For test tiles having a visible specular reflection lobe, the error in angular measurement can be determined by the angular location of the peak BRDF value. The angular position error, based on the location of the peak BRDF value, was estimated to be ± 0.5 deg.

7.0 TEST RESULTS

The test results are presented in Figs. 6 (broadband) and Fig. 7 (spectral). The plot title identifies the test sample, measurement wavelengths, and the test area on the sample, when applicable. The horizontal and vertical scales are the same for all of the plots, 0 to 40 deg and 0.001 to 1 sr^{-1}, respectively. The broadband BRDF data (Fig. 6) are presented on a single plot for each color tile. The legend identifies flown or unflown tile samples, and the left and right spot locations. Unflown left, unflown right, flown left, and flown right are abbreviated UNFL, UNFR, FL, and FR, respectively. The broadband BRDF of the white shuttle tiles was independent of the tile sample and the beam incidence location, and decreased only slightly with reflectance angle (Fig. 6a). The broadband BRDF of the unflown black tile was higher then that of the flown black tile at angles near specular reflectance, but lower at large angles from the surface normal (Fig. 6).

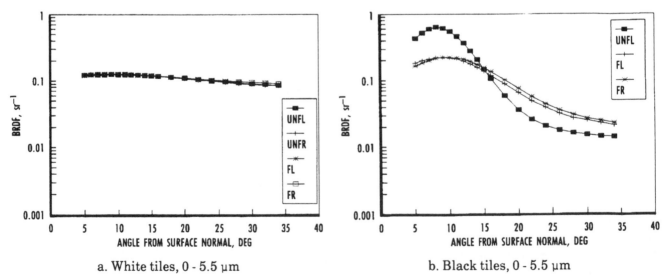

a. White tiles, 0 - 5.5 μm   b. Black tiles, 0 - 5.5 μm

Fig. 6. Broadband BRDF of shuttle tile samples.

Plots of spectral BRDF data are presented in Fig. 7. The 2.49-µm data was not presented in this plot because the software limits the user to six curves per plot. Tabulated data for the entire measurements set are included in Ref. 3. The legend includes the center filter wavelength in microns. The spectral BRDF of the flown and unflown white were almost identical (Figs. 7a - b). The spectral BRDF of the white tiles decreases from 2.25 µm to 2.8 µm, then increases at 3.46 µm, then continued to decrease with increasing wavelength. The black tile BRDF was not as dependent on wavelength, and tended to decrease monotonically with increasing wavelength (Figs. 7c - d).

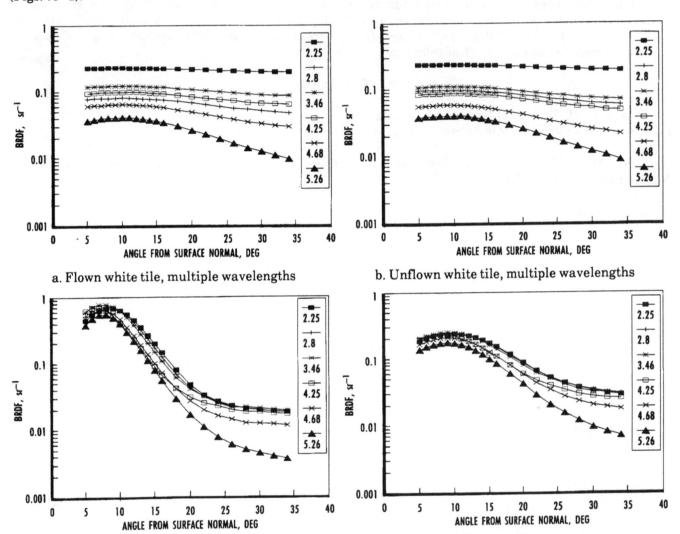

a. Flown white tile, multiple wavelengths   b. Unflown white tile, multiple wavelengths

c. Flown black tile, multiple wavelengths   b. Unflown black tile, multiple wavelengths

Fig. 7. Spectral BRDF of shuttle tile samples.

8. SUMMARY

The BRDF of both white and black, flown and unflown, NASA shuttle tiles was measured in the wavelength range between 2.0- and 5.5-µm. The flown and unflown white tile BRDF's were nearly identical. The white tile BRDF was characteristic of a good diffuse reflector, i.e., having a nearly constant BRDF as a function of reflectance angle (almost Lambertian). The white tile spectral BRDF was not a monotonic function. A significant dip was observed at 2.8 µms. This was probably a result of water absorption in the material porous structure. This spectral characteristic may not exist when the tiles are exposed to the vacuum of space. The water may outgas, thus causing an increase in the white tile BRDF at 2.8 µm.

The BRDF of the black tile was changed as a result of being flown on the space shuttle. The specular nature of the flown tile diminished as a result of space flight. The spectral BRDF of the black tiles was a monotonic decreasing function of radiation wavelength.

9. ACKNOWLEDGMENTS

The work reported herein was conducted at the request of the Geophysics Laboratory (GL), Hanscom Air Force Base, Mass. The GL project and test program manager was Dr. Edmund Murad. The authors would like to acknowledge Mr. Ray Cooper for his assistance in assembling the test hardware, and for acquiring all of the test data. Thanks to Dr. J.T. Neu of the Surface Optics corporation in San Diego for providing the spectral reflectance measurement of the diffuse reference reflector.

10. REFERENCES

1. Young, R. P. "Metal-Optics Scatter Measurements," *Proceedings of the Society of Photo-Optical Instrumentation Engineers*, Vol. 64, No. 12, August 1975.

2. Nicodemus, F. E., "Directional Reflectance and Emissivity of an Opaque Surface." *Applied Optics*, Vol. 4, No. 7, July 1965, pp.767-773.

3. Young, R. P. and Wood, B. E., "Bidirectional Reflectance Distribution Function (BRDF) of NASA Shuttle Tiles," AEDC-TSR- , February 1991.

Scattering Contribution to the Error Budget of an Emissive IR Calibration Sphere

J. Chalupa
W. K. Cobb
T. L. Murdock

General Research Corporation, 5 Cherry Hill Drive, Suite 220
Danvers, MA 01923

ABSTRACT

Because the thermal history of an isothermal metal sphere with an emissive coating can be accurately modelled, such a reference sphere is a suitable calibration object for a space-based IR sensor. To achieve high quality calibration, the uncertainty in the sphere's material parameters must be constrained by relating calibration requirements to design tolerances in the sphere parameters. A methodology for doing this will be presented and applied to an orbiting reference sphere. For clarity, the approach will be illustrated with a gray-body model of the sphere thermal behavior, but results for a non-gray-body sphere will also be given. Sources of uncertainty in the sphere signature will be identified and estimated. In particular, earth flux scatters from the sphere and contaminates the sphere's thermal signature. While the scattered earth flux constitutes a small fraction of a highly emissive sphere's IR signal, it will be shown that the uncertainty in the scattered flux is a significant fraction of the uncertainty in the signal. In sunlight the uncertainty in scattered earthflux reduces the total uncertainty by cancelling other error terms, but in darkness the total uncertainty can be increased by scattering effects.

1.   INTRODUCTION

Calibration of measurements from a space-based IR sensor involves interesting technical issues. Such calibration requires reproducible, well-characterized signatures. Stars are an obvious candidate source, but the difficulty arises of how exactly the spectral signature of a star is known a priori. Artificial reference objects have the advantage that exhaustive ground characterization is feasible before deployment, but the signature of such sources is not constant because of interaction with the time-dependent space thermal environment. This interaction must be characterized sufficiently to permit accurate prediction of the signature. This paper presents an overview of the associated modelling, with emphasis on the role of scattering in the error budget of an orbiting sphere.

The key features of a reference object's interaction with the environment are depicted in Figure 1. The incident flux impinging on the object has three main components: the thermal flux from the earth, the thermal flux from the sun (if present), and solar flux which is reflected from the earth onto the sphere. Because the temperature of an orbiting reference object is driven by a time-harmonic thermal background, the uncertainty in the signature is periodic and therefore bounded, and so the object can be used for calibration until long-term changes in material or surface properties occur. A portion of the incident flux is absorbed by the reference object and heats it, thereby contributing to the blackbody radiation emitted by the sphere. The remainder of the incident flux which is not absorbed is scattered. Thus the signal received by an IR sensor observing a spherical reference object has two components (in addition to background signals): the blackbody signature of the sphere and the background fluxes scattered off the sphere.

The paper consists of four sections of which this Introduction is the first. Section 2 summarizes the thermal model and illustrates the error analysis. Section 3 explores the importance of scattering to the error budget. Section 4 ends the paper with a brief discussion. These sections are synopsized below.

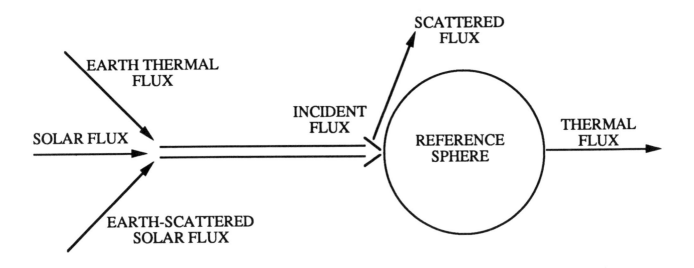

Fig. 1. Interaction of Background Radiation Fluxes with the Reference Sphere

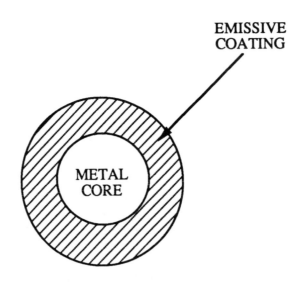

Fig. 2. Structure of Reference Sphere Not to Scale

Section 2 of this paper discusses the connection between the material characteristics of the sphere in Figure 2 (which are measured on the ground) and the calibration requirements on the sphere signature (which is measured in space). This connection is established by use of a thermal model for the sphere's temperature history: the material properties and background fluxes are inputs into the model, and the sphere signature is an output. The key to controlling the error budget is to systematically monitor how uncertainties are propagated through the model. For example, the emissivity affects the rate at which the temperature changes, and the temperature itself is therefore emissivity-dependent; an uncertainty in emissivity will appear in the model both implicitly and explicitly. Careful treatment of the error budget allows tradeoffs, for instance, between characterization of the thermal environment and characterization of the material properties. These tradeoffs will be illustrated by applying the gray-body approximation, which treats the emissivity as a constant function of wavelength, temperature, and angle, to an approximate expression for the total IR power coming from the sphere.

Section 3 is concerned with the role of scattering in the error budget. If all the background flux is absorbed and re-emitted as blackbody radiation, none is scattered. If not all the background flux is absorbed, that portion of the flux which is not absorbed is scattered. Because of this constraint linking the absorbed and scattered flux, the uncertainty in the reflected flux can be comparable to the uncertainty in the blackbody flux even if the expectation value of the blackbody flux is much larger than that of the scattered flux. In fact, at short wavelengths the scattered earth flux can assume the leading role in the nighttime error budget. An example will be constructed for the narrow-band limit.

Section 4 ends the paper with a summary of the analysis.

2. THERMAL MODELING AND ERROR BUDGET

Ideally, the configuration of the reference object has a simple design which yields a well-characterized signature. A large isothermal sphere with a specular surface is an obvious choice because its signature is independent of orientation and rotation (scattering off a specular sphere is isotropic in the short-wavelength limit [1]). Isothermal conditions are attained in a sphere with thermal conductivity κ sufficiently high that appreciable temperature gradients are not supported. The sphere interacts with its thermal environment. If the reflectivity is high, the absorption and thermal radiation are weak but the ill-characterized environmental flux, with its wavelength-dependent structure, makes the dominant contribution to the signature of the sphere. If the reflectivity is low and the emissivity high, the bulk of the environmental flux is absorbed and the signature is characterized by a blackbody spectrum. Because, in the latter case, one is concerned only with the total absorbed radiation rather than with the radiation's spectral dependence, the emissive case is preferable for characterizing the signature.

The most convenient examples of large κ materials are metals, which typically have the drawback of high reflectivity. The advantages of high κ and high emissivity ε can be secured simultaneously by covering a metallic sphere with a thin emissive coating. Such a sphere radiates and absorbs like a black body, but its thermal history is determined by the thermal properties of the metal.

The rate of change of the heat content of the sphere is the difference between the blackbody radiation rate and the rate of heat absorption from incident fluxes:

$$V\rho c_p \frac{dT}{dt} = -A\varepsilon \left[\sigma T^4 - j_{inc} \right] . \tag{1}$$

T = temperature
t = time
ε = gray-body emissivity of sphere
j_{inc} = orientationally averaged incident flux
ρ = density
c_p = specific heat
V = volume of metal
A = surface area of sphere
σ = Stefan-Boltzmann constant = 5.670 x 10^{-12} W/cm^{2}K^4.

The incident heat flux j_{inc} is the orientational average of the actual angle-dependent flux over the surface of the sphere. The dominant constituents of the incident flux are the solar flux j_s, the earth flux j_e, and the solar flux j_{ess} scattered by the earth onto the sphere:

$$j_{inc} = j_s + j_e + j_{ess}. \tag{2}$$

The earth flux and earth-scattered solar flux are respectively proportional to the earth's gray-body emissivity ε_e and the corresponding reflectivity, i.e. albedo:

$$j_e = \varepsilon_e j_e \tag{3}$$
$$j_{ess} = (1-\varepsilon_e) j_{ess}.$$

These fluxes constitute the thermal background of the sphere.

The sphere has an equilibrium temperature associated with the background flux:

$$T_{eq} = (j_{inc}/\sigma)^{1/4}. \tag{4}$$

At T_{eq} the radiated flux is equal to the absorbed flux and the temperature does not change. T_{eq} can be reached by deploying the sphere at a temperature close to its value, or by deploying a sphere designed with such low thermal inertia that a temperature close to T_{eq} is reached between crossings of the earth's shadow boundary.

The model has been solved for a reference sphere characterized by the parameters in Table 1. The resulting baseline temperature history is shown in Figure 3a, and the corresponding error envelope is shown in Figure 3b. After the first orbit or so, the sphere comes into dynamic equilibrium with its thermal environment and the temperature history locks into time-periodic behavior.

The signal received by an IR sensor is proportional to the power received from the sphere in the sensor bandwidth. A generic description, independent of bandwidth, of the sphere's IR power can be formulated as follows. The sphere's blackbody spectrum is predominantly in the IR, as is the earth's. The sun's spectrum is predominantly in the visible. Thus, a reasonable way to characterize the IR signature is to include the total blackbody power of the sphere and the total reflected earthshine while neglecting the reflected sunshine:

$$P_{IR} \equiv A\varepsilon\sigma T^4 + A(1-\varepsilon) j_e. \tag{5}$$

Figure 4a presents the baseline IR power, and Figure 4b presents the relative uncertainty in this quantity.

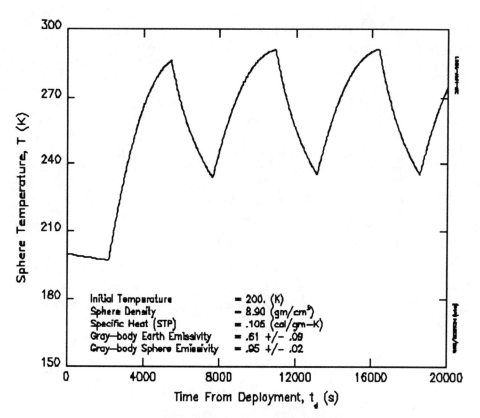

Figure 3a. Thermal History for Emissive Reference Sphere Deployed from a Satellite in a 333 km Polar Orbit at Local Sunset.

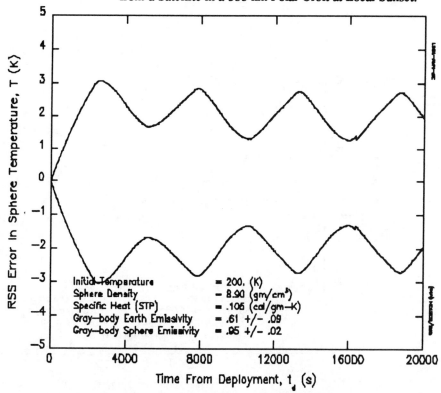

Figure 3b. RSS Error Envelope for Emissive Reference Sphere

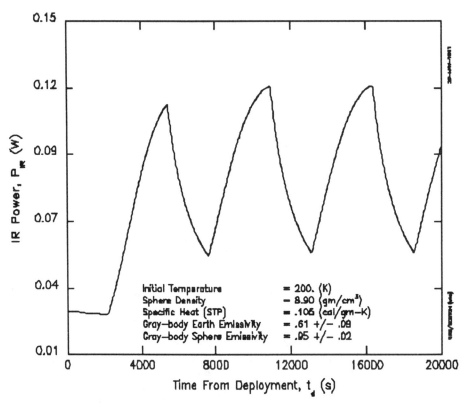

Figure 4a. Total Infrared Power Emitted by an Emissive Reference
Sphere Deployed from a Satellite in a 333 km Polar Orbit
at Local Sunset.

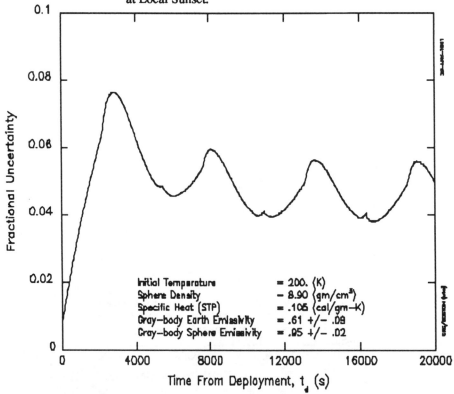

Figure 4b. Worst Case Fractional Uncertainty in Total Infrared Power

If the emissivities of the sphere and earth are imperfectly characterized and $\delta\epsilon$ and $\delta\epsilon_e$ are the corresponding uncertainties, the resulting uncertainty in the IR power is

$$\delta P_{IR} = \frac{\partial P_{IR}}{\partial \epsilon} \, \delta\epsilon \; + \; \frac{\partial P_{IR}}{\partial \epsilon_e} \, \delta\epsilon_e. \tag{6}$$

The uncertainty in power is the sum of two terms which are respectively proportional to the uncertainties $\delta\epsilon$ and $\delta\epsilon_e$. Calibration requirements determine the maximum allowed power uncertainty. Two extreme situations are embodied in (6): the uncertainty in ϵ_e can assume a maximum value if the sphere emissivity is determined exactly ($\delta\epsilon=0$), and the uncertainty in ϵ can be maximized if the earth emissivity is known exactly. In practice, both ϵ and ϵ_e have nonvanishing uncertainties which might be reduced by measurement or analysis, and a calibration experiment planner should optimally apportion effort between characterizing ϵ and characterizing ϵ_e. Because Equation 6 enables quantitative budgeting of the overall calibration requirements in terms of the uncertainties in $\delta\epsilon_e$ and $\delta\epsilon$, it can serve as a planning tool (one would actually work with the mean square of (6), which relates the variance of the power to the variances of ϵ and ϵ_e).

Clearly, (6) can be generalized to include uncertainty in other quantities affecting the sphere's thermal history (specific heat, density, sphere radius, initial temperature, etc.). Such possibilities will not be pursued further in this paper because the paper is primarily concerned with the role of scattering in the emissivity term in the error budget.

Table 1.

Characteristic Parameters of a Nickel Calibration Sphere

Sphere emissivity = ϵ = .95$\pm$ .02
Sphere density = ρ = 8.90 gm/cm^3
Sphere specific heat = c_p = .105 cal/gm-K
Sphere radius = r = .50 cm
Earth temperature = T_e = 285 K
Earth emissivity = ϵ_e = .61 $\pm$ .10
Orbital radius = R = 6711 km
Inclination angle from polar orbit = θ = 0.0 degrees
Solid angle subtended by earth = Ω_e $\simeq$ 4π/3

3.   ROLE OF SCATTERING IN THE ERROR BUDGET

The role of scattering off the sphere in the error budget for the IR power can be clarified by relating the uncertainty in power to the uncertainty in emissivity:

$$\delta P_{IR} = \left\{ \sigma T^4 + 4 \varepsilon \sigma T^3 \frac{\partial T}{\partial \varepsilon} - j_e \right\} A \delta \varepsilon. \tag{7}$$

The total uncertainty in the IR power is the sum of three contributions which need not have the same sign. First, the T^4 blackbody term in the power is proportional to the emissivity and changes accordingly. Second, the emissivity appears in the dynamical equation for the temperature T, and so T implicitly depends on the emissivity. Finally, at the gray-body level the reflected earthshine varies with sphere emissivity.

Equation 7 shows that the signs of the change in the T^4 blackbody term and in the reflected earthshine are opposite. The relative uncertainty in the IR power is comparatively small during daylight because the sphere's thermal evolution is dominated by absorption of sunlight and the magnitude of the solar flux is well characterized. As the sphere cools at night, the uncertainty in the scattered earthshine increases in importance. Nevertheless, the scattering term in the error budget always reduces the total uncertainty for IR power for the sphere of Table 1.

The situation can be different for a finite bandwidth. The inband power is[2]

$$P_{IR} (band) = 2 \pi C_1 A \int_{band} \frac{\varepsilon d\lambda}{\lambda^5 \left(e^{C_2 / \lambda T} -1 \right)} + A \int_{band} d\lambda \, (1-\varepsilon) \frac{dj_e}{d\lambda} . \tag{8}$$

$C_1 = .59544 \times 10^8$ W-μ 4/cm^2
$C_2 = 1.4388 \times 10^4$ $\mu \cdot$ K .

The emissivity has been taken inside the integral in (8) to illustrate the formalism for a wavelength-dependent emissivity, i.e. beyond the gray-body level. For a gray-body emissivity, the integrals over an infinite bandwidth reduce to the total IR power of (5). For a gray-body sphere in equilibrium ($T = T_{eq} = 195$ K), the uncertainty in inband power associated with an uncertainty in the emissivity is

$$\delta P_{IR} (band) = 2 \pi C_1 A \left\{ \int_{band} \frac{d\lambda}{\lambda^5} \left[\frac{1}{e^{C_2/\lambda T} -1} - \frac{\varepsilon_e \Omega_e}{4\pi} \frac{1}{e^{C_2/\lambda T_e} -1} \right] \right\} \delta \varepsilon . \tag{9}$$

To display how the scattering error can dominate the emissivity term in the error budget, it is convenient to consider (9) in the narrow-band limit for a gray-body sphere in shadow. The temperature of the earth is greater than the equilibrium night temperature of the calibration sphere. Thus, the blackbody spectra of the earth and sphere differ in that at sufficiently short wavelengths the sphere's is smaller in magnitude than the earth's. Accordingly, there is a critical wavelength below which the uncertainty in scattered earthflux is more than twice as large as the uncertainty in emitted power, and scattering increases the magnitude of the error. For a bandwidth $\Delta \ll \lambda$, the power and corresponding short-wavelength limit are

$$\delta P_{IR} \text{ (band)} = \frac{2\pi C_1 A \Delta}{\lambda^5} \left[\frac{1}{e^{C_2/\lambda T} - 1} - \frac{\varepsilon_e \Omega_e}{4\pi} \frac{1}{e^{C_2/\lambda T_e} - 1} \right] \delta \varepsilon \tag{10a}$$

$$\simeq \frac{2\pi C_1 A \Delta}{\lambda^5} e^{-C_2/\lambda T} \left[1 - \frac{\varepsilon_e \Omega_e}{4\pi} e^{(C_2/\lambda T)(1 - T/T_e)} \right] \delta \varepsilon. \tag{10b}$$

It is evident that for sufficiently small wavelength, the second term has more than twice the magnitude of the first and the total uncertainty is increased by scattering effects. For the temperatures of interest ($T = T_{eq} \simeq 195$ K and $T_e = 285$ K), the critical wavelength is approximately 10 microns. It is straightforward to take the long-wavelength limit of (10a) and show that at these temperatures the long-wavelength part of the uncertainty is reduced by scattering effects. This is an expected result because it has been demonstrated that the uncertainty in total power is reduced by scattering: thus, the foregoing short-wavelength enhancement must be compensated at longer wavelengths.

The arguments of this section can be extended to scattering of sunlight off the earth: there is a cancellation in the $\delta\varepsilon_e$ term in (6) between scattered sunlight and thermal radiation from the earth. However, the earth deviates from gray-body behavior[3], and presumably does so to a greater degree than a reference object which is designed to mimic a grey body. Thus, a gray-body analysis is more directly appropriate for sphere scattering than for earth scattering.

4.   SUMMARY

This paper has considered an orbiting sphere used as a calibration source for an IR sensor. The optimal sphere design is a metal interior covered with a thin coating of high emissivity. Sphere parameters can be related to the calibration requirements by construction of an error budget based on the thermal history of the sphere.

The thermal history and error budget were examined within the gray-body approximation within an orbit after deployment, the temperature of the orbiting sphere locks into a time-periodic pattern. Because the sum of the absorbed and scattered background flux is known, the uncertainty in the total IR power from the sphere is reduced by including scattering effects in the error budget. However, at short wavelengths and low temperatures the uncertainty associated with scattering can become so predominant that the total uncertainty can increase because of scattering.

5.   REFERENCES

1.   J.D. Jackson, Classical Electrodynamics, Ch. 9, Wiley, New York, 1962.

2.   R. Siegel and J.R. Howell, Thermal Radiation Heat Transfer, pp. 24-25, Hemisphere, New York, 1981.

3.   G.H. Suits, "Natural Sources" in The Infrared Handbook edited by G. J. Zissis and W. L. Wolfe, Ch. 3, ERIM, Ann Arbor, 1989.

Light Scattering Properties of New Materials for Glazing Applications

M. Bergkvist and A. Roos.
Solid State Physics Group, Department of Technology, Uppsala University
P.O. Box 534, S-751 21 Uppsala, Sweden

ABSTRACT

Several new materials are available for glazing applications, many of which require careful optical characterization, especially with regards to light scattering. Measuring scattering requires special equipment and is inherently difficult. An integrating sphere can be used for the total and diffuse components but great care must be taken in interpreting the instrument readings. Angular resolved scattering measurements are necessary for a complete characterization and this is difficult for low levels of scattering. In this paper measurements on electrically switchable NCAP materials and thick panes of aerogel are reported. The NCAP films switch reversibly from a translucent, scattering state to a transparent, clear state with the application of an ac-voltage. Airglass has a porous SiO_2 structure with a refractive index n = 1.04 and a very low heat transfer coefficient. Integrated scattering measurements were performed in the wavelength range 300 to 2500 nm on a Beckman 5240 spectrophotometer equipped with a 198851 integrating sphere. In this instrument we can measure the total and diffuse components of the reflectance or transmittance separately. The angular distribution of the scattered light was measured in a scatterometer, which can perform scattering measurements in the wavelength range 400-1100 nm in both transmittance and reflectance mode with variable angle of incidence.

1. INTRODUCTION

During the last few years several new materials for glazing applications have become available. They can, for instance, be used for energy saving, privacy glazing, glare protection or light control. Whatever the intended use the light scattering properties are of great importance, since the human eye is very sensitive even to small amounts of scattered light. In this paper we report on measurements of light scattering from materials of two different types: electrically-controllable glazing products based on the nematic droplet/polymer film (NCAP) technology[1,2], commercially available under the trade name Varilite, and silica aerogel[3,4], tradename Airglass. The NCAP-films are especially interesting since their function is based on light scattering. Small birefringent droplets are dispersed in a polymer matrix, and by applying an electric field the crystals are aligned in the field so that transmitted light sees no change in the refractive index upon film traversal (cf. Fig. 1). This of course requires matching the refractive index of the matrix to that of the crystallites. In the voltage off state the crystallites are randomly oriented and a transmitted light beam is scattered making the film translucent. By changing the birefringency of the crystals one can choose between a film with low haze in the clear state but not completely translucent in the off state, or a film which is highly translucent in the off state but with some residual scattering in the clear state. By adding a dye a film can be manufactured which will absorb more of the scattered light in the translucent state and thus allow a variation in total transmittance with applied voltage. We used ordinary 50 Hz sinus voltage (0 - 100V rms) in our measurements. A square wave is recommended by the manufacturer and may have resulted in slightly lower scattering levels in the clear state.

0-8194-0658-9/91/$4.00

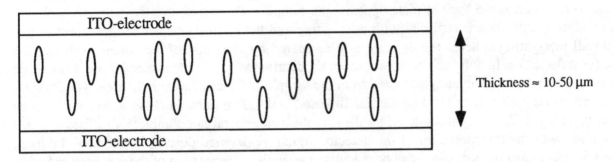

Fig.1 Schematic figure of a NCAP-sample in the clear state with the crystallites aligned.

The Airglass samples (silica aerogel) can be described as foam-glass consisting of small silica particles linked together with air filled pores, as shown in fig. 2. The material is highly transparent due to the resulting low refractive index, typically n= 1.04, and the small particle size, but the thermal conductivity is very low. It has been suggested that this material could be used in solar collectors to minimize convection losses and it has also been proposed for use in domestic glazing. The latter application requires careful optimization of the manufacturing process so that the scattering from the silica particles and/or air-pores can be minimized. If this can be achieved, a window with extremely low heat losses could be manufactured,.and, since it is transparent, it would allow solar radiation to pass into the house, thus contributing to the energy required for heating.

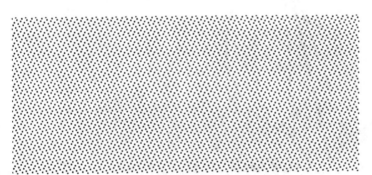

Fig. 2 Schematic figure of an Airglass™ sample

2. EXPERIMENTAL

2.1 Integrating sphere measurements.

One of the most common ways of measuring scattering is using an integrating sphere. With such an instrument only the total transmittance or reflectance can be measured, or, if the specular signal falls in a light trap, the total hemispherical scattering[5,6]. The angular distribution of the scattered radiation is in principle not possible to obtain. It is, however, necessary to know, at least approximately, the angular distribution (often described by the BSDF[7]) of the light scattered from the sample in order to obtain a correct value of the total hemispherical scattering. A common mistake when using an integrating sphere is to ignore the fact that the detector signal depends on the angular distribution of the scattered radiation. In Fig. 3 the principle for transmittance measurements with an integrating sphere is shown. In a) the total transmittance is recorded as the BaSO4-plate opposite the sample reflects the specular radiation back into the sphere, and in b) this plate is replaced with a light trap and the diffuse or scattered transmittance is recorded. In all integrating spheres the detector is illuminated by a section of the sphere wall and the transmitted (or reflected) radiation must be homogeneously distributed within the sphere before it can reach the detector, which must be completely screened from the sample and the specular light spot. The BaSO4-wall is a close approximation to an ideal lambertian diffusor, and light reflected off the sphere wall can be said to be completely diffuse. Light scattered by the sample is in general not completely diffuse and this causes problems with the interpretation of the detector signal. A detailed description of how the total, specular and diffuse transmittance values can be calculated assuming a separation of the scattered radiation into two or three components has been published recently[8]. Similar descriptions of reflected radiation and of the influence of the sphere geometry and sample texture have also been published[9,10]. At another conference within this symposium some examples of errors that can result when these effects are not properly taken into account are presented[11]. It is also obvious from Fig. 3 that some of the low angle scattering will fall within the light trap and be recorded as being specular. The division into a specular and a diffuse component is therefore always instrument specific and depends on the sizes of the sphere ports and the dimension of the light beam.

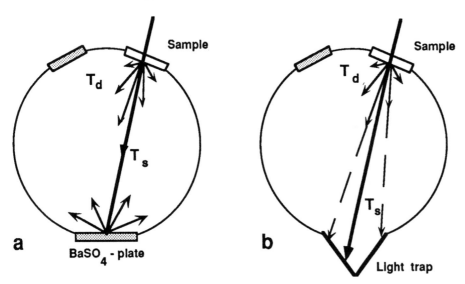

Fig.3 Integrating sphere set in mode for transmittance measurements
    a) total transmittance   b) diffuse transmittance

2.2 Scatterometer measurements.

The scatterometer consists of a light source, monochromator, sample holder and detector mounted on a movable boom. There is also a chopper, room for polarizers and different filters. A schematic diagram is shown in fig. 4. The detector is a Hamamatsu S-2386 series photodiode connected to a two stage preamplifier operating in the photovoltaic mode, i.e. no applied bias. The experimental setup is interfaced to an Apple Mac IIfx μ-computer for data logging and subsequent analysis. The detector occupies a solid angle Ω of $\approx 1.04 * 10^{-4}$ sr as seen from the sample.

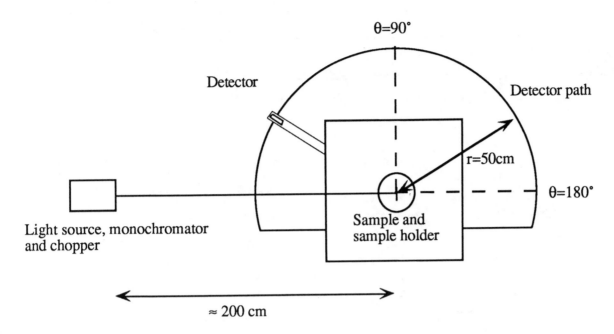

θ=90°

Detector

Detector path

r=50cm

θ=180°

Light source, monochromator
and chopper

Sample and
sample holder

≈ 200 cm

Fig. 4 Outline of the scatterometer used in the angle resolved experiments.

3. RESULTS AND DISCUSSION

3.1 Taliq samples

In Fig. 5 and 6 the total and diffuse reflectance and transmittance spectra of a dyed Taliq sample are shown for zero and 100V applied voltage respectively. It can be seen that in the zero voltage state the transmittance is completely diffuse, while in the 100V state most of the transmitted light is specular. There is also a pronounced increase of the total transmittance in the visible from around 20 to 70 percent. In the near infrared, on the other hand, the increase of the transmittance is only around 25%. The total reflectance spectra do not change so much although the scattering has almost completely disappeared. The response is fast at low voltages and saturates at 80 - 100 V. The undyed Taliq samples, Varilite 106 and 305, show a similar response to the applied voltage, but the total transmittance does not change much, only the diffuse component is modulated.

In Figs. 7 to 11 we show the results from the angularly resolved scattering measurements for the different samples. Comparing the Varilite 106 and 305 in Fig. 7 we first notice that in the clear state the large angle scattering from the 305 sample is slightly larger, consistent with the reported higher haze for this sample[2]. In the translucent state (zero voltage) in Fig. 8 and 9 sample 305 has a smaller specular peak and a different angular distribution of the scattering than sample 106. Sample 305 scatters more towards larger angles and less close to the specular beam. The total transmittance for these two samples is in principle not affected by applying the voltage, the samples switch from a scattering state to a specular state. In Fig. 8 and 9 this can be seen as a drop in the scattered signal by one order of magnitude and an increase of the specular signal by 2-3 orders of magnitude when the voltage is applied.

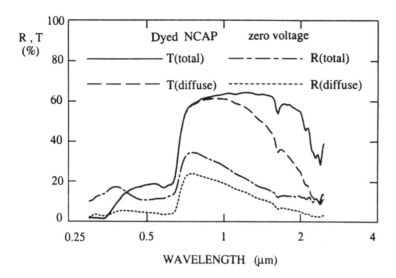

Fig. 5 Total and diffuse reflectance and transmittance spectra for a dyed NCAPsample measured by an integrating sphere. Applied voltage = 0V

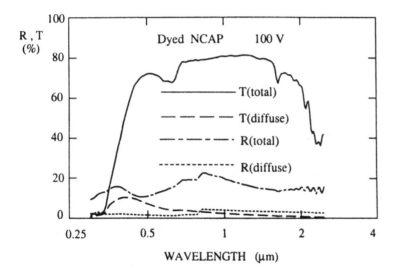

Fig. 6 Total and diffuse reflectance and transmittance spectra for a dyed NCAP sample measured by an integrating sphere. Applied voltage = 100V

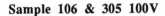

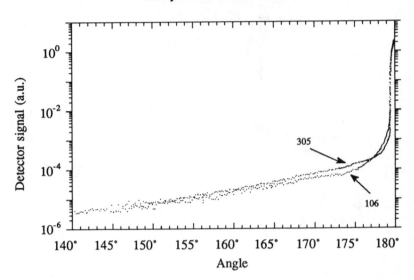

Fig. 7 Scatterometer spectra for the two NCAP films recorded with an applied voltage of 100V

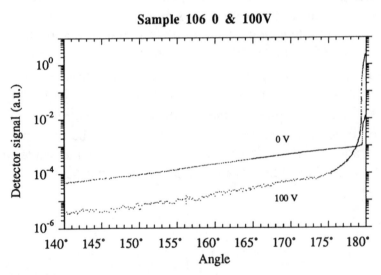

Fig. 8 Scatterometer spectra for the NCAP film #106 recorded for an applied voltage of zero and 100 V.

Sample 305 0 & 100V

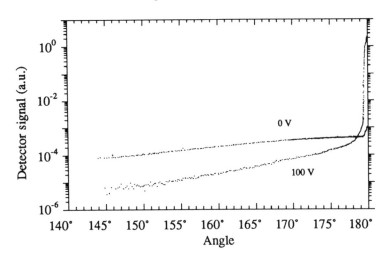

Fig. 9 Scatterometer spectra for the NCAP film #305 recorded for an applied voltage of zero and 100 V.

The situation is different for the dyed sample where the total transmittance is higher in the clear state (Fig. 6) than in the translucent (Fig. 5). The angular distribution of the transmitted radiation can be seen in Fig. 10, where the difference between the scattered signals in the two states is much less pronounced for this sample than for the undyed ones in Fig. 5 and 6. An interesting feature of these sample is demonstrated in Fig. 11, where the scattered intensity as a function of applied voltage for several fixed angles is shown, the sample reported is #106, a similar behaviour is seen for the 305. In the specular direction, $\theta=180°$, we notice a monotonous increase in detector signal as the applied voltage is increased. For larger angles, ($\theta=169.2°$), the signal steadily decreases with increased applied voltage. In the near specular region there is first an increased signal which then starts to decrease. These transient maxima can be explained by assuming a partial initial alignment of the scattering crystallites which consequently tend to scatter light preferrably in the near specular region. When fully aligned their scattering power is very low.

Dyed sample 0 & 100V

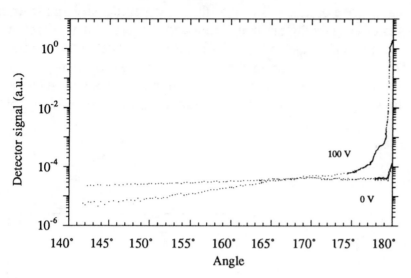

Fig. 10 Scatterometer spectra for the dyed NCAP film recorded for an applied voltage of zero and 100 V.

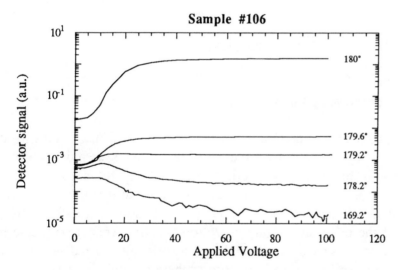

Fig. 11 Scatterometer data versus applied voltage for the NCAP 106 film recorded for different scattering angles as indicated.

3.2 Airglass samples

The airglass samples are difficult to characterize optically owing to their geometrical size and scattering properties. It is difficult to prepare sheets of airglass thinner than 5 mm and because of the low refractive index bulk scattering dominates over surface scattering. In Fig. 12 scatterometer data are shown for two airglass samples, 5 and 10 mm thick. The specular peak dominates but there is a high fraction of low angle scattering as can be seen from the smoothly rounded shoulders in the diagrams. As the bulk scattering is most likely Rayleigh scattering, this low angle scattering probably originates from the surfaces of the samples.

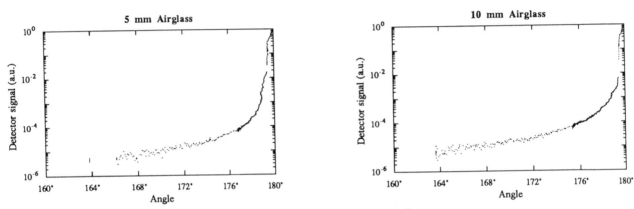

Fig. 12 Scatterometer spectra for 5 and 10 mm thick Airglass samples.

The transmittance and reflectance spectra of the same samples as in Fig. 12 as measured by the integrating sphere are shown in Fig. 13 and 14. The "absorption" spectra shown in these figures have been calculated as 1-R-T, which is the normal way to calculate absorption. This is an example of the difficulties involved in measurements of this type of sample. Clearly the true absorption levels of the samples are much lower than indicated by the 1-R-T spectra. The reason is that both for the reflectance and transmittance measurement some of the light is scattered from the bulk of the samples at such high angles that it is lost and does not enter the integrating sphere. The detected signal becomes too low and the absorption value 1-R-T too high. This can also be seen from the shape of the reflectance and absorption spectra. Apart from some true absorption bands in the near infrared, these spectra have exactly the same wavelength dependence. It can be concluded that both the R and A values as recorded by the integrating sphere are due to a large extent to bulk scattering from the airglass samples. Experimentally determining the true absorption spectra for these samples is therefore a difficult task.

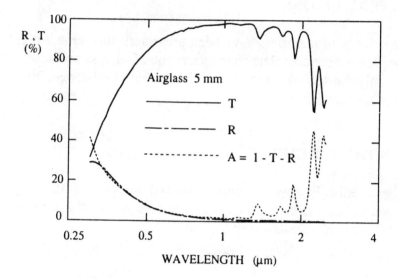

Fig. 13 Total reflectance and transmittance spectra for the 5 mm Airglass sample as recorded by the integrating sphere. Also shown is the value A=1-R-T.

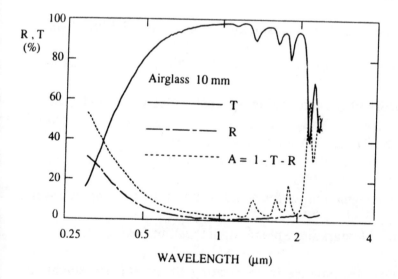

Fig. 14 Total reflectance and transmittance spectra for the 10 mm Airglass sample as recorded by the integrating sphere. Also shown is the value A=1-R-T.

4. CONCLUSIONS

In this paper some scattering data for two new types of materials have been presented. It is shown that accurate scattering measurements are of importance for the complete characterization of such samples. It is essential to be able to distinguish between not only the specular and scattered light, but also between low and high angle scattering as well as bulk and surface scattering.

5. ACKNOWLEDGEMENT

This work is part of a project sponsored by the Swedish Energy Administration and the Swedish Council for Building Research. The NCAP samples were obtained through the IEA Task 10 work. The aerogel samples were manufactured by Airglass AB P.O. Box 150 Staffanstorp Sweden.

6. REFERENCES

1. J.L. Fergason, US Patent 4,435,047 (1984) and following Patents
2. P.Van Konynenburg, R. Wipfler and J Smith, "Optical and envitomental properties of NCAP glazing products", Paper presented at SPSE/SPIE Electronic imaging Devices and Systems Symp., Los Angeles (Jan. 1989).
3. S. Henning, "Airglass-Silica Aerogel, A transparent heat insulator", Document D7:1990, Swedish Council for Building Research
4. M. Rubin and C. M. Lampert, "Transparent Silica Aerogels for Window Insulation", Solar Energy Materials 7 p. 393-400 (1983)
5. J.A.J. Jacquez and H.F. Kuppenheim, "Theory of the Integrating Sphere", J.Opt.Soc.Am. **45**, 460-70 (1955).
6. D.G. Goebel, "Generalized Integrating Sphere Theory", Appl. Opt. **6**, 125 (1967).
7. J. Stover, "Optical Scattering. Measurement and analysis", Optical and Electro-Optical Engineering Series, R. E. Fischer and W. J. Smith eds, McGraw Hill 1990
8. A. Roos, "Interpretation of integrating sphere signal output for nonideal transmitting samples", Appl. Opt. **30**, 468-74 (1991)
9. A. Roos and C-G. Ribbing, "Interpretation of Integrating Sphere Signal Output for non- Lambertian Samples", Appl. Opt. **27**, 3833-37 (1988).
10. A. Roos, C-G. Ribbing and M. Bergkvist, "Anomalies in Integrating Sphere Measurements on Structured Samples", Appl. Opt. **27**, 3828-32 (1988).
11. A. Roos, "Optical characterization of solar selective transmitting coatings", Proc. SPIE **1536** (1991)

SERS used to study the effect of Langmuir-Blodgett spacer layers on metal surface

Bingkun Yu, Yu Li and Yingtin Wang

Department of Physics
Shanghai University of Science and Technology
Shanghai, 201800, P. R. China

ABSTRACT

This paper reports the experimental results on using Langmuir-Blodgett (LB) films as spacer layers between molecules and metal surface to study electromagnetic enhancement mechanisms of Surface Enhanced Raman Scattering (SERS), and to measure the change of SERS intensity with spacer layer thickness. In pyridine + KCL/LB films/Ag films configuration, the experimental results indicate SERS effect exists when the spacer layer thickness is 5 nm, and eventually it becomes unobservable with 15 nm thickness. The experiments supported the idea that the enhancement arises from an electromagnetic mechanism.

1. INTRODUCTION

SERS has been of interest mostly because of the application value in the investigation of various surface processes and trace analysis, etc.; but the basic understanding of enhancement mechanisms is still incomplete. Much of the attention to date has been focused on the function of electromagnetic enhancement or molecular enhancement; in other words, the problem is whether SERS has a "long range" property or a "short range" property.[1.2]

In recent years, there have existed several different enhancement mechanism researches, however, much of the current interest has been directed toward accounting for the role played by surface roughness.[3] At the same time, enhancement factors of theoretical calculation are obviously associated with distance between adsorbed molecules and a surface. It is difficult to explain the "long range " property using any other mechanisms, so that experimental research about the effect of distance between molecules and a surface on SERS is an important evidence of testing electromagnetic enhancement.

In order to study the long range property of the electromagnetic enhancement mechanism, the distance between molecules and surface must be controlled. According to Murry's method to prepare multilayer configuration, it is easy to lead to an inhomogeneous structure due to complex technique.[4] This causes a great differentia between theoretical calculation and experimental measurement. When considerations are restricted to several tenths nm of distance between molecules and surface, we must consider the short range effect. This complicates the study of enhancement mechanisms.

The study of LB film has recently been receiving a great deal of attention.[5] It is appealing to researcher because of the facile manner in which molecular layer structure solid film can be transferred from a liquid surface to a substrate. This creates an ideal spacer layer. We chose to use LB films of stearic acid as spacer layers between adsorbed molecules and a silver surface to observe the electromagnetic effect on molecular SERS spectra and to understand long range enhancement mechanisms from electromagnetic model expectation. We also discussed pyridine adsorbed schemes, according to the SERS intensity ratio of two ring-breathing vibrational modes of pyridine.

2. EXPERIMENTAL

2.1 Preparation of spacer layers

In the spectra measurement, ideal spacer layers between adsorbed molecules and metal surface must be homogeneous, transparent with no or a few defects (such as holes). Also, the film must be easy to prepare and the thickness of the film must be controllable. All of these conditions are satisfied by using stearic acid LB films as the spacer layers. The instrument used for this experiment was a Mayer circle trough of Withelmy film balance, equipped with an electronically controlled dipping device (Germany). LB Film layers of stearic acid were prepared at room temperature. We dropped 3.71×10^{-2} M solution of stearic acid in chloroform onto a purified aqueous subphase (deionized, secondary distilled). After the molecules spread on the water surface, we pressurized the surface to reduce the film area of stearic acid molecules. When the pressure reached 30 mN/m, the molecules of the film arrayed together homogeneously and densely to form a stearic acid thin solid film. Then the film was transferred at 5 mm/min in a Z type to a substrate plate. With this method, we can prepare as many spacer layers as we need.

Stearic acid molecules $(CH_3(CH_2)_{16}COOH)$ consist of a hydrophilic polar head group and a hydrophobic tail. It is represented by o—— (o represented the hydrophilic head). The structure scheme used in this experiment is shown in Fig. 1.

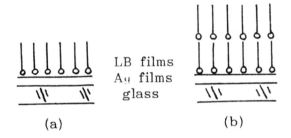

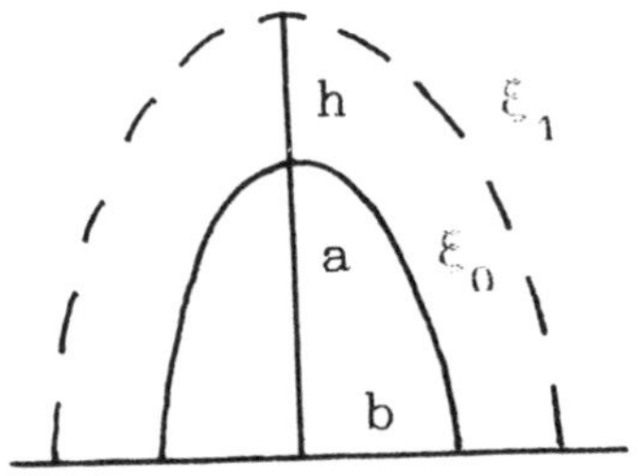

Fig. 1. Schematic structure of
      solid LB films on Ag
      island films
  (a) monolayer LB films
  (b) multilayer LB films

Fig. 2. Hemi-ellipsoid
      protrusion

2.2. Silver film and sample

There are many experimental reports about surface morphology of silver films with SERS activity. In our experiment, the silver films are not only substrates of SERS "active sites", but also LB film substrates. Because of this, silver films need to be attached to substrates to produce a characteristic SERS and LB behavior for that site.

In order to obtain sufficient adhesion, we prepared the silver films with vacuum deposit and chemical deposit on the supersonically cleaned glass slides. The pyridine of SERS sample and stearic acid of LB films wre treated strictly for purity again.

2.3. Raman spectra measurements

The Raman measurements were performed on the sample with 514.5 nm argon laser excitation. The incident angle was taken at 55° and the right–angle geometry was used for collecting scattering light. The instrument used for spectra measurements was a SPEX 1403 Raman spectrometer. We put the LB films glass slide of coating silver on the quartz sample cell. The solution of the sample consisted of 0.05 M pyridine and 0.1 M KCL. The melting point of stearic acid is not too high, so we had to control the laser power to avoid damage and carbonization of the LB film and pyridine. Through time trace research, it should be pointed out that the Raman spectra is stable for ten minutes. In this experiment, laser power was restricted to 35 mw at the sample area and only two pyridine ring-breathing modes were measured, i.e. at less scanning range of Raman spectra (980 cm^{-1}—1060 cm^{-1}). The spectra was recorded in 1 minute, so that the spectra was reliable.

3. RESULTS AND DISCUSSION

There are many experiments about SERS spectra of pyridine adsorbed on silver islands film. [6–8] Generally, the SERS spectra intensity has to be caused mainly by surface morphology of silver islands. Electromagnetic enhancement theory indicates that size and shape of extrusion randomly distributed on a roughened silver surface decides collective electron oscillations, namely, surface plasmon resonance. Therefore, local field near the surface vicinity is much greater than incident filed. We can use a spinning hemi–ellipsoid on a planar substrate to describe a small protrusion on a metal surface. Supposing that the hemi–ellipsoid size is less than the incident light wavelength and at the same time taking into account the ellipsoid electrodynamic property, we obtained an enhancement factor of SERS as follows:[9]

$$R = \left| \frac{1+(1-\varepsilon)\xi_0 Q_1'(\xi_1) \ / \ [\varepsilon Q_1(\xi_0)-\xi_0 Q_1'(\xi_0)]}{1-\Gamma} \right|^4 \qquad (1)$$

where geometry parameters related to surface morphology of metal films are

$$\xi_0 = a \ / \ (a^2-b^2)^{1/2}$$
$$\xi_1 = (a+h) \ / \ (a^2-b^2)^{1/2}$$

a, b are long and short ellipsoid axes respectively, h is the distance between molecule and surface, Q_1, Q'_1 are second kind Legendre functions and differential.

If we only consider the effect of the distance between molecules and surface, we have the expression

$$R \propto \left| \ 1-\xi_0 Q'_1(\xi_1) \ / \ Q_1(\xi_0) \right|^4 \tag{2}$$

We observed a surface morphology of silver film using a scanning electron microscope. The extrusion of chemical deposition on the silver island films surface have different sizes. Assuming that the average size was a=20 nm, b=10 nm, we can calculate the magnitude of SERS enhancement factor of different distances between molecules and surface, as shown in table 1.

Table 1. Relation between spacer thickness and
calculated values of enhancement factor

h (nm)	R	normalized intensity
0.5	7.4×10^3	1.0
2.5	8.6×10^2	0.12
5.0	2.5×10^2	0.03
10.0	1.8×10	0.002

We put sample with different numbers of LB layers and silver films into an aqueous solution of 0.1 M KCL containing 0.05 M pyridine. When the distance between pyridine molecules and silver surface changed, we measured SERS spectra of pyridine (see Fig. 3).

Added spacer layers of the modified silver surface made SERS spectra of pyridine molecules change. The scattering intensity of its ring-breathing mode (1010 cm^{-1}) was observed as a function of LB spacer layers' thickness (see Fig. 4).

LB films are organic molecular films of ordered arrays. In the process of transferring from a liquid surface to a silver film, as long as the area of the film remained constant, homogeneous crystal-like solid film are formed on the surface of the silver film. In the 980 cm^{-1}—1060 cm^{-1} range, there are no Raman virbrational peaks for stearic acid. Raman signals of adsorbed molecules on the surface in this frequency range are not affected by spacer layers between adsorbed molecules and the surface. Pyridine molecules adsorbed on LB films/Ag films complex substrate are not in direct contact with silver, thus they cannot produce a strong interaction; but Raman frequency shift of

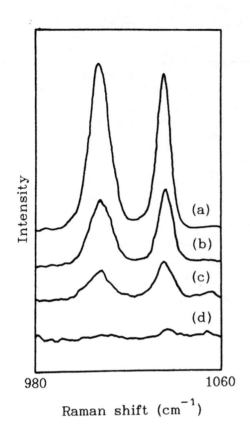
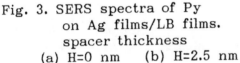

Fig. 3. SERS spectra of Py
on Ag films/LB films.
spacer thickness
(a) H=0 nm   (b) H=2.5 nm
(c) H=5 nm   (d) H=15 nm

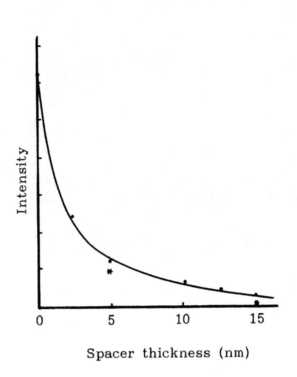

Spacer thickness (nm)

Fig. 4. Relationship between
Raman intensity of
Py (1010 cm^{-1}) and
spacer thickness
. chemically deposited
Ag films
* evaporated Ag films

ring-breathing mode is the same as that of pyridine adsorbed directly on silver films. Therefore, we deduced that LB films as spacer layers have less effect on vibration of pyridine molecules. Pyridine molecules are physisorbed on the surface.

A single spacer layer thickness of stearic acid molecule LB film is 2.5 nm. When one or two LB film spacer layers of stearic acid were transferred to the silver surface, the distance between pyridine molecules and the silver surface are 2.5 nm for one layer and 5 nm for two layers. In both cases, we still observed the SERS spectra, but the intensity was remarkably decreased. In this situation, the image field effect of the induced dipole in the metal surface can be negligible. The lightning-rod effect related to molecules adsorbed near high-curvature edges of surface irregularities is also not considered. It is quite possible that pyridine SERS was caused by electromagnetic long range enhancement mechanisms. We measured the absorption spectrum of silver films (see Fig.5). There is a prominent peak at 440 nm. It is consistent with plasmon resonance absorption of a rough silver surface to an incident light field. The absorption peak apparently broadened owing to relatively complex morphology of the silver surface. When the incident light field excited surface plasmon reaches resonance or near resonance, a local field in the silver surface vicinity increased greatly and SERS was remar- kably enhanced. Local fields decay in a certain manner with the increase of distance between molecules and surface, so that Raman signals decrease quickly. The pyridine SERS signal intensity depends strongly on LB spacer layers' thickness. The intensity was remarkably decreased when the thickness was 5 nm (the first two layers), and eventually became unobservable with 15 nm thickness (six layers).

The SERS intensity ratio of two pyridine ring-breathing modes has an important advantage as a surface spectroscopy probe, because it is sensitive to the environment. Comparing the scattering intensity ratio of pyridine ring-breathing modes in different adsorbed schemes, we found that the value of I_i/I_j in the physical adsorption was much greater than the value in the chemical adsorption. When pyridine was adsorbed on the silver electrode and silver film surface, the I_i/I_j was similar to the value obtained liquid pyridine. This suggests that most of adsorbed pyridine molecules are often physisorbed in this condition. In our experiment with LB spacer layers, the value of I_i/I_j Was 1.1. That not only clearly indicates this was physisorption, but also indicates there was interaction between stearic acid LB molecules and pyridine molecules, which changed the orientation of adsorbate molecules (see Table 2).

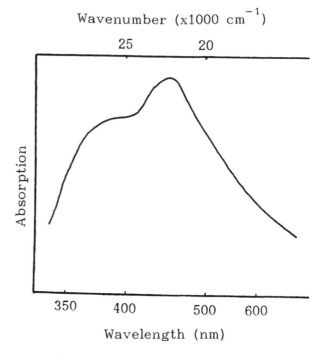

Fig. 5. Absorption spectrum of chemically deposited Ag films

Table 2. Intensity ratio of ring-breathing modes of pyridine in different environment [10]

Environment	Raman shift (cm^{-1})		I_i/I_j
	υ_i	υ_j	
Liquid pyridine (Py)	1032	992	0.80
$[(Py)_4Ag](NO_3)_2$	1075	1025	0.05
Py+CL$^-$/Ag electrode	1036	1008	0.75
Py+CL$^-$/Ag films	1036	1008	0.88
Py+CL$^-$/LB films/Ag films	1036	1010	1.10

In order to understand the effect of LB films itself on SERS of pyridine molecules, we deposited different layers (one, ten, thirty) of LB films on a glass substrate without Ag films. In these cases, we didn't observe any Raman signals. This meant that SERS of pyridine came from resonance excitation of Ag surface plasmons. LB films only have two functions here; (1) in interaction with pyridine molecules which caused the change of pyridine molecules orientation and; (2) in making spacer layers.

4. CONCLUSIONS

There has been considerable interest in enhancement mechanism research of SERS because this study will promote broad application of the SERS technique. For the first time, we have used LB films as spacer layers between adsorbed molecules and a surface to observe the long range property of SERS enhancement mechanism and to support electromagnetic enhancement theory. Compared with other researches, our method was simple and feasible, and capable of overcoming many difficulties of techniques. Using pyridine/stearic acid LB films/Ag films configuration has more advantages. It not only reveals the enhancement mechanism, but also helps us to understand the interaction between surface molecules.

5. REFERENCES

1. M. Moskovits, "Surface-Enhanced Spectroscopy," *Rev. Mod. Phys.*, Vol.57, No 3, Part 1, pp. 783-826, July 1985.
2. A. Otto, et al., "Model of Electronically Enhanced Raman Scattering from Adsorbates on Cold-Deposited Silver" *Surf. Scie.*, 210, No 3, pp 363-386, May 1989.
3. R. K. Chang and T.E. Furtak, Eds. "Surface Enhanced Raman Scattering," Plenum Press, New York 1982.
4. C. A. Murry and D. L. Allara, "Measurement of the Molecule–Silver Separation Dependence of Surface Enhanced Raman Scattering in Multilayered Structures," *J. Chem. Phys.*, Vol 76,No 3, pp 1290-1303, February 1982.
5. J. D. Swalen, et al., "Molecular Monolayers and Films," *Langmuir*, Vol. 3, No. 6, pp. 932-950, 1987.
6. Y. Mo, I. Morke and P. Wachter, "Surface Enhanced Raman Scattering of Pyridine on Silver Surface of Different Roughness," *Surf. Scie.*, Vol 133, No 1, pp 452-458, October 1983.
7. H. Seki, "Surface Enhanced Raman Scattering of Pyridine on Different Silver Surface" *J. Chem. Phys.* Vol 76, No 9, pp 4412- 4418, May 1982.
8. R. Aroca and F. Martin, "Tuning Metal Island Films for Maximum Surface- Enhanced Raman Scattering," *J. Raman Spectros.*, Vol 16, No. 3, pp 156-161, 1985.
9. J. Gersten and A. Nitzan, "Electromagnetic Theory of Enhanced Raman Scattering by Molecules Adsorbed on Rough Surfaces," *J. Chem. Phys.*, Vol.73, No 7, pp 3023-3037, Octorber 1980.
10. R. J. H. CLARK Eds. "Adv. in Infrared & Raman Spectroscopy," Vol 9, Chap. 4. Heyden, London 1982.

ADDENDUM

The following papers, which were scheduled to be presented at this conference and published in this proceedings, were canceled.

[1530-24]   **Cryovacuum BRDF measurements of MMH-nitrate**
W. Krone-Schmidt, R. C. Loveridge, Hughes Aircraft Co.

[1530-37]   **Self-heating of radiation and brightness increasing in optical scattering**
A. V. Shepelev, A. B. Shvartsburg, I. N. Sisakian, Central Design Bureau
for Unique Instrumentation (USSR)

AUTHOR INDEX

9-8-94